T0205746

Communications in Computer and Information Science 1930

Rationale
The CCIS series is devoted to the publication of proceedings of computer science conferences. Its aim is to efficiently disseminate original research results in informatics in printed and electronic form. While the focus is on publication of peer-reviewed full papers presenting mature work, inclusion of reviewed short papers reporting on work in progress is welcome, too. Besides globally relevant meetings with internationally representative program committees guaranteeing a strict peer-reviewing and paper selection process, conferences run by societies or of high regional or national relevance are also considered for publication.

Topics
The topical scope of CCIS spans the entire spectrum of informatics ranging from foundational topics in the theory of computing to information and communications science and technology and a broad variety of interdisciplinary application fields.

Information for Volume Editors and Authors
Publication in CCIS is free of charge. No royalties are paid, however, we offer registered conference participants temporary free access to the online version of the conference proceedings on SpringerLink (http://link.springer.com) by means of an http referrer from the conference website and/or a number of complimentary printed copies, as specified in the official acceptance email of the event.

CCIS proceedings can be published in time for distribution at conferences or as post-proceedings, and delivered in the form of printed books and/or electronically as USBs and/or e-content licenses for accessing proceedings at SpringerLink. Furthermore, CCIS proceedings are included in the CCIS electronic book series hosted in the SpringerLink digital library at http://link.springer.com/bookseries/7899. Conferences publishing in CCIS are allowed to use Online Conference Service (OCS) for managing the whole proceedings lifecycle (from submission and reviewing to preparing for publication) free of charge.

Publication process
The language of publication is exclusively English. Authors publishing in CCIS have to sign the Springer CCIS copyright transfer form, however, they are free to use their material published in CCIS for substantially changed, more elaborate subsequent publications elsewhere. For the preparation of the camera-ready papers/files, authors have to strictly adhere to the Springer CCIS Authors' Instructions and are strongly encouraged to use the CCIS LaTeX style files or templates.

Abstracting/Indexing
CCIS is abstracted/indexed in DBLP, Google Scholar, EI-Compendex, Mathematical Reviews, SCImago, Scopus. CCIS volumes are also submitted for the inclusion in ISI Proceedings.

How to start
To start the evaluation of your proposal for inclusion in the CCIS series, please send an e-mail to ccis@springer.com.

Rama Krishna Challa · Gagangeet Singh Aujla ·
Lini Mathew · Amod Kumar · Mala Kalra ·
S. L. Shimi · Garima Saini · Kanika Sharma
Editors

Artificial Intelligence of Things

First International Conference, ICAIoT 2023
Chandigarh, India, March 30–31, 2023
Revised Selected Papers, Part II

Editors
Rama Krishna Challa
National Institute of Technical Teachers Training and Research
Chandigarh, India

Lini Mathew
National Institute of Technical Teachers Training and Research
Chandigarh, India

Mala Kalra
National Institute of Technical Teachers Training and Research
Chandigarh, India

Garima Saini
National Institute of Technical Teachers Training and Research
Chandigarh, India

Gagangeet Singh Aujla
Durham University
Durham, UK

Amod Kumar
National Institute of Technical Teachers Training and Research
Chandigarh, India

S. L. Shimi
National Institute of Technical Teachers Training and Research
Chandigarh, India

Punjab Engineering College (Deemed to be University)
Chandigarh, India

Kanika Sharma
National Institute of Technical Teachers Training and Research
Chandigarh, India

ISSN 1865-0929 ISSN 1865-0937 (electronic)
Communications in Computer and Information Science
ISBN 978-3-031-48780-4 ISBN 978-3-031-48781-1 (eBook)
https://doi.org/10.1007/978-3-031-48781-1

This Springer imprint is published by the registered company Springer Nature Switzerland AG
The registered company address is: Gewerbestrasse 11, 6330 Cham, Switzerland

Preface

It is a great privilege for us to present the proceedings of ICAIoT-2023 which contain the papers submitted by researchers, practitioners, and educators to the International Conference on Artificial Intelligence of Things, ICAIoT 2023, held during 30th and 31st March, 2023 at National Institute of Technical Teachers Training and Research, Chandigarh, India. We hope that you will find this book educative and inspiring.

The conference attracted papers from international researchers and scholars in the field of Artificial Intelligence (AI) applications in Internet of Things (IoT). The ICAIoT 2023 conference received 401 papers from around the world out of which 60 were accepted for oral presentation. The total number of papers presented at ICAIoT 2023 was 57. All submitted papers were subjected to strict Single Blind peer review by at least three national and international reviewers who are experts in the area of the particular paper. The acceptance rate of the conference was 15%.

The aim of ICAIoT 2023 was to provide a professional platform for discussing research issues, opportunities and challenges of AI and IoT applications. ICAIoT 2023 received unexpected support and enthusiasm from the delegates. The papers presented at this premier international gathering of leading AI researchers and practitioners from all over the world are collected in two volumes which will connect advances in engineering and technology with the use of smart techniques including Artificial Intelligence (AI), Machine Learning and Internet of Things (IoT).

The book presents the most recent innovations, trends and concerns as well as practical challenges encountered and solutions adopted in the fields of AI algorithms implementation in IoT Systems. The book is divided into two volumes, covering the following topics:

Volume I:

- AI and IoT Enabling Technologies
- AI and IoT for Smart Healthcare

Volume II:

- AI and IoT for Electrical, Electronics and Communication Engineering
- AI and IoT for other Engineering Applications

This book will be especially useful for graduate students, academic researchers, scientists and professionals in the fields of Computer Science and Engineering, Electrical Engineering, Electronics and Communication Engineering and allied disciplines.

Organizing a prestigious conference such as ICAIoT 2023 with the Springer CCIS series as publication partner is a substantial endeavour. We would like to extend our sincere thanks to the organizing committee members for all their support with the conference organization. We would like to thank the authors of all submitted papers for sending high-quality contributions to ICAIoT 2023. We would like to express our sincere gratitude to our sponsors, the advisory committee members, technical program committee members and Ph.D./M.E. scholars for all their support throughout the conference. Our special thanks are due to all the external reviewers who contributed their expertise, insights, and judgment during the review process. Last but not least, we extend our heartfelt thanks to the staff of Springer for their extensive support in the publication of this book under the Springer CCIS Series.

March 2023

Rama Krishna Challa
Gagangeet Singh Aujla
Lini Mathew
Amod Kumar
Mala Kalra
S. L. Shimi
Garima Saini
Kanika Sharma

Message from the Patron

I am happy to learn that the Department of Computer Science and Engineering, Department of Electronics and Communication Engineering and Department of Electrical Engineering of National Institute of Technical Teachers Training & Research, Chandigarh has taken a timely initiative to address the emerging issues on Artificial Intelligence and Internet of Things by organizing the International Conference on Artificial Intelligence of Things (ICAIoT 2023) from March 30–31, 2023.

The organizing Committee has invited many speakers of international and national repute to make the two days' deliberations more meaningful and academically rich. I extend a warm welcome to all the dignitaries, keynote speakers, paper presenters and delegates to the conference. I am sure the conference will provide a platform to researchers, professionals, educators and students to share innovative ideas, issues, recent trends and future directions in the fields of AI, IoT and Industry 4.0 to address industrial and societal needs.

I wish the conference a grand success and all the participants a pleasant stay and great learning.

Bhola Ram Gurjar

Message from the Conference Chairs

We were delighted to welcome delegates from India and abroad to the International Conference on Artificial Intelligence of Things (ICAIoT 2023), organized by Department of Computer Science and Engineering, Department of Electronics and Communication Engineering and Department of Electrical Engineering of the National Institute of Technical Teachers Training & Research (NITTTR), Chandigarh on March 30–31, 2023.

Undoubtedly, AI and IoT technologies have transformed our society in recent times and the pace of change can only be described as revolutionary. The technology is progressing fast and new horizons are being explored. The conference aimed to provide an opportunity to researchers, engineers, academicians as well as industrial professionals from all over the world to present their research work and related development activities. This conference also provided a platform for the delegates to exchange new ideas and application experiences face to face, to establish research relations and to find global partners for future collaboration.

It is rightly said "Alone we can do so little, together we can do so much". We are thankful to all the contributors of this conference who have worked hard in planning and organizing both the academic activities and necessary social arrangements. In particular, we take this opportunity to express our gratitude to the Conference Patron and members of the Advisory Committee for their wise advice and brilliant suggestions in organizing the Conference. Also, we would like to thank the Technical Committee and Reviewers for their thorough and timely reviews of the research papers and our Ph.D./M.E. scholars for their support. We would also like to thank all the sponsors who have supported us in organizing this conference.

We hope all attendees enjoyed this event.

Rama Krishna Challa
Gagangeet Singh Aujla
Lini Mathew
Amod Kumar

Message from the Conference Co-chairs

We were pleased to extend our most sincere welcome to all the delegates to the International Conference on Artificial Intelligence of Things (ICAIoT 2023), being organized by three departments, namely, Department of Computer Science and Engineering, Department of Electronics and Communication Engineering and Department of Electrical Engineering of the National Institute of Technical Teachers Training & Research (NITTTR), Chandigarh during March 30–31, 2023 in association with DSIR, Govt. of India.

Artificial Intelligence plays an enormous role in promoting knowledge and technology which is essential for the educators, researchers, industrial and commercial houses in the present digital age. Being a core part of this conference from the beginning, we feel very enthusiastic about the event and hope that we all will benefit academically through mutual collaboration. This International Conference will provide exposure to recent advancements and innovations in the field of Internet of Things (IoT), Artificial Intelligence (AI) and Industry 4.0. It is expected to be an intellectual platform to share ideas and present the latest findings and experiences in the mentioned areas.

The successful organization of ICAIoT 2023 required the talent, dedication and time of many volunteers and strong support from sponsors. We express our sincere thanks to the paper reviewers, keynote speakers, invited speakers and authors. A special mention about our Ph.D./M.E. scholars who worked hard to make this International Conference a success, would be in order.

We hope all attendees found the event to be a grand success.

Mala Kalra
Garima Saini
S. L. Shimi
Kanika Sharma

Organization

Conference Patron

Gurjar, Bhola Ram	National Institute of Technical Teachers Training and Research, Chandigarh, India

Conference Chairs

Challa, Rama Krishna	National Institute of Technical Teachers Training and Research, Chandigarh, India
Aujla, Gagangeet Singh	Durham University, UK
Mathew, Lini	National Institute of Technical Teachers Training and Research, Chandigarh, India
Kumar, Amod	National Institute of Technical Teachers Training and Research, Chandigarh, India

Conference Co-chairs

Kalra, Mala	National Institute of Technical Teachers Training and Research, Chandigarh, India
Shimi, S. L.	National Institute of Technical Teachers Training and Research, Chandigarh, India and Punjab Engineering College (Deemed to be University), Chandigarh, India
Saini, Garima	National Institute of Technical Teachers Training and Research, Chandigarh, India
Sharma, Kanika	National Institute of Technical Teachers Training and Research, Chandigarh, India

Advisory Committee

Ahuja, Rajeev	Indian Institute of Technology (IIT), Ropar, India
Awasthi, Lalit K.	National Institute of Technology (NIT), Uttarakhand, India
Babu, K. M.	Administrative Management College Institutions, Bengaluru, India

Bollen, Math	Luleå University of Technology, Sweden
Chakrabarti, Saswat	Indian Institute of Technology (IIT), Kharagpur, India
Chatterjee, Amitava	Jadavpur University, India
Das, Debabrata	International Institute of Information Technology (IIIT), Bengaluru, India
Datta, Debasish	Indian Institute of Technology (IIT), Kharagpur, India
Fjeldly, Tor A.	Norwegian University of Science and Technology, Norway
Gaur, Singh Manoj	Indian Institute of Technology (IIT), Jammu, India
Girdhar, Anup	Sedulity Solutions & Technologies, India
Govil, M. C.	National Institute of Technology (NIT), Sikkim, India
Gupta, Savita	Panjab University, India
Hyoung, Kim Tae	Nanyang Technological University, Singapore
Jat, Dharam Singh	Namibia University of Science and Technology, Namibia
Kakde, O. G.	Indian Institute of Information Technology (IIIT), Nagpur, India
Kapur, Avichal	University Grants Commission, India
Kuruvilla, Abey	University of Wisconsin, USA
Limiti, Ernesto	University of Rome Tor Vergata, Italy
Maddara, Ramesh	Cerium Systems Pvt. Ltd., India
Madhukar, Mani	IBM, India
Maurya, U. N.	LM Healthcare, India
Mekhilef, Saad	Swinburne University of Technology, Australia
Mishra, S.	Indian Institute of Technology (IIT), Delhi, India
Naidu, K. Rama	Jawaharlal Nehru Technological University, India
Nassa, Anil Kumar	National Board of Accreditation, India
Pati, Bibudhendu	Rama Devi Women's University, India
Pillai, G. N.	Indian Institute of Technology (IIT), Roorkee, India
Poonia, M. P.	All India Council for Technical Education, India
Popov, Yu. I.	St. Petersburg National Research University of Information Technologies, Russia
Reddy, B. V. Ramana	National Institute of Technology (NIT), Kurukshetra, India
Reddy, Siva G.	Satliva.com, India
Samuel, Paulson	Motilal Nehru National Institute of Technology (MNNIT), Allahabad, India
Sarwat, Arif I.	Florida International University, USA

Sehgal, Rakesh	National Institute of Technology (NIT), Srinagar, India
Sharma, Ajay	National Institute of Technology (NIT), Delhi, India
Shukla, Anupam	Sardar Vallabhbhai National Institute of Technology (SVNIT), Surat, India
Singh, Bhim	Indian Institute of Technology (IIT), Delhi, India
Somayajulu, D. V. L. N.	Indian Institute of Information Technology, Design and Manufacturing, Kurnool, India
Subudhi, Bidyadhar	Indian Institute of Technology (IIT), Goa, India
Tariq, Mohd	Florida International University, USA
Vig, Renu	Panjab University, India
Zin, Thi Thi	University of Miyazaki, Japan

Technical Programme Committee

Agarwal, Ravinder	Thapar Institute of Engineering and Technology, India
Agarwal, Suneeta	Motilal Nehru National Institute of Technology, Allahabad, India
Aggarwal, Naveen	Panjab University, India
Agnihotri, Prashant	Indian Institute of Technology (IIT), Bhilai, India
Aujla, Gagangeet Singh	Durham University, UK
Bakhsh, Farhad Ilahi	National Institute of Technology (NIT), Srinagar, India
Bali, S. Rasmeet	Chandigarh University, India
Bali, Vikram	IMS Engineering College, Ghaziabad, India
Bansod, B. S.	CSIR-CSIO, India
Barati, Masoud	Carleton University, Canada
Bashir, Ali Kashif	Manchester Metropolitan University, UK
Behal, Sunny	Shaheed Bhagat Singh State University, India
Bhardwaj, Aditya	Bennett University, India
Bhatia, Rajesh	Punjab Engineering College (Deemed to be University), India
Bhatnagar, Vishal	Netaji Subhash University of Technology, India
Bindhu, Shoba	Jawaharlal Nehru Technological University Anantapur, India
Bindra, Naveen	Postgraduate Institute of Medical Education and Research, Chandigarh, India
Chakraborty, Chinmay	Birla Institute of Technology, Mesra, India
Chawla, Anu	Post Graduate Government College for Girls, Sector 42, Chandigarh, India

Chawla, Naveen K.	University of Oklahoma, USA
Chhabra, Indu	Panjab University, India
Chhabra, Jitender Kumar	National Institute of Technology (NIT), Kurukshetra, India
Chhabra, Rishu	Chitkara University, India
Chinthala, Ashok Praveen	Tata Consultancy Services Limited, India
Dave, Mayank	National Institute of Technology (NIT), Kurukshetra, India
Demirbaga, Umit	University of Cambridge, UK
Fjeldly, Tor A.	Norwegian University of Science and Technology, Norway
Gangadharappa, M.	Netaji Subhash University of Technology, India
Garg, Sahil	Resilient Machine Learning Institute, Canada
Gaur, Manoj	Netaji Subhash University of Technology, India
Gogna, Monika	Post Graduate Government College for Girls, Sector 42, Chandigarh, India
Gupta, Varun	Chandigarh College of Engineering and Technology, India
Gupta, Vipin	U-Net Solutions, India
Handa, Rohit	Citi Canada Technology Services, Canada
Himanshu	Punjabi University, India
Hyoung, Kim Tae	Nanyang Technological University, Singapore
Ilamparithi, T.	University of Victoria, Canada
Jat, Dharam Singh	Namibia University of Science and Technology, Namibia
Jha, Devki Nandan	University of Oxford, UK
Jindal, Anish	Durham University, UK
Karar, Vinod	CSIR-CRRI, India
Kassey, Philip B.	Petrasys Global Pvt. Ltd., India
Kaur, Amandeep	Central University of Punjab, India
Kaur, Gurpreet	SGGS College, India
Kaur, Kuljeet	École de Technologie Supérieure, Montreal, Canada
Kaur, Lakhwinder	Punjabi University, India
Krishna, P. Murali	Broadcom Inc., India
Kumar, Anil	Chandigarh College of Engineering and Technology, India
Kumar, Arvind	National Institute of Technology (NIT), Kurukshetra, India
Kumar, Gaurav	Magma Research and Consultancy Services, India
Kumar, Harish	Panjab University, India
Kumar, Krishan	Panjab University, India
Kumar, Naresh	Panjab University, India

Kumar, Nishant	Indian Institute of Technology (IIT), Jodpur, India
Kumar, Prabhat	LUT University, Sweden
Kumar, Preetam	Indian Institute of Technology (IIT), Patna, India
Kumar, Rakesh	Central University of Haryana, India
Kumar, Ritesh	CSIR-CSIO, India
Kumar, Satish	CSIR-CSIO, India
Kumar, Sunil	Meerut Institute of Engineering and Technology, India
Kuruvilla, Abey	University of Wisconsin, USA
Limiti, Ernesto	University of Rome Tor Vergata, Italy
Maheswari, Ritu	eMANTHAN Inspiring Innovation, India
Maini, Raman	Punjabi University, India
Maswood, Ali I.	Nanyang Technological University, Singapore
Mathew, Jimson	Indian Institute of Technology (IIT), Patna, India
Mehan, Vineet	Maharaja Agrasen University, India
Mekhilef, Saad	Swinburne University of Technology, Australia
Mittal, Ajay	Panjab University, India
Mittal, Meenakshi	Central University of Punjab, India
Mohapatra, Rajarshi	Indian Institute of Information Technology (IIIT), Raipur, India
Murthy, Ch. A. S.	Centre for Development of Advanced Computing, India
Nanayakkara, Samudaya	University of Moratuwa, Sri Lanka
Noor, Ayman	Taibah University, Saudi Arabia
Pal, Sujata	Indian Institute of Technology (IIT), Ropar, India
Panda, Surya Narayan	Chitkara University, India
Panigrahi, Chhabi Rani	Rama Devi Women's University, India
Patel, R. B.	Chandigarh College of Engineering and Technology, India
Patil, Nilesh Vishwasrao	Government Polytechnic, Ahmednagar, India
Pilli, E. S.	Malaviya National Institute of Technology (MNIT), Jaipur, India
Qureshi, Saalim	Quarbz Info Systems, India
Rajpurohit, Bharat Singh	Indian Institute of Technology (IIT), Mandi, India
Ramesh, K.	National Institute of Technology (NIT), Warangal, India
Ravindran, Vineeta	Scania, Sweden
Reaz, Md. Mamun Bin Ibne	Universiti Kebangsaan Malaysia, Malaysia
Reddy, S. R. N.	Indira Gandhi Delhi Technical University for Women, India
Sahu, Benudhar	Institute of Technical Education and Research, India
Saini, Surender Singh	CSIR-CSIO, India

Sangal, A. L.	Dr. B. R. Ambedkar National Institute of Technology, Jalandhar, India
Sarwat, Arif	Florida International University, USA
Sehgal, Navneet Kaur	Chandigarh University, India
Seshadrinath, Jeevanand	Indian Institute of Technology (IIT), Roorkee, India
Shanmuganantham, T.	Pondicherry University, India
Sharma, Shweta	Punjab Engineering College (Deemed to be University), India
Sharmeela, C.	Anna University, India
Singh, A. K.	National Institute of Technology (NIT), Kurukshetra, India
Singh, Amandeep	National Institute of Technology (NIT), Srinagar, India
Singh, Amardeep	Punjabi University, India
Singh, Baljit	Central Scientific Instruments Organization, India
Singh, Balwinder	Centre for Development of Advanced Computing, India
Singh, Brahmjit	National Institute of Technology (NIT), Kurukshetra, India
Singh, Dheerendra	Chandigarh College of Engineering and Technology, India
Singh, Jeetendra	National Institute of Technology (NIT), Sikkim, India
Singh, Paramjeet	Giani Zail Singh Campus College of Engineering & Technology, Maharaja Ranjit Singh Punjab Technical University, India
Singh, Rajvir	Chitkara University, India
Singh, Satwinder	Central University of Punjab, India
Singh, Shailendra	Punjab Engineering College (Deemed to be University), India
Singh, Sukhwinder	Panjab University, India
Singh, Sunil K.	Chandigarh College of Engineering and Technology, India
Sofat, Sanjeev	Punjab Engineering College (Deemed to be University), India
Srinivasa, K. G.	Indian Institute of Information Technology (IIIT), Raipur, India
Sunkaria, Ramesh Kumar	National Institute of Technology (NIT), Jalandhar, India
Tariq, Mohd	Florida International University, USA
Tiwari, Anil Kumar	Indian Institute of Technology (IIT), Jodhpur, India

Toreini, Ehsan	University of Surrey, UK
V. S., Ananthanarayana	National Institute of Technology (NIT), Surathkal, India
Vasantham, Thiru	Durham University, UK
Verma, Harsh K.	Dr. B. R. Ambedkar National Institute of Technology, Jalandhar, India
Verma, Yajvender Pal	Panjab University, India
Vuppala, Anil	Indian Institute of Information Technology (IIIT), Hyderabad
Yadav, Yashveer	Datum Analysis, Jordan
Zin, Thi Thi	University of Miyazaki, Japan

Organizing Committee

Dhaliwal, Balwinder S.	National Institute of Technical Teachers Training and Research, Chandigarh, India
Doegar, Amit	National Institute of Technical Teachers Training and Research, Chandigarh, India
Solanki, Shano	National Institute of Technical Teachers Training and Research, Chandigarh, India
Thakur, Ritula	National Institute of Technical Teachers Training and Research, Chandigarh, India
Sharan, Amrendra	National Institute of Technical Teachers Training and Research, Chandigarh, India

Student Volunteers

Kaur, Amandeep	National Institute of Technical Teachers Training and Research, Chandigarh, India
Agrawal, Deepika Vikas	National Institute of Technical Teachers Training and Research, Chandigarh, India
Agrawal, Lucky Kumar Dwarkadas	National Institute of Technical Teachers Training and Research, Chandigarh, India
Bala, Anju	National Institute of Technical Teachers Training and Research, Chandigarh, India
Bhardwaj, Neha	National Institute of Technical Teachers Training and Research, Chandigarh, India
N., Raj Chithra	National Institute of Technical Teachers Training and Research, Chandigarh, India
Gupta, Mahendra	National Institute of Technical Teachers Training and Research, Chandigarh, India

Gupta, Vinita	National Institute of Technical Teachers Training and Research, Chandigarh, India
Krishna, Banoth	National Institute of Technical Teachers Training and Research, Chandigarh, India
Kulkarni, Atul M.	National Institute of Technical Teachers Training and Research, Chandigarh, India
M. Soujanya	National Institute of Technical Teachers Training and Research, Chandigarh, India
Mahajan, Palvi	National Institute of Technical Teachers Training and Research, Chandigarh, India
M. K., Shajila Beegam	National Institute of Technical Teachers Training and Research, Chandigarh, India
Patel, Anwesha	National Institute of Technical Teachers Training and Research, Chandigarh, India
Pushparaj	National Institute of Technical Teachers Training and Research, Chandigarh, India
Rani, Puja	National Institute of Technical Teachers Training and Research, Chandigarh, India
Shukla, Praveen	National Institute of Technical Teachers Training and Research, Chandigarh, India
Siddiqui, Mohd. Ahsan	National Institute of Technical Teachers Training and Research, Chandigarh, India

Sponsors

Department of Scientific and Industrial Research, Ministry of Science and Technology, Government of India, New Delhi, India

Syngient Technologies, India

IETE Chandigarh Centre, India

Alakh Infotech, India

Design Tech Systems, India

Gigabyte Networks, India

Fore Solutions, India

Hitech Solutions, India

Pinnacle Enterprises, India

Luxmi Enterprises, India

Contents – Part II

Contents – Part I

AI and IoT Enabling Technologies

AI and IoT for Electrical, Electronics and Communication Engineering

Amalgamation of Machine Learning Techniques with Optical Systems: A Futuristic Approach

Alka Jindal[1](✉) and Shilpa Jindal[2]

[1] PEC University of Technology, Chandigarh 160012, India
alkajindal@pec.edu.in
[2] CCET (Degree Wing), Sector 26, Chandigarh 160012, India

Abstract. With the availability of huge bandwidth backbone networks in terms of optical fiber links, there is a surge in data at the user end. It has been boosted further by the deployment of 5G services and IoT. Hence to incorporate this big data, demand for flexible and scalable networks has risen thereby increasing complexity. So, there is a need of new techniques that can monitor their performances and accordingly adjust the parameters as per the real time requirements and reduce the cost/bit/sec. In this direction, we have reviewed various machine learning algorithms that have the potential to be incorporated with optical systems that can further optimize the efficiency across many dimensions. Hence, the complex network can be self-reconfigured by managing failure and estimating Quality of Transmission (QoT) using machine learning algorithms. Further, input to machine learning algorithms is the dataset. In the present scenario, dataset to this scientific domain of optical communication and networks is not abundantly available and hence can either be generated through simulations, experimentation or can be synthetically provided by advanced machine learning algorithms. The dataset can, thus, be taken from field trials and testbeds, lab trials and testbeds, open-source platforms and from some government funded networks etc. As per the type of data (image data, sequential data or augmented data), machine learning algorithms are classified. In this paper, we have generated data for optical mesh network connecting four Indian cities (Delhi, Mumbai, Bangalore and Kolkata) using different transceivers on a public and open-source platform – GNpy model – a QoT estimation tool and optimized the data rates and modulation format for distance and generalized signal to noise ratio (GSNR) values. Data similar to this can further be generated for different networks and can be applied as input to machine learning algorithms in some standard format and ML algorithms can be developed in optical domain.

Keywords: QoT · quality of transmission · GSNR · Generalized Signal Noise Ratio · MFR · Modulation Format Recognition · IM-DD · Intensity Modulation–Direct Detection · GAN · Generative Adversarial Network

R. K. Challa et al. (Eds.): ICAIoT 2023, CCIS 1930, pp. 3–12, 2024.
https://doi.org/10.1007/978-3-031-48781-1_1

1 Introduction

Many areas of engineering are heading towards involving artificial intelligence, machine learning approaches in them, due to its applied advantages despite being complex. Many supervised and unsupervised machine learning algorithms work on the dataset available either online or on real time data like face recognition, face verification, speech processing, image processing, handwriting recognition etc. But in some technological domains like optical system, antenna design, renewable energy etc. much data is not available and hence one has to rely on synthetic data either generated from simulations or through expensive hardware or experimental setup. This is the major challenge faced by such technologies which hinders the implementation of available machine learning and further deep learning algorithms in these areas. In this work, we have investigated various machine learning algorithms as a powerful interdisciplinary tool that can act as viable candidate for optical system analysis. It can help in realizing an autonomous, self-learning, self-driving, self-healing network that can auto configure and predict traffic and optimize to lower the capital expenditure as well as operational expenditure thereby reducing cost and making technology cheaper. Hence complex optical communication systems and network problems can be solved by utilizing the capacity of ML and DL algorithms both at physical layer and at network layer. CNN is used for extracting parameters from image while RNN is used for sequential data analysis. CNN can estimate channel length, channel impair-ments, modulation format and hence monitor performance of optical systems while sequential data like optical and electrical signals, network traffic data, inter symbol interference can be realized through RNN.

2 Past Work

Wang, Danshi et al. [1] used 1600 constellation diagrams of six different modulation format i.e., QPSK, 8PSK, 8QAM, 16QAM, 32QAM and 64QAM of IM-DD optical system over OSNR range (15–30 dB and 20–35 dB) to develop an intelligent constellation diagram analyzer that estimates OSNR and recognizes modulation format using CNN with better accuracy over other four traditional machine learning algorithms like Decision tree, back propagation artificial neural network, distance weighted K-Nearest Neighbors, SVM. The images used were generated via simulation on a popular commercial software VPI transmission maker version 9.0. The results show that CNN achieved better accuracy for modulation format recognition and OSNR estimation over other 4 traditional machine learning approaches. Tizikara et al. [2] reviewed various machine learning algorithms and candidate features for monitoring the optical performance for direct detection and coherent detection systems. Wang et al. [3] reviewed various machine learning algorithms like CNN for image data that can be used for channel estimation, mode demodulation, optical signal analysis, impairment diagnosis, optical performance monitoring, digital signal processing, and spectral analysis, RNN for sequential data, to execute signal pre-distortion and post-compensation, network traffic forecasting, and fault alarming analysis, Deep learning (auto encoder i.e., interpreting the entire communication systems), GAN for augmenting data that is diverse and sufficient to conduct experimentation, Deep Reinforcement Learning for Network Automation that adds more intelligence and adaptivity to optical network. D. Wang et al. [4] used simulated images of eye diagrams of IM-DD optical system for 10 Gb/s on off keying and 20 Gb/s pulse amplitude modulation-4 obtained from popular software VPI transmission maker version 8.6 to feed transfer learning-based eye diagram analyzer. TL based approach can extract varied information and characteristics from eye diagram data only and hence can execute large number of tasks (6) like estimation of fibre length, Q factor, essential eye diagram parameters, impairment characteristics, eye height and width, levels of "0" and "1" from source task learning to target tasks. Results showed that training times for both the modulation formats were reduced using TL approach. Esteves et al. [5] monitored optical performance that predicted BER and analysed eye-pattern over short distance communication with PAM4 signals, intensity modulation-direct detection modulation in inter data centres optical connections using CNN. The transmission fibre used was multi core fiber. The synthetic dataset used was created by Monte Carlo simulations and the results showed that CNN was able to predict BER without surpassing Root Mean Square Error (RMSE) limit of 0.1. Wang et al. [6] reviewed various machine learning algorithms available in literature that can manage failure for fault prediction, detection, diagnosis and location in optical layer. D. Lippiatt et al. [7] demonstrated transmitter dispersion eye closure quaternary (TDECQ) estimation and identified impairment simultaneously using convolutional neural networks. Model trained with some impairments like limited bandwidth, signal compression and SNR degradation resulted in Identification Accuracy of 100% for eye closure value TDECQ $>$ 2.6 dB. Xu et al. [8] used transfer learning technique for feed forward and recurrent neural networks on nonlinear equalizer to overcome the problem of varied channel environment impairment on short-reach direct detection systems and accordingly shift to new equalizers depending on channel conditions faster. An optical link of varied distances and bit rates was used as source task and 50Gb/s having 20 km PAM4 optical link is used as target task. Liu et al. [9] proposed equalizer for coherent optical communication system based on transfer learning aided perturbation theory convolutional neural network (PT-CNN) and demonstrated experimentally that the proposed method can accelerate the elasticity of backbone communication network. Xie et al. [10] provided comprehensive survey of machine learning techniques for short-reach optical communications having less than 100 km of coverage i.e., optical access networks, inter- and intra-data centre interconnects, mobile front haul, and in-building and indoor communications that can monitor optical performance, identify modulation format and process signal in-building/indoor. Further, the need was also felt to provide raw data for multi-layer machine learning techniques like CNN (convolution neural network), DNN (Deep Neural Network), SVM, K-Means clustering, learning the k in k means, PCA (Principle Component Analysis) , transfer learning and LSTM (Long Short-Term Memory) [14, 16–22] in some standard format. In this work, optical mesh network was designed for Indian cities and was optimized with the parameters like modulation format, data rate and OSNR for typical connectivity between four cities. The data thus generated can act as source to further develop ML based models for solving problems of prediction, classification or regression in optical networking.

3 Classification of Algorithm based on Optical Data

Machine learning algorithms can be classified as shown in Fig. 1 based on data type input. If the algorithm is working on image data like eye diagrams, constellation diagrams, linear polarized modes, strokes space etc., the algorithm used is convolution neural network. If the algorithm is working on sequential data like digital signal waveform, network traffic data etc. that has some correlation in time, then the algorithm used is Recurrent Neural Network i.e. with memory. If both the image data and sequential data is generated by augmentation, then the algorithm is called as Generative Adversarial Network that works on competition between real data and synthetic data generation.

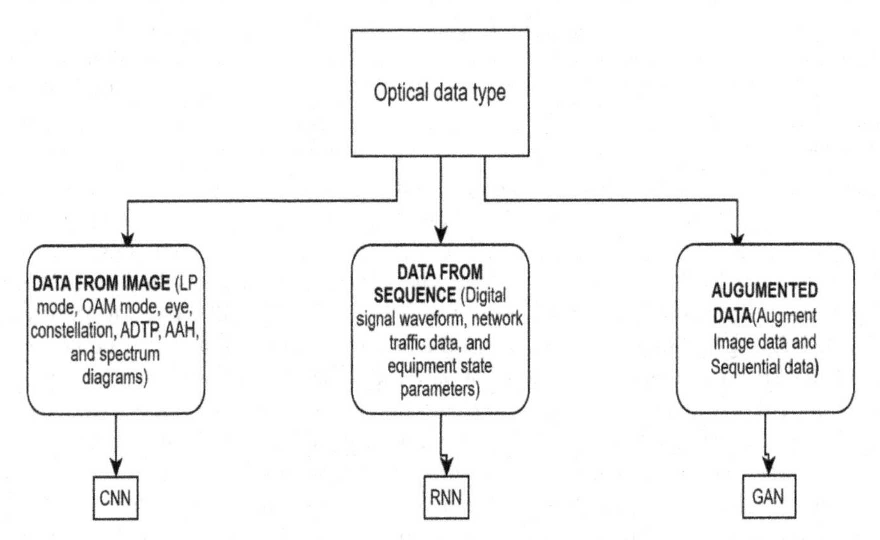

Fig. 1. Optical Data Type for Machine Learning Algorithms [3].

3.1 Feature Extraction from Data in Optical Communication for CNN [Image]

For monitoring the performance of optical system through convolution neural network, it takes data from the signal representation that is given to a network which will extract some features, learn from them and outputs the impairment type and its amount. The image data can be in any form that can give some useful information like eye diagram, constellation diagram, linear polarized mode diagram, optical power extracted either from simulations (commercial optical software) or from experimental data (lab setup, field setup). A lot of images are available in literature for various optical systems for different modulation for-mats and bit rate that diagnose impairments like OSNR, PMD, CD, non-linearity, crosstalk caused because of different reasons from spectrum, strokes-space con-stellation, eye diagrams etc. Figure 2 shows some of such potential image data or features that can act as candidate for CNN Networks for monitoring performance [1, 4, 11].

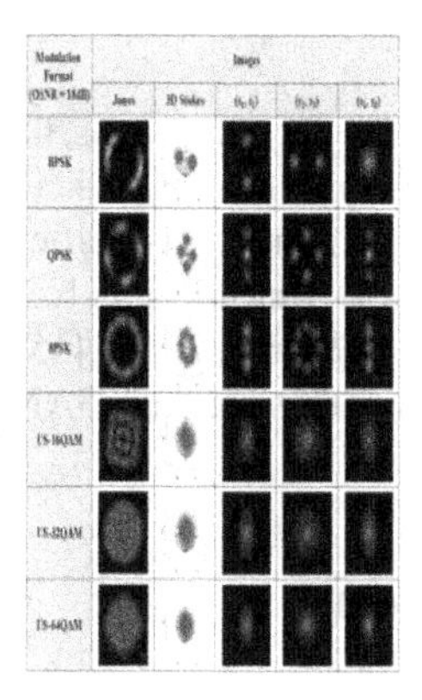

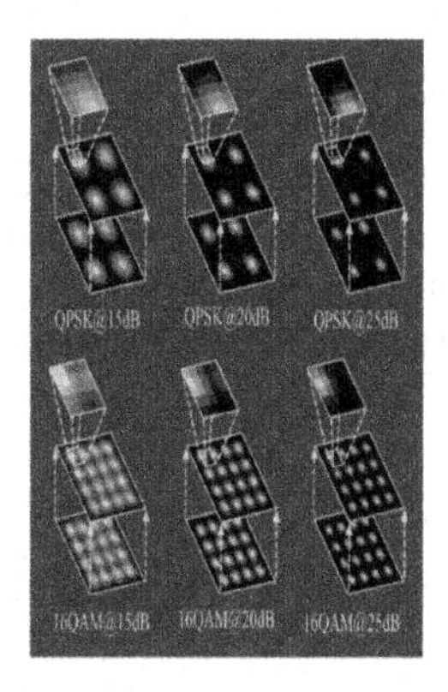

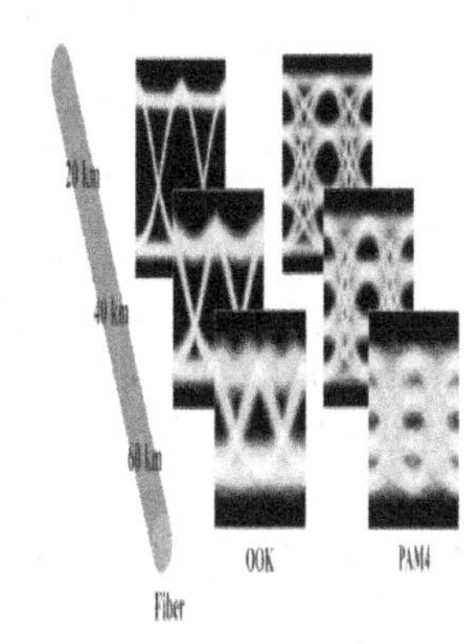

Fig. 2. Potential Image data for Optical Performance Monitoring using CNN a) Images of 20,000 symbols of BPSK, QPSK, 8PSK, US-16QAM, US-32QAM, US-64QAM signals in 2D Stokes planes with their corresponding 3D Stokes space constellation [11] b) The constellation diagram images of QPSK and 16QAM signals at OSNR of 15, 20, 25 dB [1]. c) Fiber transmission distance as reflected by eye diagrams for three OOK and PAM4 signals captured following transmission through 20-, 40-, and 60-km fibers [4].

3.2 Sequential Data for Recurrent Neural Network

Sequential data is a data that has correlation of time at different time scale. It includes data of network traffic, optical and electrical signals, inter-symbol interference (ISI) cancellation, digital signal waveforms etc. To analyze this kind of data RNN are used similar to speech recognition (used by Apple's Siri and Google's voice search.), handwriting recognition etc. These are the neural networks with memory i.e. current output is dependent on past outputs as well as present inputs. RNN network architecture is classified based on number of inputs and outputs as One to Many architecture, Many to One Architecture, Many to Many Architecture (two types as first type is when the input length equals to the output length, second type of many to many architecture is when input length does not equal to the output length) as shown in Fig. 3 [13].

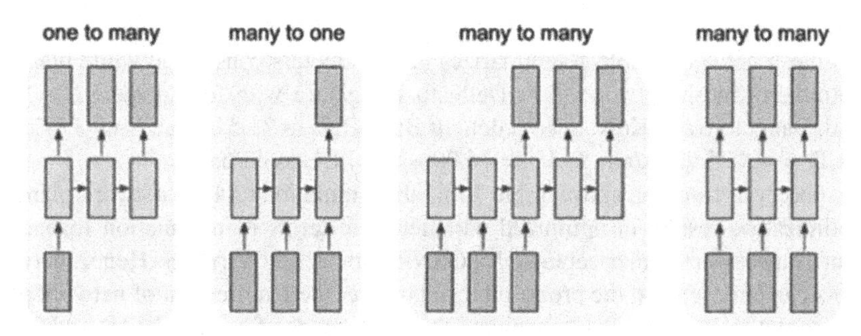

Fig. 3. RNN architectures [13].

3.3 Augmented Data for GAN

As the dataset for machine learning algorithms is not widely available for many areas, the same can be artificially created using Generative Adversarial Network. In GAN, either a new synthetic data is generated or available data is modified by changing its properties. Chen et al. designed variable length Data Augmentation for Optical Transmission Networks under Multi-condition constraint (MVOTNDA) based on GAN architecture [15]. As per our knowledge gained from literature, GAN can be classified as original GAN, LSTM and MVOTNDA.

4 Machine Learning Applications in Optical Systems

Machine learning is a technology that has entered in almost all domains because of its inbuilt advantages. There are various algorithms [13] that can be incorporated in optical communication system for monitoring the performance of networks and managing network traffic. Accordingly, machine leaning algorithms find applications both at the physical layer (Quality of Traffic estimation, Digital twins for optical networks, short reach equalization and Fiber nonlinear noise mitigation) and at the network layer (Traffic prediction and generation, Core network parameter optimization) [12]. Machine learning models can be trained from image data, sequential data or the augmented data. Many applications are consolidated in Fig. 4 for further work.

5 Present Work

In this paper, optical performance is monitored and potential optimized data is generated from public open source gaussian noise python library (GNpy) plat-form for real-world mesh optical networks for Indian scenario. The optical net-work is planned between four metropolitan cities of India. Frequency from 191.30 THz to 196.10 THz with 50GHz spacing is used. EDFA and Reconfigurable Add Drop Multiplexers (ROADM) are used to boost the power level and add or drop the channel as desired between the cities for network planning with de-sired power levels. Chromatic Dispersion (CD), Polarization mode dispersion (PMD), polarization dependent loss (PDL), gaussian noise and non-linear impairments are the impairments optical signal faces while propagating through mesh network. Table 1 summarizes the parameters considered while planning and optimizing mesh network from Delhi to Bangalore with four nodes i.e., Delhi, Mumbai, Bangalore and Kolkata with default Baud Rate as 31.57GBaud and Tx OSNR as 35, CD = 40075.92 ps/nm, PMD = 16.00ps and PDL as 4.36dB.

It is observed from the above Table 1 that the simulation tool for network planning and optimization results in optimized parameters in terms of modulation format for different Transceivers with acceptable Rx OSNR without any warnings. Hence, network operators can choose from the proposed results as per the requirement of network plan. Further, such tool can be used to evaluate the performance for other nodes and routes using different networks i.e., govt networks, field testbeds, lab testbeds etc. It is proposed that the data generated from this tool can be employed to machine learning models as discussed above to add flexibility to optical mesh networks at the backbone/ core network

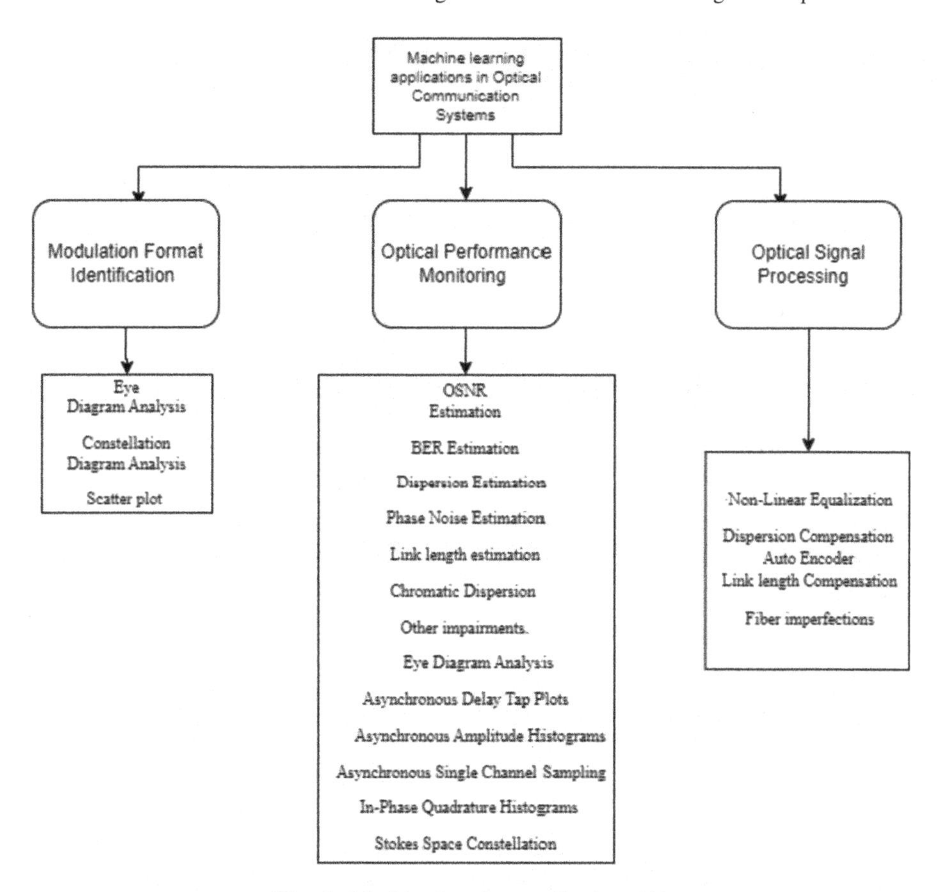

Fig. 4. Machine learning applications [2]

infrastructure in some standard form for classification and regression type of problems. Figure 5 shows the optimized route plan from Delhi to Bangalore with four nodes i.e., Delhi, Mumbai, Bangalore and Kolkata for Indian national network. Similar routes can be planned and optimized results can be achieved using more nodes. Other data like fiber type, distances, optical power levels, optical signal to noise ratio etc. can be used for developing machine learning algorithms for estimating Quality of transmission (QoT) by planning other different type of networks. Fur-ther, deep learning models can be generated and thereby performance can be enhanced depending on the type of data collected as categorical data, sequential data, image data.

Table 1. Parameters considered for mesh network

Delhi to Bangalore (Optical Mesh Network Planning and Optimization)			
Feature 1 Independent variable	Feature 2 Independent variable	Feature 3 Independent variable	Feature Dependent variable
Transceiver parameters	Mode	Modulation format	Rx OSNR (Avg)
CFP2-DCO-T-WDM-HG	100 Gbit/s	DP-QPSK (37.5 GHz)	18.30
OIF 400 ZR	400 Gbit/s,	DP-16QAM (75.0 GHz)	21.02
OPEN ROADM MSA V5.0	200 Gbit/s,	DP-QPSK (87.5 GHz)	21.70
CISCO NCS 1004	200 Gbit/s	DP-QPSK (87.5 GHz)	21.66
Juniper QFX10000	100 Gbit/s,	DP-QPSK (50.0 GHz)	19.48

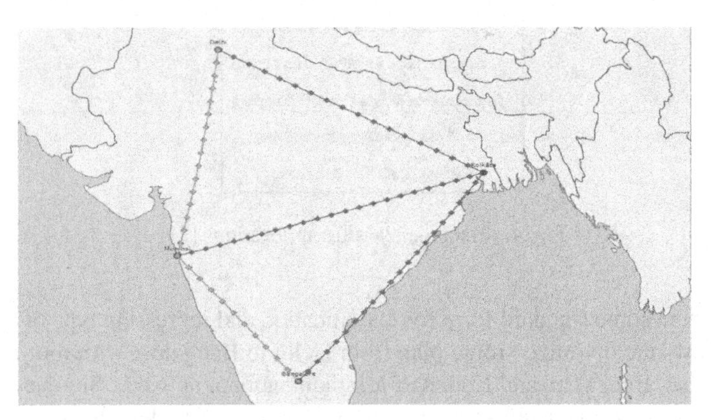

Fig. 5. Optimized route planned from Delhi to Bangalore with four nodes i.e., Delhi, Mumbai, Bangalore and Kolkata shown in marked line with given parameters.

6 Conclusion and Future Scope

In this paper, we have reviewed the work in literature regarding dataset that have been used as input to various machine learning algorithms and potential algorithms for analyzing optical systems and networks. It is observed that in this scientific domain, no such standard dataset is available and the researchers are generating data either through simulations, experimentations or augmentation. In this direction, we have demonstrated optimized parameters of Indian network based on different trans-receivers and estimated the quality of transmission using public open access optimization tool GNpy for optical mesh networks for four nodes for Indian scenario. Similar results can be analyzed with more nodes and networks and the dataset thus generated can be applied in some standard format to machine learning algorithms to make optical networks reconfigurable, scalable and cost effective. Hence, in future, customized ML optical communication-based algorithms can be developed to enhance the capability and scalability of optical systems and networks in real time.

References

1. Wang, D., et al.: Intelligent constellation diagram analyzer using convolutional neural network-based deep learning. Opt. Express **25**(15), 17150–17166 (2017)
2. Tizikara, D.K., et al.: Machine learning-aided optical performance monitoring techniques: a review. Front. Comms. Net. **2**, 756513 (2022). https://doi.org/10.3389/frcmn.2021.756513
3. Wang, D., Zhang, M.: Artificial intelligence in optical communications: from machine learning to deep learning. Front. Commun. Net. **2**, 656786 (2021). https://doi.org/10.3389/frcmn.2021.656786
4. Wang, D., et al.: Comprehensive eye diagram analysis: a transfer learning approach. IEEE Photonics J. **11**(6), 1–19 (2019)
5. Esteves, S., Rebola, J., Santana, P.: Deep learning for BER prediction in optical connections impaired by inter-core crosstalk. In: 13th International Symposium on Communication Systems, Networks and Digital Signal Processing (CSNDSP), pp. 440–445. Porto, Portugal (2022)
6. Wang, D., Wang, D., Zhang, C., Wang, L., Liu, S., Zhang, M.: Machine learning for optical layer failure management. In: Opto-Electronics and Communications Conference (OECC), pp. 1–3. Hong Kong (2021)
7. Lippiatt, D., et al: Impairment identification for PAM-4 transceivers and links using machine learning. In: Optical Fiber Communications Conference and Exhibition (OFC), pp. 1–3. San Francisco, CA, USA (2021)
8. Xu, Z., Sun, C., Ji, T., Manton, J.H., Shieh, W.: Feedforward and recurrent neural network-based transfer learning for nonlinear equalization in short-reach optical links. J. Lightwave Technol. **39**(2), 475–480 (2021)
9. Liu, J., Wang, Y., Yang, H., Huang, X., Zhang, Q., Tian, Q.: Transfer learning aided PT-CNN in coherent optical communication systems. In: 20th International Conference on Optical Communications and Networks (ICOCN), pp. 1–3. Shenzhen, China (2022)
10. Xie, Y., Wang, Y., Kandeepan, S., Wang, K.: Machine learning applications for short reach optical communication. Photonics **9**(1), 30 (2022)
11. Zhang, W., Zhu, D., Zhang, N., Xu, H., Zhang, X., Zhang, H., et al.: Identifying probabilistically shaped modulation formats through 2d stokes planes with two-stage deep neural networks. IEEE Access **8**, 6742–6750 (2020)
12. Nevin, J.W., Nallaperuma, S., Shevchenko, N.A., Li, X., Faruk, M.S., Savory, S.J.: Machine learning for optical fiber communication systems: an introduction and overview. APL Photonics **6**, 121101 (2021). https://doi.org/10.1063/5.0070838
13. https://towardsdatascience.com/sequence-models-and-recurrent-neural-networks-rnns-62cadeb4f1e1. 27 Jul 2020
14. https://calvinfeng.gitbook.io/machine-learning-notebook (2013)
15. Chen, L., et al.: Data augmentation algorithm based on generative antagonism networks (GAN) model for optical transmission networks (OTN). In: Tallón-Ballesteros, A.J. (ed.) Proceedings of CECNet 2021: The 11th International Conference on Electronics, Communications and Networks (CECNet), November 18–21, 2021. IOS Press (2021). https://doi.org/10.3233/FAIA210453

16. Boser, B.E., Guyon, I.M., Vapnik, V.N.: A training algorithm for optimal margin classifiers. In: Proceedings of the fifth annual workshop on Computational learning theory, pp. 144–152 (1992)
17. Cortes, C., Vapnik, V.: Support-vector networks. Mach. Learn. **20**, 273–297 (1995)
18. Cover, T., Hart, P.: Nearest neighbor pattern classification. IEEE Trans. Inform. Theory **13**(1), 21–27 (1967). https://doi.org/10.1109/TIT.1967.1053964
19. Hamerly, G., Elkan, C.: Learning the k in k-means. Advances in Neural Information Processing Systems, vol. 16 (2003)
20. Olliffe, I.T.: Principal Component Analysis. Springer Series in Statistics, 2nd edn. Springer-Verlag, New York Inc (2002)
21. O'Shea, K., Nash, R.: An Introduction to Convolutional Neural Networks. CoRR, abs/1511.08458 (2015)
22. Torrey, L., Shavlik, J.: Transfer Learning. Handbook of Research on Machine Learning Applications and Trends: Algorithms, Methods, and Techniques, pp. 242–264. IGI global, Hershey (2010)

Antenna Array Fault Detection Using Logistic Regression Technique

Atul M. Kulkarni(✉), Garima Saini, and Shyam S. Pattnaik

National Institute of Technical Teacher Training and Research, Chandigarh, India
amkulkarni.ece22@nitttrchd.ac.in

Abstract. Array antenna is widely used in 5G wireless networks. The main issue with the array antenna is that, failure or fault in one or more elements disturbs its radiation pattern, and its directivity pattern gets disturbed. Fault in elements of the array antenna enhances the side lobe levels; To tackle this kind of issue, pattern regeneration techniques can be utilized, but for that, the detection of the faulty element is needed. Hence to cater this kind of problem exemplary machine learning technique, i.e., logistic regression classifier, is used in this paper. The eight-element planar antenna is simulated using the Ansys-HFSS tool. To build training and testing datasets, a discontinuity is formed in the array's feed network to simulate various fault conditions. The array is designed using RT Duriod 5880 substrate with relative permittivity 2.2 and thickness 1.6 mm; the simulated result shows a high gain of 12 dB and S_{11} of −32 dB for 3.67 GHz frequency. Before applying a logistic regression machine learning algorithm for fault detection in an antenna array, a review of various techniques applied by researchers is carried out. The logistic regression multiclass model with a liblinear solver obtained 95% accuracy over 105 test samples.

Keywords: Antenna Array · Fault Detection · Logistic Regression · Machine Learning · One-vs-rest (OVR)

1 Introduction

Nowadays, smart and machine learning is become a buzzword in the engineering. In the case of 5G or next-generation wireless communication networks, smart antenna arrays play an important role, as apparent in the literature [1–5]. In the wireless network, that may be deployed indoors, outdoors, in industrial, medical, or vehicular applications, array antennas are used because of their high gain and bandwidth feature. Addressing beamforming effectively happens if the antenna array is faultless. Fault in an antenna element disturbs the radiation pattern with respect to the main lobe, side lobe, back lobe, and nulls, and to tackle this issue, various reshaping techniques are reported in the literature [6–8]. As discussed, a fault in the antenna array disturbs the radiation pattern. So, reshaping techniques can be used to reshape the disturbed radiation pattern, but to use these techniques location of the fault needs to be detected first. Hence, various linear and planar antenna array fault diagnosis techniques using the heuristics approach as well as machine learning techniques reported in the literature.

R. K. Challa et al. (Eds.): ICAIoT 2023, CCIS 1930, pp. 13–29, 2024.
https://doi.org/10.1007/978-3-031-48781-1_2

Logistic regression is an exemplary algorithm in machine learning, primarily used in the detection, classification, and segmentation applications. Some of the selected literature related to this are mentioned next. Tadelo et al. used logistic regression ML for predicting disease in tomato plants with respect to the association of climate change [9]. Vincent F et al. showed the application of logistic regression ML to detect polyps in computed tomography (CT) colonography [10]. Harikrishnan et al. reported the application of logistic regression ML to detect disease on dry beans [11]. Poreba et al., and Del Rosso et al., both literature presented the application of the logistic regression ML model to detect a fault in the motor by using a vibrational acoustic signal [12, 13]. Landstrom et al. presented the use of logistic regression ML for morphology-based detection of cracks in the steel slabs. Wthout human intervention, non-contact identification of cracks reported in this literature [14]. Khurshid et al. reported the segmentation and categorical prediction of multi-temporal multi-spectral SPOT 5 satellite images using logistic regression [15]. Recently Sanchez et al. presented a logistic regression model combined with feature extraction methodology for identifying phishing sites or URL [16]. Considering these diverse domain applications of logistic regression, the author is motivated to use the same for fault diagnosis of the antenna array.

In this paper, the method is formulated to diagnose the fault in an antenna array. A single or many element failure causes the array's radiation pattern to deviate. This encourages the development of a fault-finding strategy. The suggested technique uses a shift in the radiation pattern to find defects. It is practical to get the far-field radiation pattern without disconnecting or moving the antenna from its deployed location or significantly altering its regular operating circumstances. It is simple to identify problems in applications like, satellite-borne antennas by measuring the far-field radiation pattern. This paper provides a review of the fault detection techniques in an antenna array. Also, it provides experimental results of the application of logistic regression machine learning (ML) to detect a fault in a planar eight-element antenna array. This paper also discusses the feasibility of the logistic regression ML model for real-time fault detection.

The remaining structure of this paper is outlined as follows; Sect. 2 provides a literature survey, Sect. 3 discusses the problem statement and methodology flow, Sect. 4 discuss design of antenna array and train-test data preparation, Sect. 5 presents logistics regression ML technique and implementation, Sect. 6 provides results and discussion, and Sect. 7 concludes the work.

2 Literature Survey

Peters et al. presented a methodology to correct the shape of radiation pattern in existence of failed element in an antenna array. The suggested approach helps to reduce the side lobe levels that happened due to failure of element in an antenna array. A conjugate gradient based algorithm is suggested to recover the radiation pattern of the antenna array when element failure arises, where prediction of element failure needs to be accurate [6]. Appasani et al. presented pseudo measurement technique so as to detect and correct the errors occurred due to element failure in linear antenna arrays. The radiation features of the single antenna elements and the array factor are utilized to mathematically formulate errors in an element for a particular radiation pattern. Author used MATLAB for the simulations [7].

Guoliang Zheng et al. presented a deep learning based deep residual shrinkage network (DRSN) methodology to dignose fault in antenna array. The reported methadology showed improved accuracy considering presence of noise in the signal. Author also compared the DRSN with the examplarary ML technique i.e. SVM. Prediction accuracy of the DSRN and SVM is compared for noisy input radiation pattern data also [17]. Grewal et al. proposed a methodology to correct or resahpe radiation patterns, when element failure occurs. The correction of radiation pattern is achieved by adjusting the excitation. The comparative performance of the proposed method i.e. improved BAT algorithm with other soft computing techniques is reported [8].

Balamati et al. utilized bacteria foraging optimization (BFO) to locate the fault in an antenna array. The proposed method is tested on a 24-element linear broadside Chebyshev array. The linear antenna array inter element spacing is considered as $\lambda/2$. The developed methodology used to find different types of fault situations like combination of partial and complete fault, complete single fault, and more than one fault [18]. Shafqat et al. discussed the cuckoo search soft computing technique to identification of failure element in a array antenna. A linear array (Classical Dolph Chebyshev) of fourty elements used while implementing the antenna array [19]. Mukherjee et al. used a GA to detect failed element in 8x8 planar array. The method was applied for planar arrays with two or more than two elements failed. The number of iterations used clearly notifies the computationally intensive nature of this algorithm [20]. Khan et al. demonstrated how to use the Firefly algorithm to find an issue with a linear antenna array.

The suggested approach is effectively employed for both the partial and total faulty element position detection. 34 element Chebyshev array is referred detect the fault using firefly algorithm [21]. Grewal et al. proposed BAT algorithm to detect faulty element in a symmetric linear antenna array of 32 elements. BAT algorithm has been compared with other soft computing techniques for confirmation of the performance of the proposed method [22]. Acharya et al. presented a soft computing technique to detect fault in an antenna array. Author reported a particle swarm optimization (PSO) based method to identify the whole or partial faulty element for 10 elements linear antenna arrays [23]. Amalendu et al. developed artificial neural networks (ANN) to locate fault finding in 16 elements linear microstrip antenna array. Author used IE3D tool to simulate the antenna array, train test dataset is generated with respect to creation of various faults, by means of introducing the discontinuity in the feed network. The demerit of various soft computing techniques utilised in past to detect fault in antenna array is reported by author. Hence, to avoid the computational load because of iterative process used in soft computing technique, ANN is proposed in this literature. The author also reported that the same methodology can be applied to the planar array [24]. Vakula et al. used ANN to detect a fault in a planar array with 5×5 and 8×8 isotropic antenna elements with identical excitation and equal distance between consecutive elements. Varients of ANN are compared and analysed. It has been observed that, PNN network performance is better than RBF network. Performance measures of ML model like train time, test time and peak memory usage is not reported in this paper [25].

Srikanth et al. showed that support vector machine (SVM) prediction for defective elements in a planar array (5 × 5) is better as compared to linear array (11 elements). Measured powers for varied Signal to Noise Ratios made up the training vectors. Less than 5 dB of signal-to-noise ratio produced detection accuracy that was only about 30%, whereas more than 10 dB of SNR produced accuracy of greater than 90%. The author used variants of SVM classifier just by changing the kernels, but not used the optimized SVM [26]. Nan Xu et al. referred SVM multiclass classifier as a machine learning tool to diagnose faulty elements in a 4-element dipole antenna array. The performance of classifier is compared over different SNRs using RBF and polynomial kernel, but SVM is not optimized using any soft computing algorithm. The model's suitability for real time applications is reported but the performance measures of ML model like train time, test time and peak memory usage is not reported in this paper [27].

Alzubaidi et al. presented a review of deep learning concepts; the author stated that ML techniques utilized by many researchers to detect fault require pre-processing of the data, whereas the introduction of deep learning for various fields in the past decade reduced human supervision subsequently. Among the different deep learning methods, CNN is the maximum popular and largely employed algorithm where the convolutional layer extracts the features of data [28]. Sunita rani et al. proposed a virtual instrument for failure detection in antenna array. Four element linear antenna array is designed and radiation pattern data set corresponding to the introduction of fault situations is created. Fault are created by introducing the discontinuity in the feed network. Any machine learning, model should be trained and test, with balanced dataset to confirm the generalization of the model. Whereas use of stratified sampling for splitting the train test is not mentioned and total sample set is also 93. Hence, probability that proposed model is biased is more. The best thing about this paper is that Matlab implemented ANN model is embedded with Lab-VIEW to get visualization of fault in antenna array if any based on aforementioned literature review, the research gaps are identified as below.

There are some highly efficient and popular ML/DL techniques which, are not yet utilized to address the fault detection in antenna array. Various soft computing, and ML methodologies for detection of fault in antenna array are reported in literature, however approach to optimize the architectural parameter of machine learning model using soft computing techniques to diagnose fault in antenna array is not reported. The literature of fault finding in antenna array using AI/ML techniques lacks deployment approach of the model. Hence, to make array antenna adaptive and reconfigurable or to automatically reshape the disturbed radiation pattern (because of element failure) in antenna array there is a need to detect and locate fault in antenna array using radiation pattern far field data, hence efficient AI/ML techniques needs to be developed to handle large antenna array. Figure 1 reflects the various AI/ML techniques reported for fault detection of antenna array in literature.

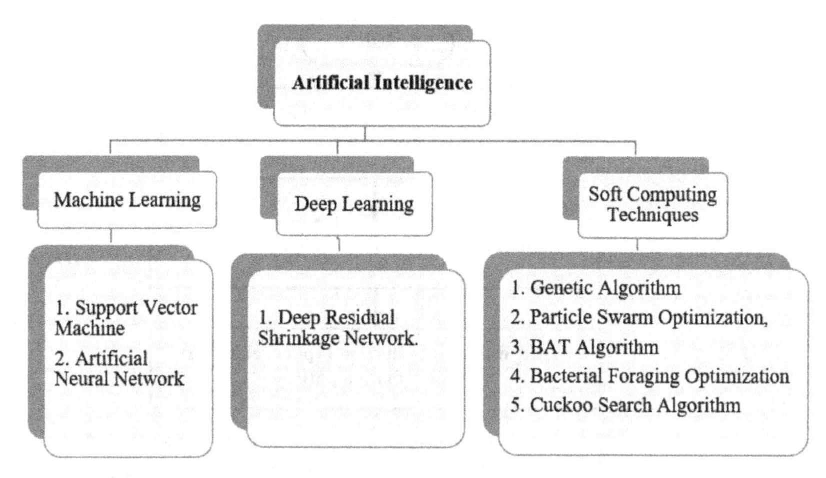

Fig. 1. Various types of AI techniques reported in literature for fault detection in antenna array.

3 Problem Statement and Methodology Flow

Earlier array antennas are used in high end applications, like satellite, military RADAR, traffic control and collision avoidance system (TCAS) and astronomical observations. Recently, noticing the vast use of array antenna in other domains of wireless networks, raised the attention of addressing issues related to array antenna. The main issue of the array antenna is disturbed radiation pattern due to fault in one or many elements of an array. To reshape or regenerate the pattern in the existence of defective element in the antenna array, the essential requirement is to locate the fault. Hence, this paper presents logistics regression ML technique-based detection of failure of an antenna element. The methodology flow to attain the problem statement is depicted in Fig. 2.

4 Design of Antenna Array and Train Test Dataset Preparation

A planar array of 8 elements is designed and simulated in the Ansys HFSS tool. The design equations of the patch and corporate feed network i.e. from Eq. (1) to Eq. (10) are referred from [29, 30]. The antenna is designed using RT/Duriod −5880 substrate having relative permittivity 2.2, and a thickness of 1.6 mm. The design frequency of the array antenna is 3.67 GHz. The design steps of microstrip array antenna with design equations are further mentioned.

The width of patch (W) is calculated from Eq. (1).

$$W = \frac{C}{2f_r}\sqrt{\frac{2}{\varepsilon_r + 1}} \tag{1}$$

where W is patch width, f_r is resonant frequency, C light velocity, and the ε_r is the dielectric constant of substrate. Now the effective dielectric constant is calculated from

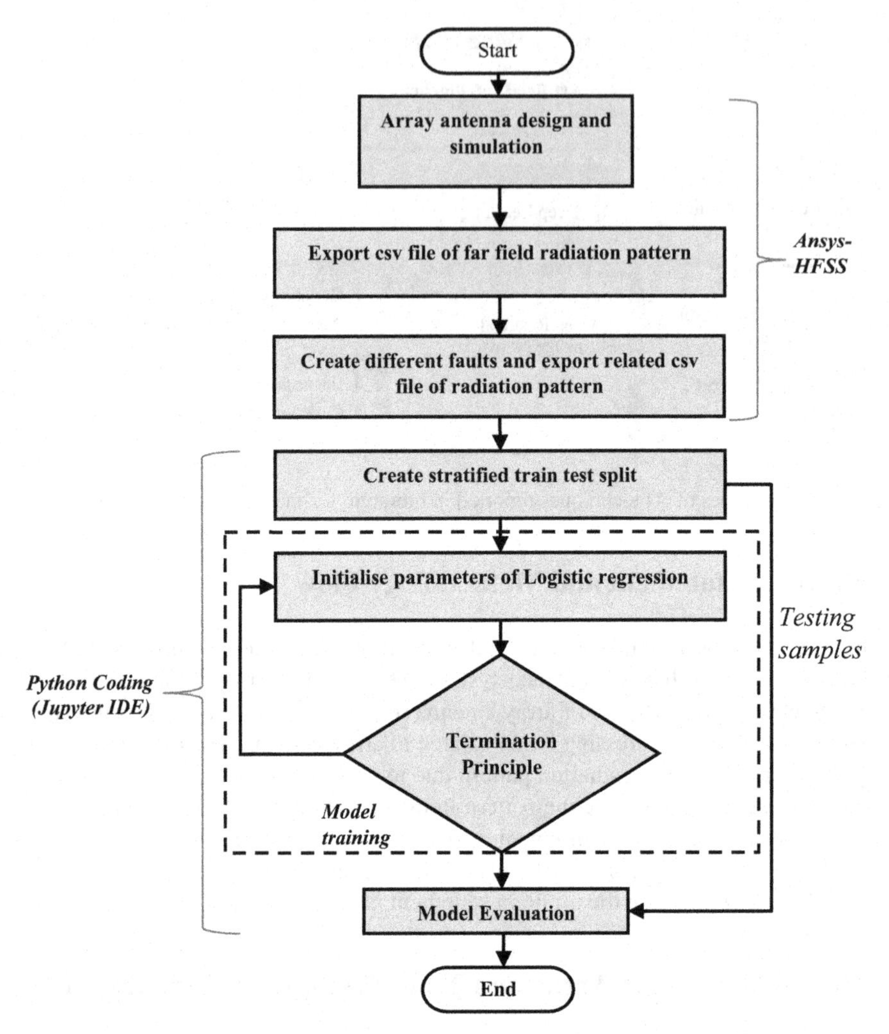

Fig. 2. Methodology flow to detect faulty element in an antenna array.

Eq. (2).

$$\varepsilon_{reff} = \frac{\varepsilon_r + 1}{2} + \frac{\varepsilon_r - 1}{2}\left(1 + 12\frac{h}{w}\right)^{-\frac{1}{2}} \tag{2}$$

Here, h is thickness of substrate and effective dielectric constant is ε_{reff}.

The length of the patch is calculated from Eq. (3).

$$L = \frac{C}{2f_r\sqrt{\varepsilon_{reff}}} - 2\nabla L \tag{3}$$

where, ∇L is the extension of length due to fringing, calculated using Eq. (4).

$$\nabla L = 0.412h\frac{(\varepsilon_{reff} + 3)(\frac{W}{h} + 0.264)}{(\varepsilon_{reff} - 0.258)(\frac{W}{h} + 0.8)} \tag{4}$$

Now, the determination of inset feed position is done by calculating the edge resistance (R_{in}) of the patch, which is given by Eq. (5).

$$R_{in} = \frac{1}{2(G_1 \pm G_{12})} \tag{5}$$

where, G_1 is input edge conductance of the single slot, and G_{12} is the mutual conductance between the two slots, which is given by Eq. (6) and (7) respectively.

$$G_1 = \frac{I_1}{120\pi^2} \tag{6}$$

$$G_{12} = \frac{1}{120\pi^2}\int_0^{\pi}\left[\frac{sin\left(\frac{K_0W}{2}Cos\theta\right)}{Cos\theta}\right]^2 J_0(K_0LSin\theta)Sin^3\theta d\theta \tag{7}$$

where, J_0 is a Bessel function of the first kind of order zero, and I_1 is the integral defined by Eq. (8).

$$I_1 = \int_0^{\pi}\left[\frac{sin\left(\frac{K_0W}{2}Cos\theta\right)}{Cos\theta}\right]^2 Sin^3\theta d\theta \tag{8}$$

Equation (8) is further simplified as,

$$I_1 = -2 + cos(X) + S_i(X) + \frac{sin(X)}{X} \tag{9}$$

where, $X = K_0W$, K_0 is a free space wave propagation constant, W is width of patch, and S_i is sine integral.

Now the location of the inset-feed from edge of the rectangular patch is determined by Eq. (10).

$$R_{in}(y = y_0) = R_{in}(y = 0)Cos^2\left(\frac{\pi}{L}y_0\right) \tag{10}$$

where, $R_{in}(y = y_0)$ is the impedance of the feed point, which is generally 50Ω, $R_{in}(y = 0)$ is edge impedance of the patch. y_0 is the position of inset feed point or inset feed depth which is given by Eq. (11).

$$y_0 = \frac{L}{\pi}Cos^{-1}\left[\frac{R_{in}(y = y_0)}{R_{in}(y = 0)}\right]^{\frac{1}{2}} \tag{11}$$

Now, the inset feed gap is calculated by using Eq. (12) given in [31].

$$N_g = \frac{4.65 \times 10^{-18}cf_r}{\sqrt{2\varepsilon_{reff}}} \tag{12}$$

where, N_g is a notch gap or inset feed gap. Hence, single inset feed patch element of array antenna is designed using Eqs. (1), (2), (3), (4), (5), (6), (7), (8), (9), (10), (11) and (12). Further corporate feed network is designed using microstrip lines having impedances 50 Ω, 100 Ω and 70.7 Ω. The width of the microstrip line for different impedance value is calculated from Eq. (13).

$$Z_0 = \begin{cases} \frac{60}{\sqrt{\varepsilon_{reff}}}\left[\frac{8h}{W_f} + \frac{W_f}{4h}\right] & for\ {}^{W_f}/_{h} \leq 1 \\ \frac{120\pi}{\sqrt{\varepsilon_{reff}}\left[1.393+\frac{W_f}{h}+\frac{2}{4}ln\left(\frac{W_f}{h}+1.444\right)\right]} & for\ {}^{W_f}/_{h} > 1 \end{cases} \tag{13}$$

where, Z_0 is the characteristic impedance of the microstrip line. $\frac{W_f}{h}$ ratio is determined by Eq. (14) [30].

$$\frac{W_f}{h} = \frac{8e^A}{e^{2A} - 2} \quad for\ {}^{W_f}/_{h} < 2 \tag{14}$$

and, $for\,{}^{W_f}/_{h} > 2$

$$\frac{W_f}{h} = \frac{2}{h}\left[B - 1 - ln(2B - 1) + \frac{\varepsilon_r - 1}{2\varepsilon_r}\left\{ln(B - 1) + 0.39 - \frac{0.61}{\varepsilon_R}\right\}\right]$$

where, A and B is given by Eq. (15) and (16) respectively.

$$A = \frac{Z_0}{60}\sqrt{\frac{\varepsilon_r + 1}{2}} + \frac{\varepsilon_r - 1}{\varepsilon_r + 1}\left(0.23 + \frac{0.11}{\varepsilon_r}\right) \tag{15}$$

$$B = \frac{377\pi}{2Z_0\sqrt{\varepsilon_r}} \tag{16}$$

Simple power divider network is used to feed the array with 50 Ω impedance of excitation feed. The quarter wave transformer is used to match the impedance of 100 Ω line with 50 Ω microstrip line of the corporate feed network. The designed array antenna is shown in Fig. 3.

The effect of the introduction of fault in S11, voltage standing wave ratio (VSWR), and radiation pattern can be observed in Figs. 4, 5, and 6, respectively.

Enhancement of side-lobes and alteration or shifting of the main-lobe, with reduced gain can be observed from the radiation pattern plot when a fault is introduced at element 2. Variation in S_{11} and VSWR due to fault is reflected in Figs. 4 and 5, respectively.

By creating the discontinuity near every patch, the fault is created, and after simulation csv file of the radiation pattern is exported to plot the radiation pattern. By varying the theta steps, 39 samples of the radiation pattern are created. Hence, total 351 number of the samples in the dataset is created then a balanced split is performed to create a training and testing set. 70% samples are considered for training the model, while the remaining 30% are considered to evaluate the model.

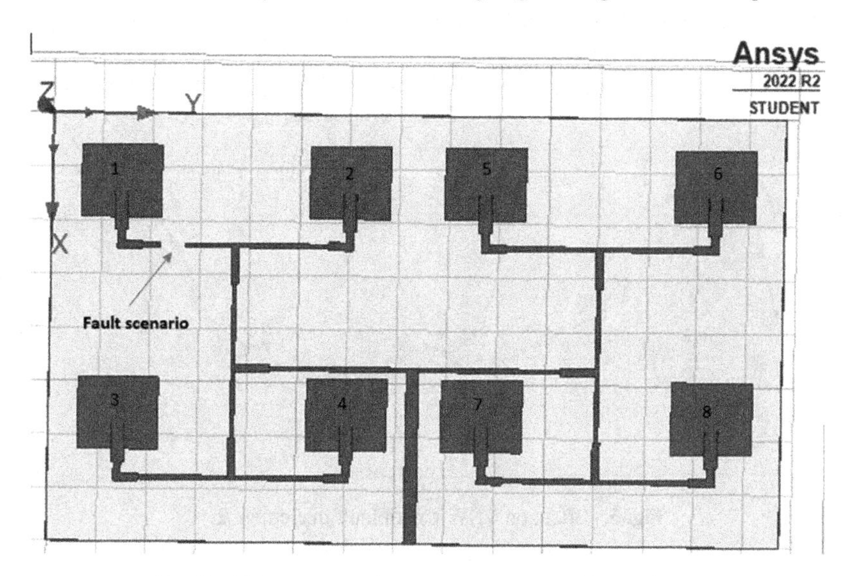

Fig. 3. Eight element planar array antenna with one fault.

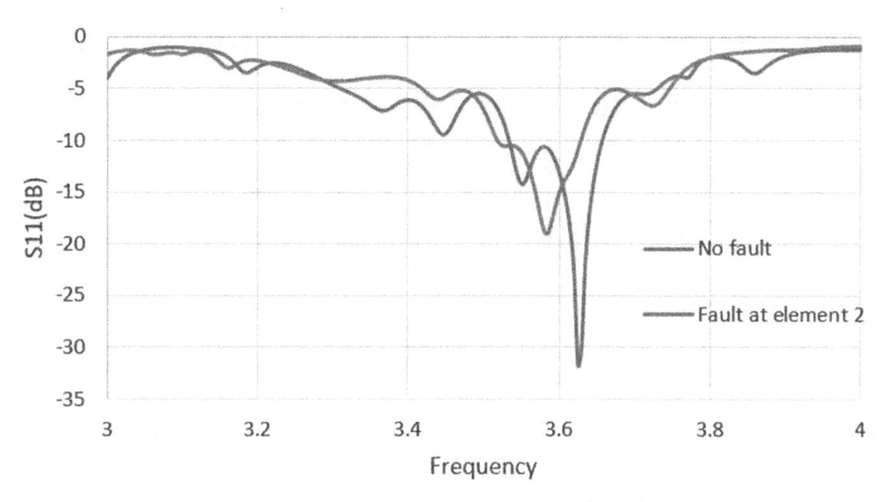

Fig. 4. Effect on S_{11} with fault in element 2.

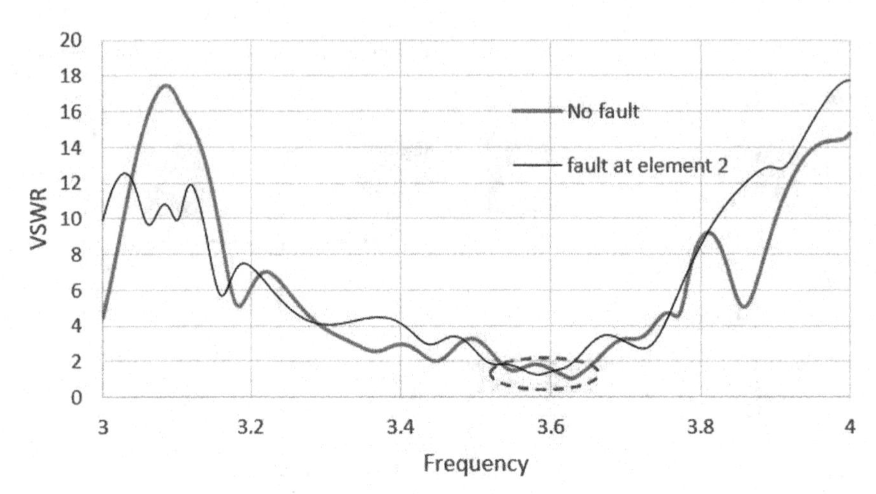

Fig. 5. Effect on VSWR with fault in element 2.

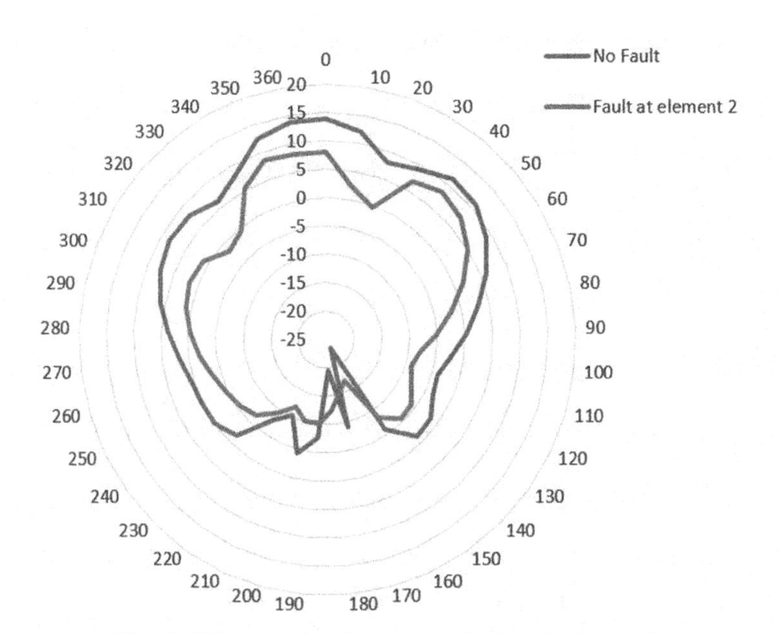

Fig. 6. Effect on Radiation pattern with fault in element 2.

5 Logistic Regression ML Technique and Implementation

Logistic regression is basically applied for binary classification problem statement. In linear regression by using single best fit line it can attain binary classification but error rate increases drastically just by inclusion of single outlier. To tackle this issue sigmoid function is used to discriminate between data points or classes instead best fit line. The cost function is given by Eq. (17) [32, 33].

$$cost\ function = J(m) = max \sum_{i=1}^{n} f\left(y_i \times w_i^T x_i\right) \tag{17}$$

where x_i are data points and y_i classification labels + 1 or -1 and the product $w_i^T x_i$ represents distance between data point and best fit line, and f is considered to be sigmoid function. i.e. multiplication of $y_i \times w_i^T x_i$ is passed through the sigmoid function. The sigmoid function is given as $\frac{1}{1+e^{-z}}$ hence, $z = y_i \times w_i^T x_i$ and as sigmoid function transforms value between 0 to 1 which helps to reduce the impact of outlier. Here optimizer task is to update the w_i^T such that it will maximize the cost function.

In case of multi-categorical problem statement logistic regression is extended and is refereed as one vs rest (OVR). One vs. all algorithm is another name for this. According to the name of the algorithm, it picks one class and place all other classes into a second virtual class before applying binary logistic regression to that class. This process is repeated for each class in the dataset. In the end, binary classifiers have been used that can identify each class in the dataset. The problem under consideration is detection of faulty element from 8 element antenna array, which is considered here as a multi class predictor problem. So, in this case of OVR, for 9 categories 9 models gets created in each iteration while training (No fault, element 1 fault......element 8 fault hence, total 9). While evaluating the model it takes input test features and provides 9 probabilities related to 9 models. Test data belongs to that category; which model has highest probability. The OVR is implemented using SciKit learn library of python.

The flow of the experimentation for implementing tuned multiclass logistic regression model i.e. OVR using cross validation technique is represented in Fig. 7. The solvers liblinear, lbfgs, newton-cg, sag, and saga are used while cross validating the model. The random state parameter used while splitting the train test data set is varied every time, to ensure the change of sample subset in train-test dataset. Termination count for cross validation is considered as five, i.e. five times stratified split is performed. The best solver is determined and then final logistic regression model is created. The trained model is saved in h5 file format to use the same on any machine afterword's.

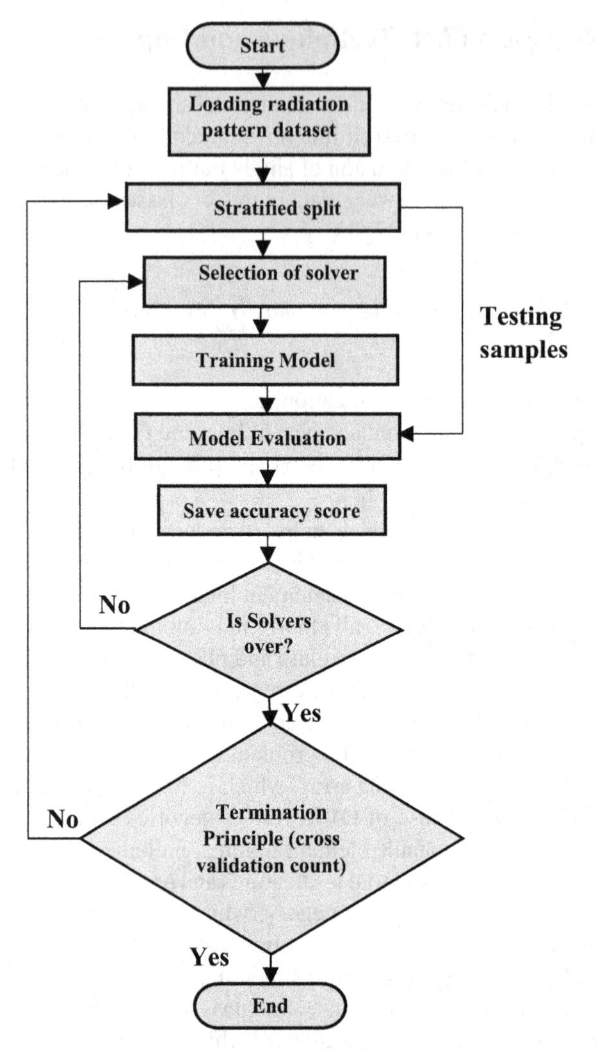

Fig. 7. Implementation flow of OVR Model.

6 Results and Discussion

One vs. rest framework of logistic regression classifier is used to multi-class the radiation pattern categories corresponding to the faults. L2 regularization, also known as ridge regression, amends the loss function by including the "squared magnitude" of the coefficient as a penalty term to improve the model's generalization. The various solvers.

used in this experimentation are liblinear, lbfgs, newton-cg, sag, and saga. All the solvers support L2 regularization. 5 times, cross-validation is performed to select the best solver for the OVR model. Table 1 shows the accuracy score for various solvers. The precision, recall, and F1 score for the optimized model are shown in Table 2. The confusion matrix for the optimized model is then plotted, as shown in Fig. 8.

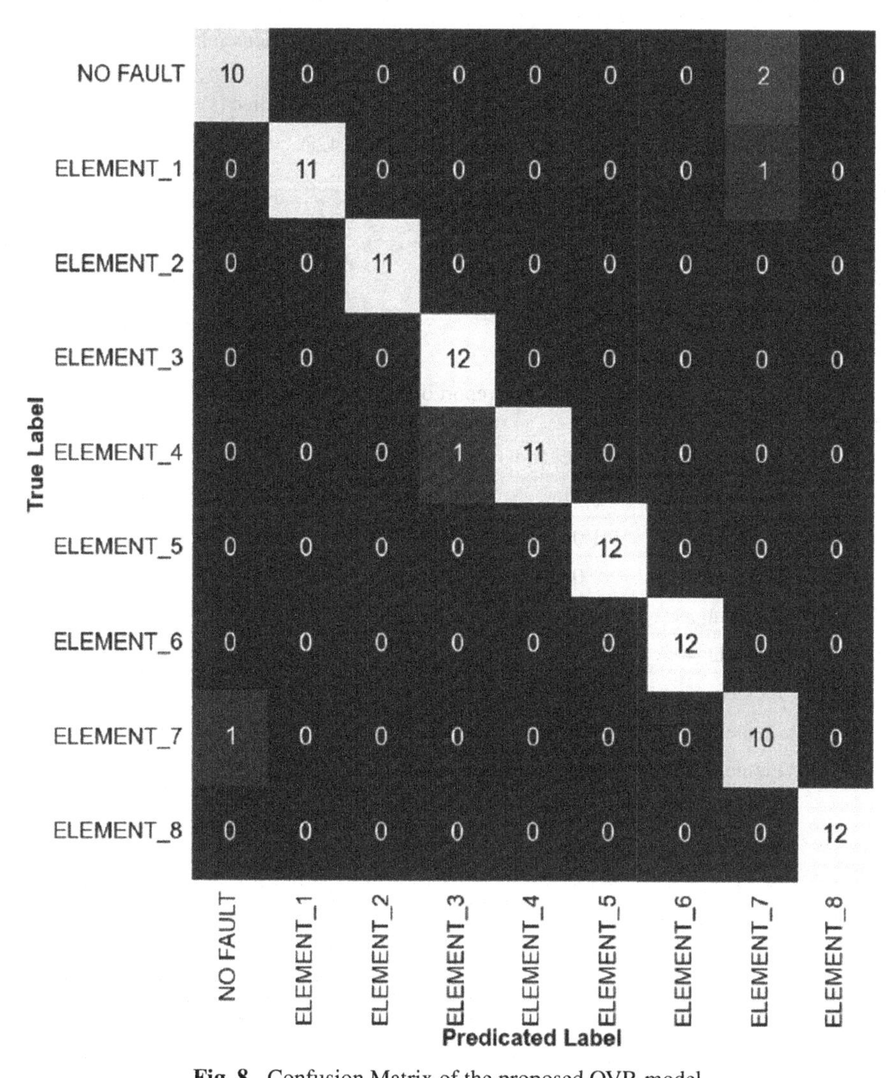

Fig. 8. Confusion Matrix of the proposed OVR model.

In the confusion matrix, the notations used for eight elements of 2×4 array are Element_1, Element_2, Element_3, and so on till Element_8 as shown in Fig. 8. The output of the trained model is made available on local machine, on which the model is deployed. The test sample fault identification report visualization appears on the display as shown in Fig. 9. Figure 9 (a) shows the input test sample given to the detection model, and corresponding fault detected is shown in Fig. 9 (b). Blue filled circle is appeared on 6, indicates sixth element is faulty. The conventions used in the plot is; 0 on the x axis is the indication for no fault, 1 for element 1 fault and so on. The visualization of the output of model can be modified as per the requirement.

Table 1. Performance metrics of OVR model for various solvers

Solver	Accuracy score(%)	Training time (s)
liblinear	95.4	0.428
lbfgs	95.4	0.568
newton-cg	95.4	0.956
sag	92	7
saga	90	8

Table 2. Prediction report of tuned OVR model

Fault	Precision	Recall	F1 Score
No Fault	0.91	0.83	0.87
Element_1	1.00	0.92	0.96
Element_2	1.00	1.00	1.00
Element_3	0.92	1.00	0.96
Element_4	1.00	0.92	0.96
Element_5	1.00	1.00	1.00
Element_6	1.00	1.00	1.00
Element_7	0.77	0.91	0.83
Element_8	1.00	1.00	1.00

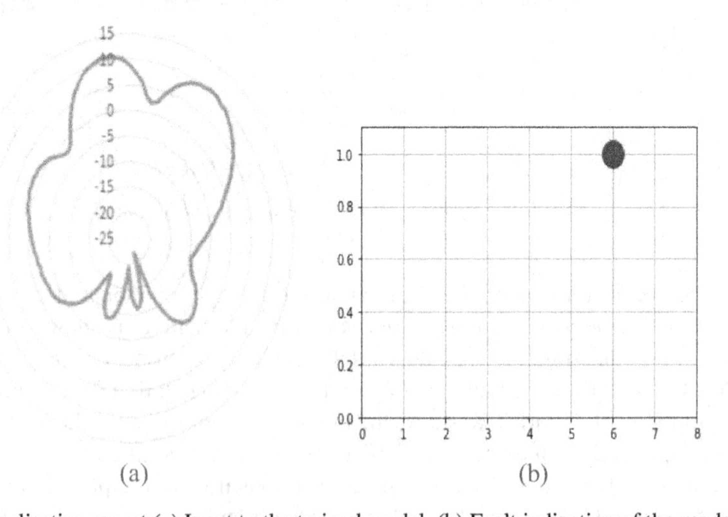

Fig. 9. Visualisation report (a) Input to the trained model, (b) Fault indication of the model.

It can be observed that liblinear solver outperforms the other solvers. The accuracy rate for the tuned OVR model is 95%, and the time required to train the model is 0.4 s. It may be noted that as the number of elements of the antenna increases, the train-test time required for the model will increase, and other performance metrics need to be evaluated. As the time required to test the sample is in microseconds, the above model can be used in real-time applications. Upcoming work will be to use ensemble learning to improvise the performance-metrics of the model.

7 Conclusion

The performance of the OVR framework of the logistic regression machine learning technique is evaluated in this paper for the identification of failure elements in an antenna array. Various solvers' performance has been evaluated using a cross-validation technique. Liblinear solver performance is better as compared to other solvers. For more than 8 element antenna array fault detection performance of OVR needs to be evaluated. The accuracy of the proposed tuned OVR model is 95% over the 105 test samples.

References

1. Valkonen, R.: Compact 28-GHz phased array antenna for 5G access. In: IEEE MTT-S International Microwave Symposium Digital, vol. June, pp. 1334–1337 (2018)
2. Diawuo, H.A., Jung, Y.B.: Broadband proximity-coupled Microstrip planar antenna array for 5G cellular applications. IEEE Antennas Wirel. Propag. Lett. **17**(7), 1286–1290 (2018)
3. Li, Y., Wang, C., Yuan, H., Liu, N., Zhao, H., Li, X.: A 5G MIMO antenna manufactured by 3-D printing method. IEEE Antennas Wirel. Propag. Lett. **16**, 657–660 (2017)
4. Hong, W., et al.: The role of millimeter-wave technologies in 5G/6G wireless communications. IEEE J. Microw. **1**(1), 101–122 (2021)
5. Kumar, S., Dixit, A.S., Malekar, R.R., Raut, H.D., Shevada, L.K.: Fifth generation antennas: a comprehensive review of design and performance enhancement techniques. IEEE Access **8**, 163568–163593 (2020)
6. Peters, T.J.: A conjugate gradient-based algorithm to minimize the sidelobe level of planar arrays with element failures. In: International Symposium on Antennas and Propagation Society, Merging Technologies for the 90's, Dallas, TX, USA, vol. 2, pp. 848–851 (1990)
7. Appasani, B., Pelluri, R.: Detection and correction of errors in linear antenna arrays. Int. J. Numer. Model.: Electro. Netw. Devices Fields **31**(5), 1–12 (2018)
8. Singh Grewal, N., Rattan, M., Singh Patterh, M.: A linear antenna array failure correction using improved bat algorithm. Int. J. RF Microw. Comput. Aided Eng. **27**(7), e21119 (2017). https://doi.org/10.1002/mmce.21119
9. Wato, T., Tadelo, M., Negash, T.: Application of logistic regression model for predicting the association of climate change resilient cultural practices with early blight of tomato (Alternaria solani) epidemics in the East Shewa. Central Ethiopia. J. Plant Interact. **17**(1), 43–49 (2022)
10. Van Ravesteijn VF, van Wijk C, Vos FM, Truyen R, Peters JF, Stoker J, van Vliet.: Computer-aided detection of polyps in CT colonography using logistic regression. IEEE transactions on medical imaging. 29(1), pp. 120–131. 2009
11. Harikrishnan, R., del Río, L.E.: A logistic regression model for predicting risk of white mold incidence on dry bean in North Dakota. Plant Dis. **92**(1), 42–46 (2008). https://doi.org/10.1094/PDIS-92-1-0042

12. Poręba, J., Baranowski, J.: Functional logistic regression for motor fault classification using acoustic data in frequency domain. Energies **15**(15), 5535 (2022)
13. Del Rosso, V., Andreucci, A., Boria, S., Corradini, M.L., Ranalli, A.: Mechanical fault detection for induction motors based on vibration analysis: A case study. In: IECON Proceedings (Industrial Electronic Conference, vol. 2021-Octob (2021)
14. Landstrom, A., Thurley, M.J.: Morphology-based crack detection for steel slabs. IEEE J. Sel. Top. Signal Process. **6**(7), 866–875 (2012)
15. Khurshid, H., Khan, M.F.: Segmentation and classification using logistic regression in remote sensing imagery. IEEE J. Sel. Top. Appl. Earth Observations Remote Sensing **8**(1), 224–232 (2015). https://doi.org/10.1109/JSTARS.2014.2362769
16. Sanchez-Paniagua, M., Fernandez, E.F., Alegre, E., Al-Nabki, W., Gonzalez-Castro, V.: Phishing URL detection: a real-case scenario through login URLs. IEEE Access **10**, 42949–42960 (2022)
17. Zheng, G., Zhang, Q., Li, S.: Failure diagnosis of linear arrays based on deep residual shrinkage network. Microw. Opt. Technol. Lett. **64**(9), 1627–163 (2022)
18. Choudhury, B., Acharya, O.P., Patnaik, A.: Bacteria foraging optimization in antenna engineering: an application to array fault finding. Int. J. RF Microw. Comput. Eng. **23**(2), 141–148 (2013)
19. Khan, S.U., Rahim, M.K.A., Aminu-Baba, M., Khalil, A.E.K., Ali, S.: Diagnosis of faulty elements in array antenna using nature inspired Cuckoo search algorithm. Int. J. Electr. Comput. Eng. **8**(3), 1870–1874 (2018)
20. Gupta, B., Pinaki Mukherjee, K.Y.: Element Failure Detection in Antenna Arrays Using Genetic Algorithm. ISAP2007, Niigata, Japan, no. 1, pp. 330–333 (2007)
21. Khan, S.U., Qureshi, I.M., Zaman, F., Basit, A., Khan, W.: Application of firefly algorithm to fault finding in linear arrays antenna. World Appl. Sci. J. **26**(2), 232–238 (2013)
22. Grewal, N.S., Rattan, M., Patterh, M.S.: A linear antenna array failure detection using Bat Algorithm. In: 8th International Conference on Contemporary Computation IC3, pp. 202–207 (2015)
23. Acharya, O.P., Patnaik, A., Choudhury, B.: A PSO application for locating defective elements in antenna arrays. In: World Congress National Biologically Inspired Computation NABIC 2009 – Proceedings, vol. 2(2), pp. 1094–1098 (2009)
24. Patnaik, A., Choudhury, B., Pradhan, P.: An ANN application for fault finding in antenna arrays. IEEE Trans. Antennas Propagat. **55**(3), 775–777 (2007)
25. Vakula, D., Sarma, N.V.S.N.: Using neural networks for fault detection in planar antenna arrays. Prog. Electromagnet. Res. Lett. **14**, 21–30 (2010)
26. Rajagopalan, S., Joshi, M., Gudla, V.: Detection of faults in antenna arrays using SVM. Stanford University (2009). http://cs229.stanford.edu/proj2009/GudlaJoshiRajagopalan.pdf. Last accessed 22 June 2022
27. Xu, N., Christodoulou, C.G., Barbin. S.E., Martinez-Ramon, M.: Detecting failure of antenna array elements using machine learning optimization. In: IEEE Antennas Propagation Society International Symposium, pp. 5753–5756 (2007)
28. Alzubaidi, L., et al.: Review of deep learning: concepts, CNN architectures, challenges, applications, future directions. J. Big Data **8**(1), 1–74 (2021)
29. Balanis, C.A.: Antenna Theory: Analysis and Design, 3rd edn. John Wiley, New York (2006)
30. Pozar, D.M.: Microwave Engineering, 4th edn. John Wiley & Sons, New York (2012)

31. Sayeed, I., Matin, M.A.: A design rule for inset-fed rectangular Microstrip patch antenna. WSEAS Trans. Commun. **9**(1), 63–72 (2010)
32. Makalic, E., Schmidt, D.F.: Review of modern logistic regression methods with application to small and medium sample size problems. In: Li, J. (ed.) AI 2010. LNCS (LNAI), vol. 6464, pp. 213–222. Springer, Heidelberg (2010). https://doi.org/10.1007/978-3-642-17432-2_22
33. Maalouf, M.: Logistic regression in data analysis: an overview. Int. J. Data Anal. Tech. Strat. **3**(3), 281–299 (2011)

Internet Based Routing in Vehicular Named Data Networking

Purva Paroch[1(✉)], Simmi Dutta[1], Vivek Mahajan[2], and Niraj Dubey[1]

[1] Goverment College of Engineering and Technology, Jammu, India
purva.paroch03@gmail.com
[2] Government Polytechnic College, Ramban, Jammu, India

Abstract. In vehicular environments, the connection between data consumers and producers is not as stable as in the traditional IP architecture. This is due to the mobile nature of the vehicular nodes that causes a change in the location of consumers, producers, and intermediate nodes. Named Data Networking (NDN), when implemented on vehicular networks provides support for efficient data retrieval and mobility. This new paradigm, dubbed Vehicular Named Data Networking (V-NDN) significantly improves content access and dissemination by making a minor change in the conventional IP architecture. The focus is drawn away from the addresses of the consumers and producers and instead, data chunks are made primary citizens in the model, by allowing consumers to refer to data chunks by name. In this paper, V-NDN is introduced along with a survey of the existing forwarding schemes developed so far. An experiment is also conducted where the mobility is checked along with other important parameters.

Keywords: IP architecture · Named Data Networking · Vehicular Named Data Networking

1 Introduction

When Named Data Networking (NDN) and Vehicular ad hoc Networks (VANETs) are combined to fulfill the purpose of vehicular nodes communicating with each other and various other entities, Vehicular Named Data Networking (V-NDN) is born. The nodes in such a network can specify the data they need by sharing the name of the said Data packet with the entire network. The node that has this Data will then send it to the requesting vehicular node. Due to VANETs, it is possible for vehicles to wirelessly communicate with roadside units (RSU), access points and base stations. By using V-NDN, researchers aim to monitor and manage traffic, reduce the possibility of accidents and smartly navigate the vehicles on the road.

But direct implementation of VANETs in V-NDN is not feasible due to some features of VANETs like high mobility requirements, security, adaptability to changing topology and intermittent connectivity. The host-centric architecture of VANETs also adds to this dilemma and the results recorded are unsatisfactory. In order to achieve the desired results by implementing V-NDN, NDN is studied first and then efficient routing protocols are developed in V-NDN [1].

R. K. Challa et al. (Eds.): ICAIoT 2023, CCIS 1930, pp. 30–40, 2024.
https://doi.org/10.1007/978-3-031-48781-1_3

2 Literature Survey

Zhang et al. [1] proposed the NDN model that is compatible with the current Internet architecture. The initial blueprint of NDN architecture has been sketched and a core set of research problems have been identified. This includes scalability of routing names, fast forwarding based on variable-length hierarchical names, efficient verification by signatures, trust models for data-centric security, content protection and privacy. They claimed that NDN has a universal overlay and provides advantages in terms of content distribution, application-friendly communication and naming, robust security, mobility and broadcast.

Zhang et al. [2] described the motivation and vision behind the new NDN architecture, its basic components and operations. A snapshot of the current design, development status and research challenges are also provided. The naming approaches for pilot NDN applications, NDN routing protocols supporting traditional link-state and hyperbolic routing are explained. The team hopes to bring broader community participation to support and explore further research on the technology.

Grassi et al. [3] applied NDN, which was a newly proposed architecture, to networking vehicles on the run. Their initial design is called V-NDN, that is, Vehicular NDN, illustrated the promising potential of NDN that enables networking among all computing devices independent of whether they are connected through wired infrastructure, ad hoc, or intermittent DTN. Their work described a prototype implementation of V-NDN and its preliminary performance assessment. They depicted the results obtained graphically and the conclusion was: when the number of nodes interested in the same set of information increases, the resulting satisfaction time and overhead show great improvements.

After substantial research was done in NDN based VANET, Khelifi et al. [4] researched and reviewed the NDN-driven VANETs. They investigated the role of NDN in VANET and discussed the feasibility of NDN architecture in VANET. They provide detailed review of NDN-based naming, routing and forwarding, caching, mobility and security mechanisms for VANET. They discussed the existing standards, solutions and simulation tools used in NDN-based VANET. They also identified the open challenges and issues faced by NDN-driven VANET and highlighted future research directions to be addressed.

Yi et al. [5] demonstrated the role and need of routing protocols in NDN despite the ability of NDN routers to handle network failures locally without relying on global routing convergence. NDN has a unique feature which allows the forwarding plane to detect and recover from network faults on its own. The impact of the intelligent forwarding plane on the design and operation of NDN routing protocols is analyzed through extensive simulations. Routing protocols compute routing tables to guide forwarding as well as spread information about initial topology and policy information along with long-term changes in the parameters. But since NDN forwarding planes are capable of failure detection and quick recovery, the handling of short-term churns need not be handled by routing protocols. This improves the scalability and stability of the network and enables NDN to implement routing protocols that were deemed unsuitable for real networks.

The traditional IP-based Internet architecture cannot establish a stable end-to-end connection between a source and destination node in a vehicular environment due to the mobility of vehicles, causing location change. NDN-based VANETs have been introduced by Ahed et al. [6] along with their benefits and limitations. The classification of NDN-based VANETs forwarding strategies and detailed review of the representative schemes has been discussed. Various important attributes like transmission mode, forwarding strategies, application scenarios, evaluation metrics and simulation platform have been considered to review and compare the existing forwarding strategies. The main open research challenges have been pointed out in this survey. Their research provides the understanding of forwarding in vehicular environment and provides inspiration to design new forwarding protocols to improve Vehicular-NDN (V-NDN) networks.

In the dynamic V-NDN environment, the presence of redundant broadcast messages leads to resources being wasted and impacts the performance of the network. A wide variety of information dissemination protocols have been formulated, like counter-based protocol, having different threshold parameters that help in rebroadcasting and discarding decisions. But, in a dynamic VANET, one cannot have a prior knowledge of these threshold values. It is a difficult feat to alter and update these values in order to maintain good latency and reachability. Bakhouya et al. [7] present a new approach for information dissemination that is decentralized and adaptive for VANETs. The simulations and results compiled clearly show that such adaptive approaches perform better as opposed to statistical-based approaches.

It is a possibility that due to no end-to-end connection between the nodes, there may be a broadcast storm of Interest packets in the network. The existing proposed solutions of this problem further lead to issues related disconnect link which prevents the consumer from accessing Data packets and isolated network among vehicles which makes vehicles unavailable to broadcast Interest packets. Burhan et al. [8] proposed the Velocity-based forwarding strategy (VRFS) that tackles the disconnect link and isolated network problems by making use of the speed and location information of the vehicles. They also evaluate VRFS along with providing a comparison with the traditional V-NDN implementation.

In the thesis presented by Duarte [9] the solutions to the problems raised by high vehicle mobility and wireless communication in the V-NDN environment are addressed. A geographic routing protocol is proposed that is receiver-based and supports multiple hops. The impacts of common VANET issues like broadcast storms, redundancy and transmission resynchronization are highlighted. There is a thorough study on the effects of consumer mobility in V-NDN. The Reverse Path Partitioning (RPP) is addressed and Auxiliary Forwarding Set (AFS) is proposed to determine the RPP probability along with making the decisions regarding when to choose additional sets of suitable vehicles to act as intermediate nodes while forwarding.

Duarte et al. [10] performed simulations that study two V-NDN scenarios. In the first scenario, a 3 km, two lane, one way road is considered. A road-side unit (RSU) is designated to be the content provider and is 2 km away from the vehicle starting point. In the second scenario, the simulation is adjusted to reflect a real world 10 km long road in Erlangen, Germany, which has 2 ways and 4 lanes. There are multiple street junctions and the RSU that is designated to be the content provider is situated at a roundabout. The results for these simulations depicted that applying AFS to address the RPP problem provided an efficient and scalable solution while also enabling the high performance of the V-NDN applications.

3 NDN Overview

Zhang et al. [2] introduced the concept of NDN along with developing a prototype for the same in 2010. The prototype was software based and existing protocols were run on the traditional packet transmission networks. The NDN communication is all about one party that is requesting for a specific named data, called the consumer. The other entity in this communication is the party that has the requested data and fulfills this request, called the producer. An NDN network has multiple nodes that can operate as consumers and producers depending upon the network topology and policy implemented. The consumer broadcasts Interest packets to transmit its Data requests and the producers respond by sending back Data packets to fulfill these requests. The exact formats and contents of both Interest and Data packets are as shown in Fig. 1 [2].

The consumer waits for a reasonable period of time to get its requests fulfilled and after that either a negative acknowledgement (NACK) is received or the session is said to have been timed out. If the consumer faces such a situation, it will then send the Interest to other nodes, through other available paths to explore other nodes for the specifically named Data packet. The NDN routers carry out this failure detection, recovery and independently re-route the Interests on their own. This role is delegated to the forwarding plane which also handles other issues like prefix hijacking and congestion control [3].

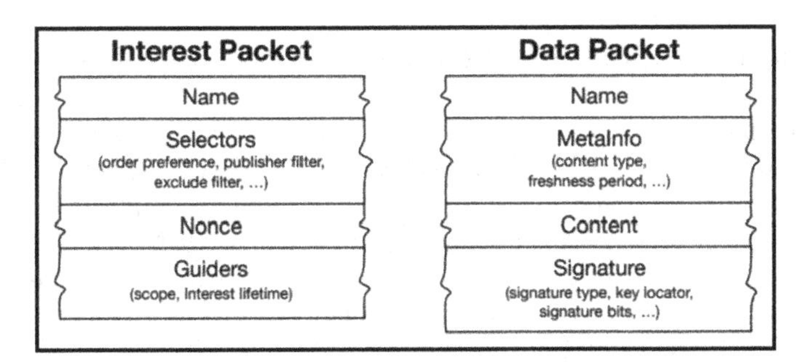

Fig. 1. Packets in the NDN Model.

All the nodes in an NDN network are tasked with managing three special data structures. These are Pending Interest Table (PIT), Forwarding Information Base (FIB), and Content Store (CS). Details about the forwarding decisions are recorded and stored in the FIB. In the FIB, PIT and CS, the indexing of the packets is done according to the name prefixes of the Data packets. This is one of the differentiating features of NDN from IP, where we make use of IP addresses of destination nodes to route the packets through the network. For each entry in the FIB, there are multiple possible interface choices listed in the table. The NDN router can make this decision dynamically. After an interface has been selected from the FIB, the PIT will then get updated and the selected interface will be recorded as having received the specific Interest. It is important to record the interfaces being used by the Interests because the Data packets will also follow the same path back to the consumer node. The CS is a cache storage that temporarily stores the Data packets that pass through each node on the way to the consumer node. The CS maintained by each node allows nodes other than the producer to satisfy Interests and this makes the communication more efficient and fast [3].

3.1 Forwarding Process

A request for a particularly named Data is made by a node in the NDN network by sending out Interest packets. Once that packet is received by a router, it will first check its CS to look for the Data. If the said data is found, it will be routed back to the consumer node. If not, the PIT is referenced and it is checked if the same Interest has been broadcasted earlier. If a matching entry is found in PIT, the incoming interface is added against the Interest. If no such Interest is in PIT, a new entry is made and the incoming interface is recorded. The Interest is then checked in the FIB. This is done by using the longest name prefix of the Data. If match is found, the Interest is routed according to the strategy mentioned in the FIB. If not, then this means that the Interest request of the consumer cannot be satisfied and the router will send back a negative acknowledgement message to let the consumer know. Once the Data packet is located at the producer node, the PIT is checked. If request for that Data packet is found, it is first stored in CS and then forwarded to the interface mentioned in PIT. If a matching Data request is not in the PIT, the Data packet is discarded since it is no longer requested.

4 Role of Routing in Named Data Networks

It has been established by Zhang et al. [2] that NDN is compatible with the existing Internet model. The aim is to evolve from IP model and adapt to the NDN model. This means that the routing protocols that have been designed to be implemented in an IP-based environment can be modified to be implemented in an NDN-based environment. So far, the IP protocols have a few shortcomings that have been highlighted, like poor scalability and slow convergence. With NDN's highly capable forwarding plane, smart decisions regarding forwarding are made.

The forwarding plane in IP architecture is stateless and does not allow for dynamic routing in case of unexpected changes in the network topology and policy. The main role of routing by creating and handling routing tables is taken care of by the routing plane. Any unexpected modification in the topology or network policy for long or short-term is managed by the routing plane and routing protocols. IP-based routing strategies have a long convergence period. Due to a change in the network, the routers need to inform each other about the change and update their routing tables accordingly. The goal here is to reach global consistency as fast as possible. The time it takes to reach global consistency after the change has occurred is called the routing convergence period. The implementation of IP-based routing protocols causes packet loss and delayed delivery due to their slow convergence.

A fast converging, scalable and stable protocol is difficult to attain. There can be trade-off between the three important features and they cannot be achieved simultaneously [5]. If we compromise on the fast convergence, we can achieve stability and scalability in routing protocols. This can be done by making it so that any network failures are handled without having to achieve global convergence. The NDN routers are required to detach from the conventional routing practices like handling short-term failures and focusing on global convergence. It becomes the NDN router's job to make sure that the network is stable for data retrieval and handle any link failures and recover from them. It is important to note that routing still plays a big role in bootstrapping forwarding process and maintaining coordination between routing and forwarding planes. This is done by integrating steps like interface ranking and probing to improve the routing stability and scalability. Due to the wireless nature of V-NDNs, researchers have developed multi-hop routing approaches like Position-Based Routing (PBR) that make it possible to reduce the retransmission of messages and wastage of network resources. For the different PBR schemes, there are different techniques that are used to determine how the best next hop is calculated [10]. A particular scheme is selected depending on the scenario. In the urban scenario, a better performance is recorded when area-based approaches are employed. Similarly, in highway scenario, it is found that distance-based approach causes less delays and high Interest Satisfaction Rates (ISR).

5 Forwarding Strategy

According to the various V-NDN forwarding strategies surveyed [6], a classification is proposed. Depending on the transmission mode implemented, forwarding can be unicast or broadcast. Furthermore, broadcast mode can be categorized as flooding, deferred broadcast and selective broadcast. Depending on the information taken into consideration while making the routing decisions, selective broadcast is further subcategorized. The existing routing schemes are studied and a comparative and statistical analysis is performed. A majority of the surveyed work was focused on broadcast mode and approximately 12% of the research was done on unicast mode. The most commonly used simulator to carry out the research is ndnSIM and the urban scenario has been researched most extensively so far.

A distributed and adaptive mechanism is presented by Bakhouya et al. [7] which perform information dissemination in VANETs. The objective is to disseminate content to the highest number of vehicles within a network. A counter-based scheme is implemented where a fixed threshold value is set to obstruct the rebroadcasting of messages. A relationship is established between the number of required retransmissions and the coverage area. This is done to make sure that in cases where a vehicle receives the same message more than the set threshold, then it is unlikely to rebroadcast that message due to an insignificant coverage area. Once a vehicle receives a broadcast message, a random timer is started and a counter is set for the received message. As soon as the vehicle receives another copy of the same message, the counter will be increased in the presence of the running timer. After the timer expires, if the counter is at less than a threshold, then the vehicle will broadcast the message, otherwise, the message will be discarded.

6 Experimental Setup and Result

Routing in V-NDN has been demonstrated through simulations using ndnSIM. An AdHoc network is established which eliminates the need for infrastructure and other complex equipment. Each node will forward the packets not intended for it to its neighbors. Initially, two nodes are created for the sake of simplicity. Next, an object is created and WiFiPhyHelper class is installed on it. This allows each node to have access to Wi-Fi. The mobility model is to be installed on each node next. The constant velocity model is selected for which the current speed is constant once it has been set until it is explicitly updated. As the next step, the NDN stack is installed on each node. The routing of all the packets in the network is done by selecting the best route strategy. As the final step, the applications are set up in each node with the help of the consumer and producer helper classes.

For this experiment, two nodes are created as a start and the above-mentioned steps are implemented on each node as more nodes are added to the network. Figure 2 depicts the network when the number of nodes added is 6. The nodes are dynamic and the speed, at which the communication in the form of packet exchange is taking place, is mentioned on the routes. A wide variety of performance metrics can be considered like average delay, Interest satisfaction delay (ISD), Interest satisfaction ratio (ISR), number of forwarded Interest packets (FIP), overhead and average hops, end-to-end-delay, successful content delivery rate (SCDR), number of forwarded data packets (FDP), throughput and so on.

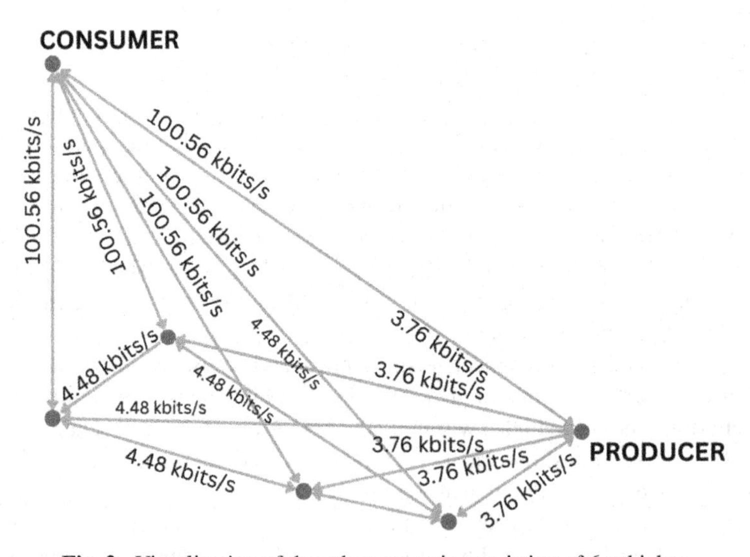

Fig. 2. Visualization of the urban scenario consisting of 6 vehicles.

The urban scenario is simulated in this experiment. This is done due to the high vehicle density in an urban scenario which is an important factor in this experiment. The speed of vehicles in this scenario range from 40 km/h to 60 km/h. The parameters considered to evaluate the performance of the system are Interest satisfaction ratio, network transmission overhead, throughput, and average delay. The graphs in the section below depict the changes in the parameters as the vehicle density changes with time.

ISR is one of the most used metrics in V-NDN as the major goal of every NDN-based VANET protocol is to achieve a higher satisfaction rate in terms of Interests transmitted [6]. Figure 3 is a graphical representation of the impact of vehicle density on ISR. Another factor that contributes to the result is the speed of the vehicles in the scenario. Three-speed limits of 40 km/h, 50 km/h, and 60 km/h are simulated here. At a range of 40 km/h, starting with 20 vehicles, the ISR increases drastically. When the vehicle density reaches 140, the increase in ISR becomes somewhat steady at 0.84–0.86%. The curves for speeds of 50 km/h and 60 km/h show similar progression when the vehicle density reaches a high number like 140. It can be concluded that with an increase in speed and average vehicle density, the ISR will also increase.

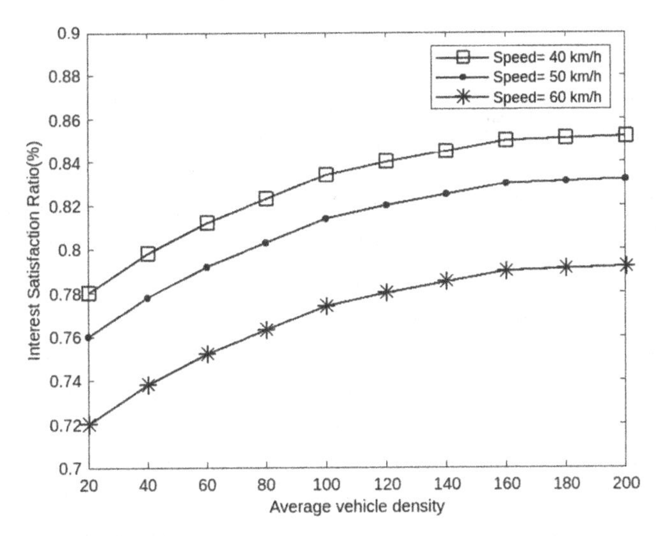

Fig. 3. Impact of vehicle density and speed on Interest Satisfaction Ratio (ISR).

For network transmission overhead, average vehicle density and the number of consumers are considered. With the increase in the number of vehicles and consumers, the network transmission overhead gets reduced considerably. As shown in Fig. 4, for 15 consumers, as the number of vehicles increase, the network transmission overhead decreases from 120 to approximately 70. The same pattern is exhibited by networks with 5 and 10 consumer nodes.

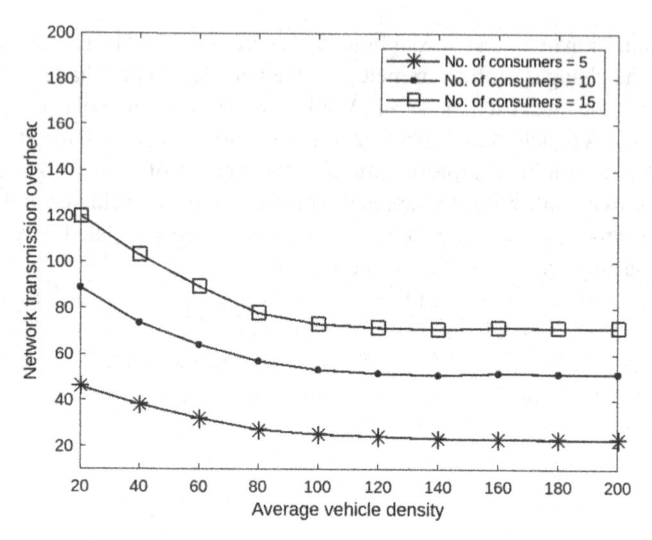

Fig. 4. Impact of vehicle density and number of consumers on network transmission overhead.

Figure 5 depicts the changes in the throughput when vehicle density and speed are moderated steadily. It can be observed that with an increase in vehicle density, there is a slight improvement in the throughput with the curves at different speeds showing similar patterns.

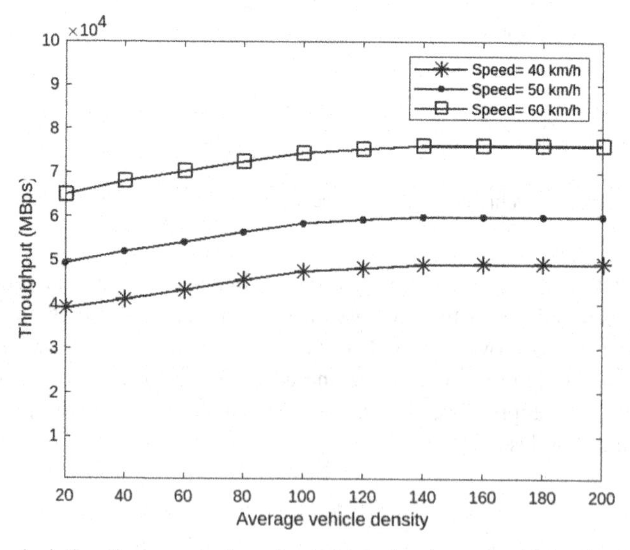

Fig. 5. Curves depicting the impact of varying vehicle density and vehicle speed on throughput of the system.

The next parameter observed is the average delay. The speed of the vehicles and the vehicle density are considered to determine the impact on the average delay in the delivery of packets from the source to the destination node. With the increase in vehicle density, the average delay gets reduced and maintains a steady value once the vehicle density has reached approximately 150 (Fig. 6).

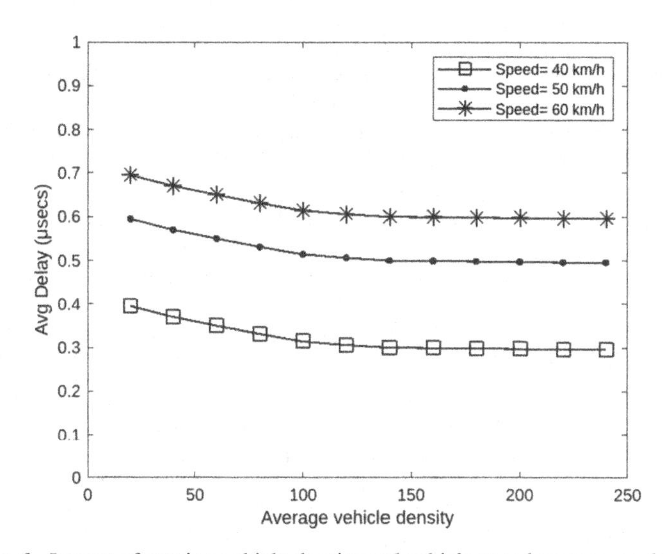

Fig. 6. Impact of varying vehicle density and vehicle speed on average delay.

7 Conclusion

The experiment conducted shows the impact of vehicle density, number of consumer nodes, and speed of the vehicle on the four main parameters namely Interest satisfaction ratio (ISR), network overhead, throughput, and average delay. The urban scenario is helpful as it allows the researchers to observe how the performance of the setup changes with a change in the number and speed of the vehicular nodes. It can be concluded from the observations above that the performance of the system increases significantly in terms of ISR and network overhead. However, in terms of throughput and average delay, after a specific increase in the vehicle density, the change in the curve remains steady which indicates that a threshold has been achieved and the parameters maintain their value after that point.

There is still a need for improvement in this area as a lot of mobility problems can be caused at the sender and receiver ends. Moreover, network partitions have not been researched in terms of V-NDN which serves as a limitation. A general-purpose V-NDN framework is needed that can support on-demand and advertised content delivery while addressing the negative impact of mobility and the unreliability of wireless mediums for communication to provide mobility support in V-NDN with high performance.

References

1. Zhang, L., Estrin, D., Burke, J., Jacobson, V., Thornton, J.D., Smetters, D.K.: Named Data Networking (NDN) Project. In: Technical Reports – Named Data Networking NDN-0001 (2010)
2. Zhang, L., Burke, J., Crowley, P., Wang, L.: Named data networking. ACM SIGCOMM Comput. Commun. Rev. **44**(3), 66–73 (2014)
3. Grassi, G., Pau, G., Wakikawa, R., Zhang, L.: Vehicular Inter-Networking via Named Data. https://www.researchgate.net/publication/258082586_Vehicular_Inter-Networking_via_Named_Data. Last accessed 21 Feb 2023
4. Khelifi, H., et al.: Named data networking in vehicular ad hoc networks: state-of-the-art and challenges. IEEE Commun. Surv. Tutorials **22**(1), 320–351 (2020)
5. Yi, C., Abraham, J., Afanasyev, A., Wang, L., Zhang, B., Zhang, L.: On the Role of Routing in Named Data Networking. In: NDN, Technical Report NDN-0016 (2013)
6. Ahed, K., Benamar, M., Ait Lahcen, A., El Ouazzani, R.: Forwarding strategies in vehicular named data networks: A survey. J. King Saud Univ. – Comput. Inform. Sci. **34**(5), 1819–1835 (2022)
7. Bakhouya, M., Gaber, J., Lorenz, P.: An adaptive approach for information dissemination in Vehicular Ad hoc Networks. J. Netw. Comput. Appl. **34**(6), 1971–1978 (2011)
8. Burhan, M., Rehman, R.A., Kim, B.S.: Velocity based reliable forwarding strategy towards disconnect link avoidance in NDN-VANETs. In: IEEE Wireless Communications and Networking Conference Workshops (WCNCW), pp. 1–6. Seoul, Korea (South) (2020)
9. do Monte Gomes Duarte, J.: Mobility Support in Vehicular Named-Data Networking. Inaugural Dissertation of the Faculty of Science of the University of Bern and the Institute of Computing of the University of Campinas, pp. 1–115 (2018)
10. Duarte, J.M., Braun, T., Villas, L.A.: Receiver mobility in vehicular named data networking. In: Proceedings of MobiArch'17, pp. 43–48 (2017)

A Very Deep Adaptive Convolutional Neural Network (VDACNN) for Image Dehazing

Balla Pavan Kumar(✉), Arvind Kumar, and Rajoo Pandey

National Institute of Technology, Kurukshetra, Haryana 136118, India
pavan_6180103@nitkkr.ac.in

Abstract. The main challenge faced by the existing methods is that they cannot efficiently eliminate the haze from the dense hazy or foggy images. The haze features of dense hazy images are not effectively learnt by the Deep Neural Networks. To resolve this drawback, a very deep adaptive convolutional neural network model is proposed for efficient image dehazing. The hazy images are first categorized into two categories viz., Less-Haze and High-Haze. Two Very Deep Convolutional Neural Networks (VDCNNs), viz., Less Haze-VDCNN and High Haze-VDCNN are developed separately for the classified images. Then the Less Haze-VDCNN is trained using the input hazy images that are less haze and their transmission maps as output. Similarly, the High Haze-VDCNN is trained separately with high-hazy images and their transmission maps. After the training process, a hazy image can be adaptively dehazed from one of the two trained VDCNNs based on the hazy image categorization. The proposed VDACNN exhibits better results for dense hazy images in comparison to existing approaches.

Keywords: High-haze · Low-haze · VDCNN · Image dehazing

1 Introduction

The pictures or images captured in bad environments such as fog are heavily deteriorated. This problem creates great challenges for real-time image processing systems viz., robot vision, automated driver-assessment systems, Closed Circuit Television (CCTV) monitoring systems, etc. To overcome these challenges, it is required to build a dehazing algorithm to eliminate or minimize the haze effect.

There are several methods implemented in the past to dehaze the hazy or foggy images. Most of these dehazing algorithms can be classified under two categories, (i) Image restoration and (ii) Deep learning. The image restoration-based techniques [1–5] use the atmospheric model where the atmospheric light and transmission map are evaluated to obtain the haze-free image. Although they produce natural results, they appear over-degraded due to inappropriate estimation of the transmission map. To overcome this problem, deep learning approaches are implemented for the effective extraction of the haze-relevant features of the transmission map. Most of the deep learning techniques [6, 7] produce natural results and also do not exhibit over-degradation, unlike image

The original version of this chapter was revised: The references section has been updated. A correction to this chapter can be found at
https://doi.org/10.1007/978-3-031-48781-1_30

R. K. Challa et al. (Eds.): ICAIoT 2023, CCIS 1930, pp. 41–47, 2024.
https://doi.org/10.1007/978-3-031-48781-1_4

restoration methods. In [6], the transmission map (T) is evaluated from the supervised Convolutional Neural Network (CNN) with the hazy image as the input image. Then using atmospheric light (A) and the refined transmission map, the dehazed image is recovered.

As mentioned in [7], the direct learning of haze-free images from the input hazy images may not yield natural outcomes. As the hazy image works on the principle of the atmospheric model, this model must be considered for image recovery.

In most of the existing techniques, the haze is not efficiently eliminated for the dense-hazy images. To overcome this drawback, an adaptive model called a very-deep adaptive convolutional neural network (VDACNN) is proposed in this paper. In this work, the hazy images are first categorized into low-hazy and high-hazy images using the No-Reference Image Quality Assessment metric called Haziness Degree Evaluator (HDE) [8]. Then two VDCNNs are trained separately for two categories of hazy images. The Low haze-VDCNN is developed by training the input hazy images of low haze and their corresponding synthetic transmission maps as output. Similarly, High haze-VDCNN is implemented using high-hazy images. After the training, the VDACNN model can be used to adaptively dehaze the hazy images. The T is obtained from one of the two trained deep neural networks (DNNs) (Low haze-VDCNN /High haze-VDCNN). The A is evaluated using the max-median approach as mentioned in [4]. The dehazed image is recovered with the evaluated T and A from the atmospheric model.

The rest of the article is arranged in the following order. The background of the atmospheric model is presented in Sect. 2. The procedure of the VDACNN is elaborated in Sect. 3. The experimental results are presented in Sect. 4. The conclusion of the proposed VDACNN is stated in Sect. 5.

2 Atmospheric Scattering model

The haze effect for an image can be expressed by Koschmieder's law [9] as:

$$I(x) = J(x).T(x) + A(1 - T(x)) \tag{1}$$

where I denotes the hazy or foggy image, x represents the 2D location of an image, and J denotes the haze-free images. T and A are the vital parameters that affect the image with the haze. These parameters have to be determined to restore the dehazed image from the corresponding foggy image. The dehazed image is determined by:

$$J(x) = \frac{I(x) - A}{T(x)} + A \tag{2}$$

Although A can be evaluated accurately by statistical approach, the main challenge lies in calculating the T. These drawbacks of existing methods can be resolved by the proposed VDACNN framework.

3 Methodology

The proposed methodology can be mainly implemented in three steps: (i) Categorization of Hazy images, (ii) Training of VDCNNs, and (iii) Image Restoration.

3.1 Categorization of Hazy Images

The hazy images have to be categorized for efficient dehazing purposes. The previous techniques of image dehazing do not efficiently eliminate the haze for dense hazy images. The DNNs are trained for all kinds of hazy images where the features of dense hazy images are not effectively learnt. Also, the transmission maps of these images exhibit different properties when compared with other kinds of foggy images as displayed in Fig. 1. By considering these factors, all the hazy images are categorized into Low-Hazy and High-Hazy images. The process of categorization is done using the Haziness Degree Evaluator (HDE) [8]. In the range of (0,1), the larger the HDE, the more the haze density for a given image. Empirically, the hazy images are categorized as:

$$\text{Hazy Image} = \begin{cases} \text{Low - Hazy, if HDE} < 0.6 \\ \text{High - Hazy, if HDE} > 0.6 \end{cases} \tag{3}$$

The categorized images are used to adaptively train the VDCNNs.

Fig. 1. Comparison between different transmission maps for the hazy images of unique haze levels. The images in the top row contain hazy images - *53_8_0.78862* and *53_9_0.93321* of RESIDE dataset [10] with HDE 0.48 and 0.65, respectively. The images of the bottom row show the transmission maps of *53_8_0.78862* and *53_9_0.93321*, respectively.

3.2 Training of VDCNNs

The VDCNNs have to be constructed before the actual training process. A VDCNN consists of 20 layers i.e., one input layer, eighteen middle layers, and one output layer.

A patch of the hazy image of size 15 × 15 is provided to the input layer. Each middle layer contains a Convolutional Neural Network (CNN) with 64 number of 3 × 3 filters and a Rectified Linear Unit (ReLU) layer. The output layer contains a CNN with a 3 × 3 filter followed by a regression layer. By this setup, a VDCNN is constructed with 20 layers (Fig. 2).

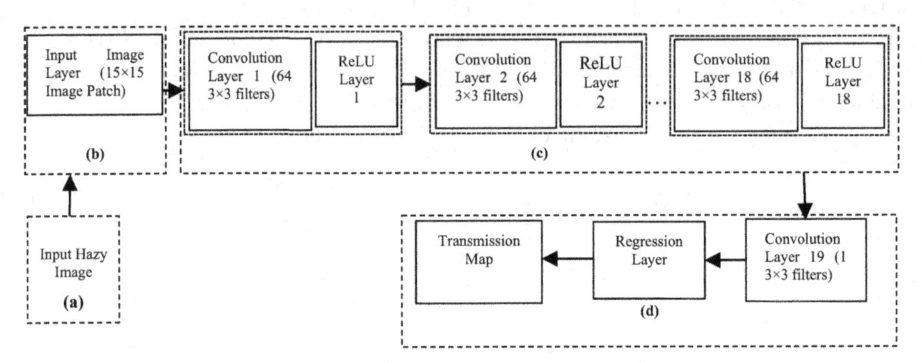

Fig. 2. Architecture of VDCNN. (a) represents the hazy image, (b) denotes the input layer, (c) represents the middle layers, and (d) denotes the output layer.

A patch of the transmission map is trained with the same patch of the corresponding hazy image. The size of the patch is empirically set as 15 × 15. Two VDCNNs are designed in our proposed work i.e., Low haze-VDCNN and High haze-VDCNN. The Stochastic Gradient Descent with momentum algorithm is used to update the weights after each batch with a batch size of 64. The Low haze-VDCNN is constructed using only the low-haze images and its transmission maps. Similarly, the High haze-VDCNN is constructed using high-haze images. The two VDCNNs i.e., Low haze-VDCNN and High haze-VDCNN are used to develop the proposed VDACNN framework.

After the VDCNNs are trained, these networks can be tested using sample hazy images. After a hazy image is categorized, a VDCNN network is chosen based on its categorization. For example, if a hazy image is categorized as high haze, then that hazy image is passed through high haze-VDCNN to obtain the T.

3.3 Image Restoration

After the T is determined using the VDACNN framework, the haze-free image has to be recovered. A can be evaluated using the dark channel approach as mentioned in [1] which is needed for image restoration. The dehazed image can be recovered using T and A as mentioned in Eq. (2). The obtained transmission is in raw form and may produce halo artefacts. Hence the T is refined using the guided filter [11] to avoid the halo artefacts.

The proposed VDACNN framework is tested for the hazy images with RESIDE dataset [10]. The proposed VDACNN framework adaptively dehazes an image based on its level of haze. The haze in the high-haze image is effectively eliminated using the VDACNN model.

4 Experimental Results

The proposed algorithm performs better than existing techniques as mentioned in Table 1, Fig. 3, and Fig. 4. The Image Quality Assessment (IQA) metrics called Structure Similarity Index Metric (SSIM), Peak Signal to Noise Ratio (PSNR), and Naturalness Image Quality Evaluator (NIQE) are used to assess the quality of dehazed images as shown in Table 1. The SSIM is applied for calculating the similarity among the dehazed image and its corresponding ground truth (GT) image. The PSNR of an image is inversely related to Mean Square Error (MSE) amongst the GT and dehazed images. NIQE denotes the naturalness of a given image. When compared to the existing methods, the proposed model produces better IQA as shown in Table 1.

Table 1. Quantitative Results of existing techniques along with the proposed method

Hazy Image - *53_8_0.78862* of RESIDE dataset [10]						
	DCP [1]	NLD [2]	CEP [3]	RBAD [4]	Dnet [6]	VDACNN
SSIM	0.8663	0.7943	0.6876	0.7477	0.9313	0.913
PSNR	15.8661	16.771	11.563	17.1909	22.6408	24.774
NIQE	3.3681	2.8599	3.3448	3.0658	3.7679	3.7704
Hazy Image - *53_9_0.93321* of RESIDE dataset [10]						
	DCP [1]	NLD [2]	CEP [3]	RBAD [4]	Dnet [6]	VDACNN
SSIM	0.848	0.8593	0.805	0.7974	0.8387	0.8515
PSNR	18.309	20.5245	16.635	19.2755	18.1938	19.6659
NIQE	3.7678	3.8515	3.8913	4.1844	4.2327	4.4977

The proposed VDACNN works well for all kinds of hazy images i.e., low-haze and high-haze images as mentioned in Fig. 3 (a), and Fig. 4 (a). The foggy image – *53_8_0.78862* of RESIDE dataset [10] exhibits low-haze with HDE $= 0.428$ and the hazy image – *53_9_0.93321* exhibits high-haze with HDE $= 0.6436$. For the images that exhibit low haze with HDE < 0.6, the low-haze VDCNN is applied; similarly, high-haze VDCNN is applied for the high-hazy images. In this manner, the optimum results are obtained using VDACNN as shown in Fig. 3 and Fig. 4.

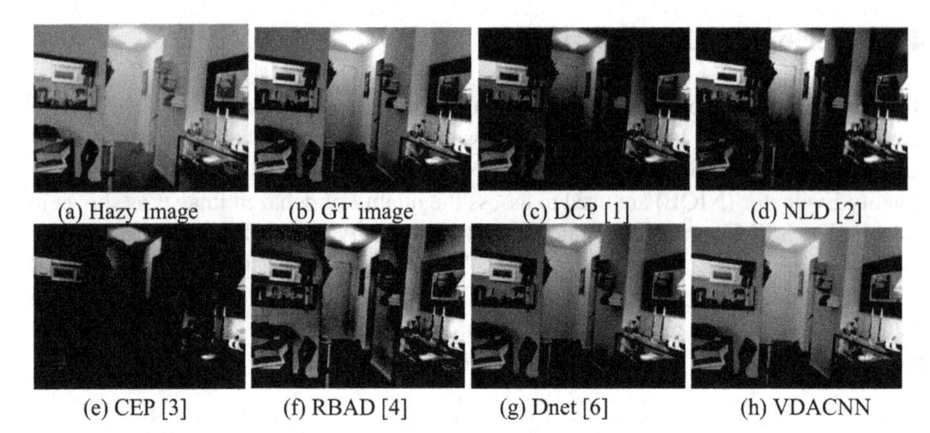

Fig. 3. Subjective assessment of previous techniques and VDACNN technique for the hazy image – *53_8_0.78862* of RESIDE dataset [10].

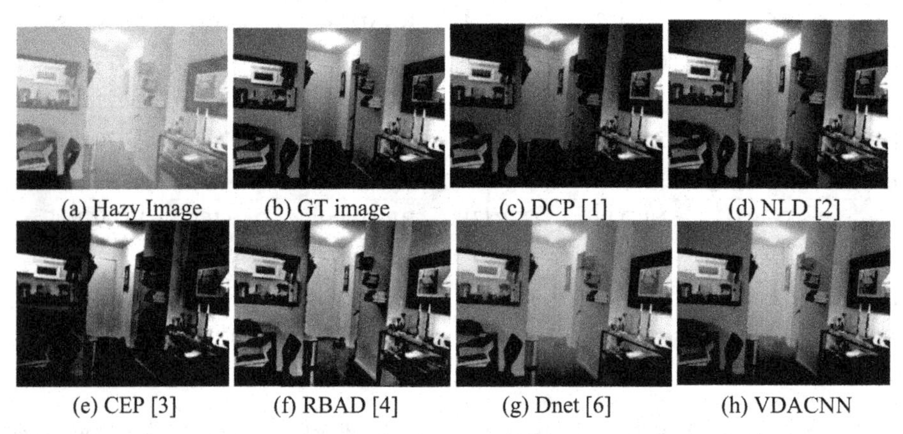

Fig. 4. Subjective assessment of previous techniques and VDACNN approach for the hazy image – *53_9_0.93321* of RESIDE dataset [10].

5 Conclusion

The VDACNN model is presented in this work which adaptively dehazes the hazy images based upon their haze levels. Two VDCNNs called low-haze VDCNN and high-haze VDCNN are developed for adaptive dehazing. A hazy image is first categorized into low-hazy and high-hazy using HDE, before the application of the VDACNN model. The proposed VDACNN produces better results and also eliminates the haze well for the high-hazy images in comparison to the previous approaches.

References

1. He, K., Sun, J., Tang, X.: Single image haze removal using dark channel prior. IEEE Trans. Pattern Anal. Mach. Intell. **33**(12), 2341–2353 (2010)
2. Berman, D., Avidan, S.: Non-local image dehazing. In: Proceedings of the IEEE conference on computer vision and pattern recognition, pp. 1674–1682 (2016)
3. Bui, T.M., Kim, W.: Single image dehazing using color ellipsoid prior. IEEE Trans. Image Process. **27**(2), 999–1009 (2017)
4. Kumar, B.P., Kumar, A., Pandey, R.: Region-based adaptive single image dehazing, detail enhancement and pre-processing using auto-colour transfer method. Signal Process.: Image Commun. **100**, 116532 (2022)
5. Kumar, B. P., Kumar, A., Pandey, R.: A generic post-processing framework for image dehazing. Sig. Image Video Process. **17**, 3183–3191 (2023)
6. Cai, B., Xu, X., Jia, K., Qing, C., Tao, D.: Dehazenet: an end-to-end system for single image haze removal. IEEE Trans. Image Process. **25**(11), 5187–5198 (2016)
7. Haouassi, S., Wu, D.: Image dehazing based on (CMT net) cascaded multi-scale convolutional neural networks and efficient light estimation algorithm. Appl. Sci. **10**(3), 1190 (2020)
8. Ngo, D., Lee, G.D., Kang, B.: Haziness degree evaluator: a knowledge-driven approach for haze density estimation. Sensors **21**(11), 3896 (2021)
9. Koschmieder, H.: Theorie der horizontalen Sichtweite. Beitrage zur Physik der freien Atmosphare **12**, 33–53 (1924)
10. Li, B., et al.: Benchmarking single-image dehazing and beyond. IEEE Trans. Image Process. **28**(1), 492–505 (2018)
11. He, K., Sun, J., Tang, X.: Guided image filtering. In: Daniilidis, K., Maragos, P., Paragios, N. (eds.) ECCV 2010. LNCS, vol. 6311, pp. 1–14. Springer, Heidelberg (2010). https://doi.org/10.1007/978-3-642-15549-9_1

Automatic Door Unlocking System Using Facemask Detection and Thermal Screening

Arun Kumar Katkoori[1(✉)], Burra Raju[2], Koteswararao Seelam[3], and Bhanu Prakash Dudi[1]

[1] Electronics and Communication Engineering Department, CVR College of Engineering, Hyderabad, India
arun.katkoori@gmail.com, pbhanududi@gmail.com

[2] Electronics and Communication Engineering Department, Jyothishmathi Institute of Technology and Science, Karimnagar, India

[3] Electronics and Communication Engineering Department, Kallam Haranath Reddy Institute of Technology, Guntur, India

Abstract. The pandemic of the coronavirus disease 2019 (COVID-19) negatively affected the lives and businesses of millions of people in 2022. Everyone is more afraid now, especially those who wish to continue their own activities while the world recovers from the epidemic and gets ready to go back to normal. Studies show that wearing a face mask significantly reduces the danger of spreading the virus and offers the user a sense of security. However, it is difficult to physically check to see whether this rule is being followed. The creation of a face mask detector that can identify face masks is the main objective of this research. Open CV and Tensor flow are used to identify face masks. Face mask recognition is included into both the Raspberry Pi and the USB camera. This encourages the use of face masks, helps catch safety violations, and maintains the workplace safe. MLX90614 is used to determine body temperature without making physical contact. The door is opened if the temperature is below the threshold, or 40 degrees Celsius, and someone is wearing a mask; otherwise, the door must be locked to stop the buzzer from going off.

Keywords: OpenCV · TensorFlow · Machine Learning · Raspberry Pi 4

1 Introduction

India, with a population of over 134 billion, is the second-most populous nation in the world after China, where the COVID-19 coronavirus infection has been rapidly escalating [1]. India would have a difficult time controlling the coronavirus due to its enormous population. The best techniques to inhibit transmission are to use face masks and maintain a constant body temperature. It has been effective in halting the spread of illness. A coronavirus infection often causes a fever, sore throat, fatigue, loss of taste and smell, and nasal congestion [2]. It is often unintentionally transferred to surfaces. In extreme cases, the virus may take up to 14 days to become completely infectious

R. K. Challa et al. (Eds.): ICAIoT 2023, CCIS 1930, pp. 48–59, 2024.
https://doi.org/10.1007/978-3-031-48781-1_5

and is mostly transmitted by respiratory droplets. To prevent the spread of the illness, governments have implemented several safety and security measures, including social exclusion, indoor masking requirements, quarantine, travel restrictions for residents travelling both domestically and internationally, self-isolation, and widespread exclusion as well as the cancellation of social events [3, 4] The COVID-19 pandemic influences all sports, off-screen and on-screen entertainment, interpersonal connections, and professional activities [5, 6]. A person's increased body temperature increases the danger of infection and dissemination; hence it is advised that masks be worn. Every city demands a temperature and mask check at every entrance to a workplace, store, hospital, and mall. Therefore, a smart access device that can also detect masks on door opening systems and analyses body temperature was developed [7]. This systems-based method makes use of temperature monitoring and face mask identification.

This is accomplished with the use of the Raspberry Pi 4 CPU, a Universal Serial Bus (USB) camera, an InfraRed (IR) sensor, a camera, a servo motor, and the MLX90614 temperature sensor. A model is created by combining these components.

2 Literature Survey

Deora et al. [8] proposed a condensed method for face mask detection which looks at a face's posture to identify whether it is covered. The issue is somewhat connected to basic object recognition to distinguish between object types. Face recognition categorizes a certain collection of items by identifying them. It may be used for a variety of things, including autonomous driving, teaching, and spying. This paper presented a condensed method for achieving the goal utilizing common machine learning (ML) tools including TensorFlow, Keras, OpenCV, and Scikit-Learn.

The approach developed by Baskaran et al. [9] for real-time applications for face recognition, however, has a high detection speed and accuracy. It could be possible to locate the face mask using an object detection method. You Only Look Once: A sophisticated, high-performing item detection technique (YOLO). A humanoid robot soccer player employs the You Only Look Once (YOLO) deep learning method to recognize the white ball and the target, according to Susanto et al. This method was created using the NVIDIA JETSON TX1 controller board. In their study, Liu et al. also used the YOLO principle. They conducted their investigation using traditional image processing to extract noise, blur, and rotation filters in the real world. Then, to improve traffic sign identification, they trained a powerful model using the YOLO approach. They improved the YOLO algorithm and compared the recognition performance with the conventional approach for face detection in a video series. Additionally, they used the Face Detection Data Set and Benchmark (FDDB) dataset to train and test the model. The YOLO model has been enhanced, as Zhao et al. have shown. They improved the YOLO model, which tackles two problems to recognize pedestrians.

Lim, & Chuah [10] suggested employing Convolution Neural Networks (CNN), a kind of Deep Neural Network (DNN) often used in image classification and identification, to create a real-time face mask detection model for this study. The recommended model might be installed in security cameras at places like malls, multiplexes, schools, and colleges to help detect and report people who are not wearing face masks automatically. Additionally, authorities should directly educate them via messaging. With the help of this model, it was possible to slow the rapid increase in the number of positive cases, control the hopeless loss rate, and break the close-contact chain of viral transmission.

Swain et al. [11] proposed a face mask detection technology that provides information to the government so it may put preventive measures in place, reduce risks, and assess its programmers. This article also serves as a warning to authorities on the need of studying local individuals' behaviors. There is a need to provide more face masks as people get used to using them. On the other hand, based on people's propensity of using face masks, the industry may use this answer to create a face mask.

Vishwesh et al. [12] proposed a technology being developed to identify whether a person is wearing a mask or not and alert the necessary authorities. First, live video is recorded utilizing CCTV cameras at a variety of metropolitan public spaces. The location of the mask on the face is then determined using the facial pictures received from this video clip. The features are first extracted from the photos using a convolutional neural network (CNN) learning method, and they are then learnt from several hidden layers. The proper authorities are alerted through the city network if the architecture notices individuals without face masks so that the required action may be taken. By incorporating information from many sources, the suggested approach forecasts favorable results. The authors provide a framework for this pandemic condition that might guarantee the correct application of the law for those who disregard crucial health precautions.

3 Proposed Technique

It is impossible to exaggerate the value of detecting body temperature in clinical diagnosis and treatment. Lack of measurement precision and a drawn-out measuring process are two drawbacks. It is challenging to automate and accurately monitor a patient's body temperature over time using artificial measuring techniques. A distributed surveillance system is suggested that might be used to assess body temperature to address the issue.

A person's body temperature signal is recorded by the system using MLX90614 temperature sensors. After it has been gathered, the data is sent to the Raspberry Pi 4. A USB camera is also used to take pictures of the person. OpenCV and TensorFlow are used to recognize the person wearing the mask. If the person is wearing a mask and their body temperature is below the threshold, the doors will open; if not, a buzzer will sound, and they won't.

To reduce the need for labor while improving accuracy, we provide an automatic door unlocking system with face mask recognition and thermal screening. Live streaming is being utilized to track people's body temperatures and check whether they are wearing face masks to address the current problem.

The fundamental components of the suggested system are face mask detection and thermal screening. In this paper, the Raspberry Pi 4 acts as the system's brain. MLX90614 was used as the temperature sensor system's input to determine body temperature. Camera and IR sensor inputs are used for people and face mask user detection.

The Liquid Crystal Display (LCD), which also serves as an output, displays the output. The output, the door open/close function, and the buzzer will sound if the temperature exceeds the cutoff point for removing the mask. Figure 1 displays the system's block diagram.

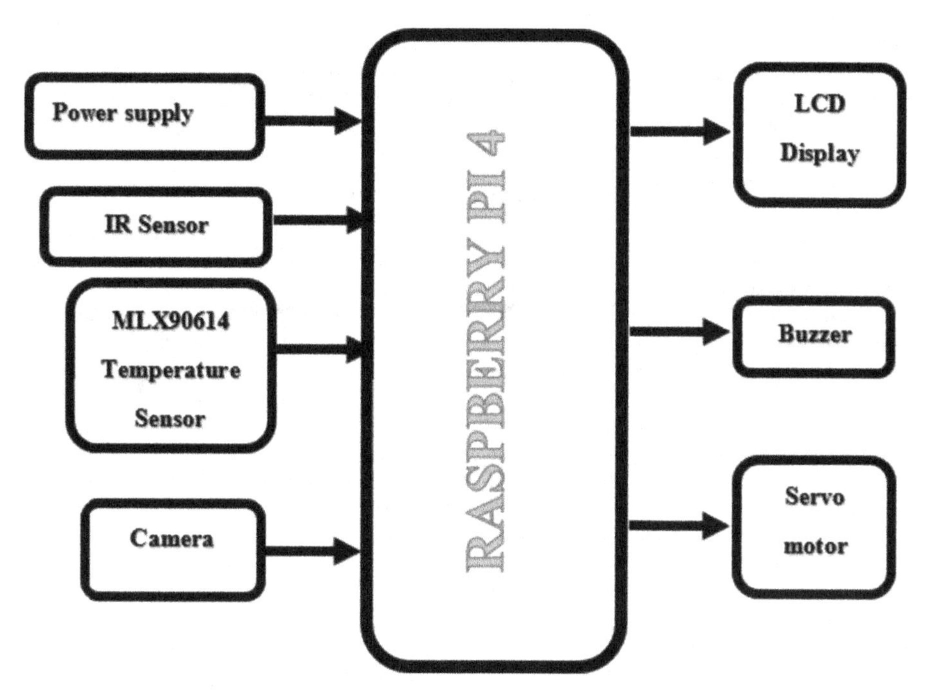

Fig. 1. Block Diagram of the System

There are three steps involved in this system:

- Face Mask Detection
- Temperature check
- System Alert

3.1 Facemask Detection

There are mainly two steps involved in Facemask Detection:

- Training
- Deployment

Training. At this step, our focus will be on loading our face mask detection dataset from disc, using it to train a model (using Keras/Tensor Flow), and serializing the model to disc.

Deployment. Deployment of a ML model is a process that involves placing a finished model in a live environment that's designed to be used for its purpose. These models can be used in various environments, and they can be integrated with various apps through an API. After training the face mask detector, load it, identify each face, and then classify each face as 'with mask' or 'with no mask'. Figure 2 shows the block diagram of facemask detector.

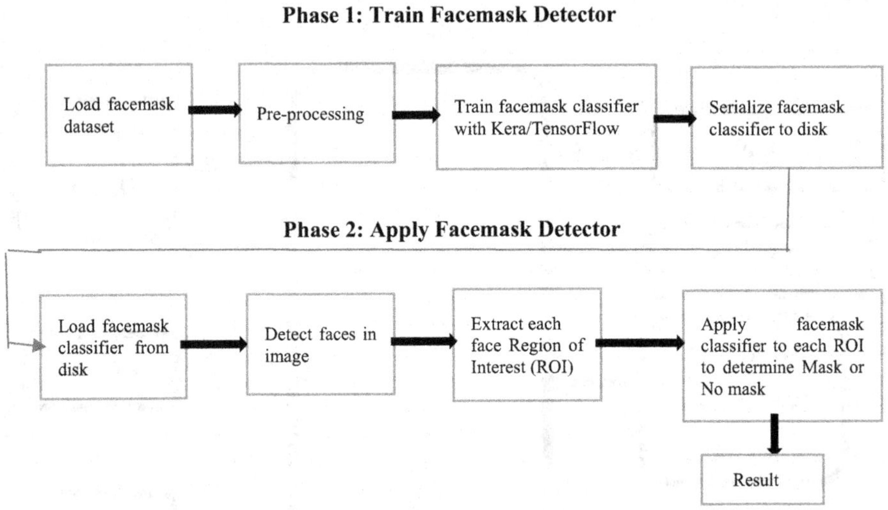

Phase 1: Train Facemask Detector

Load facemask dataset
Pre-processing
Train facemask classifier with Kera/TensorFlow
Serialize facemask classifier to disk

Phase 2: Apply Facemask Detector

Load facemask classifier from disk
Detect faces in image
Extract each face Region of Interest (ROI)
Apply facemask classifier to each ROI to determine Mask or No mask
Result

Fig. 2. Facemask Detector

3.2 Temperature Check

Thermal screening is performed using MLX90614 Temperature Sensor, which is coupled with Raspberry Pi 4.

3.3 Alert System

Servo motor and buzzer are attached to Raspberry Pi 4. The buzzer will sound, and the doors won't open if the temperature threshold is surpassed, and no face mask is found. The actuator, i.e., H., is turned on when the temperature and face mask are both adjusted to normal. There are open doorways.

The 40-pin Raspberry Pi 4 features General purpose Input/ Output (GPIO) pins, a power supply, Serial Data (SDA), and Serial Clock (SCA). The four pins of the MLX90614 Temperature Sensor, which serves as an input, are linked to the Serial Communication Board through pins 1, SDA, 3, and 5, respectively.

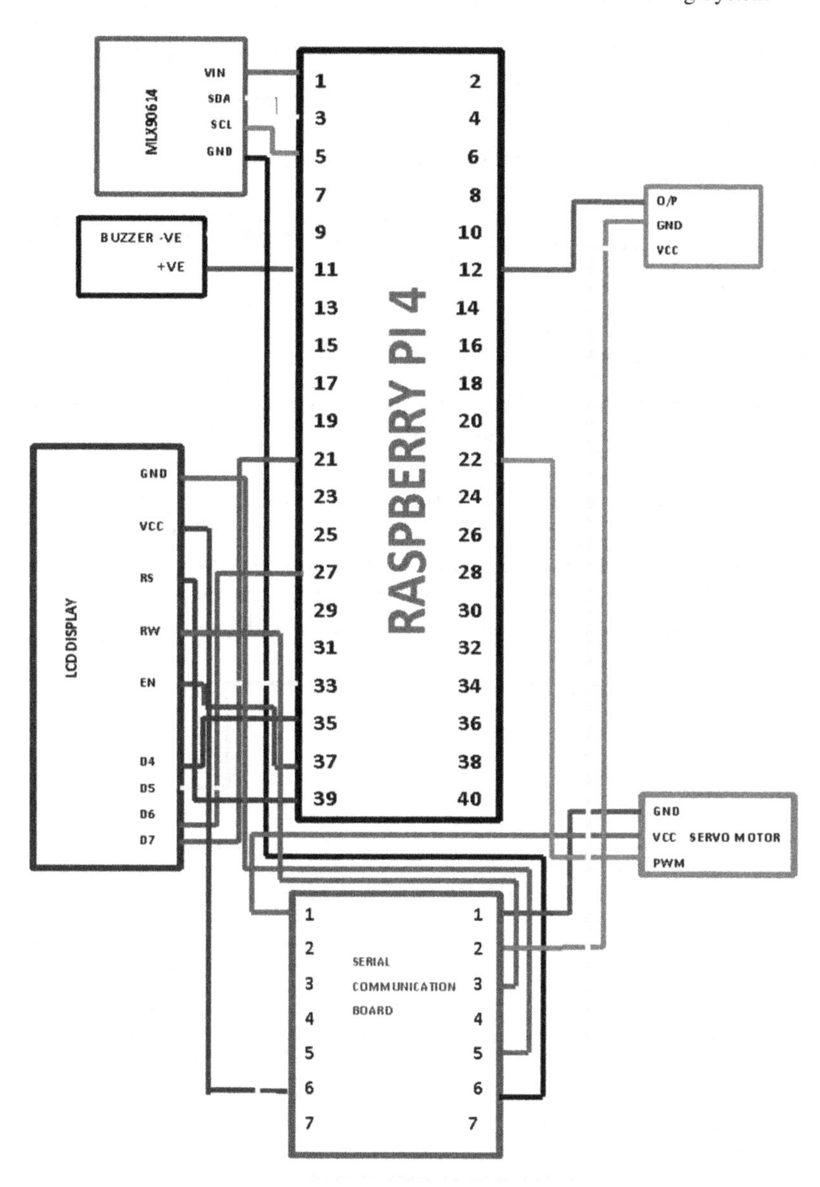

Fig. 3. Schematic Diagram

The three-pin IR sensor, which serves as both an input and an output, is linked to the serial communication board's VCC, GND, and pin 12 on the Raspberry Pi 4. The Raspberry Pi 4's pins 9 and 11 are linked to the buzzer's Positive and Negative pins, which serve as the device's output. The Raspberry Pi 4's pin 22 is linked to the servo motor's three pins, pins GND and VCC, which also serve as an output. GND, VCC, RW, RS, EN, D4 to pin 35, D5 to pin 33, D6 to pin 27 and D7 to pin 21 are linked to the Serial

Communication Board through the LCD Display's 16 pins. The schematic diagram is shown in Fig. 3.

4 Software and Hardware Interfacing

Here, two tools-Advanced IP Scanner and VNC Viewer are used. The 'Advanced IP Scanner' is used to scan the IP Address of the Raspberry Pi and 'VNC Viewer' is used to view the Rasbian OS Screen.

4.1 Advanced IP Scanner

Advanced IP Scanner is a thorough network scanning tool with more capabilities than its name. You could assume that an application dubbed "Advanced IP Scanner" is limited to advanced IP scanning.

This utility can really perform a wide variety of additional network-related tasks. For instance, you may browse distant shared files and see all the devices that are connected to your network. However, the parts that follow will go into further depth.

Advance IP Scanner can be downloaded from: https://www.advanced-ip-scanner.com/download/. The screenshot of advanced IP scanner is shown in Fig. 4.

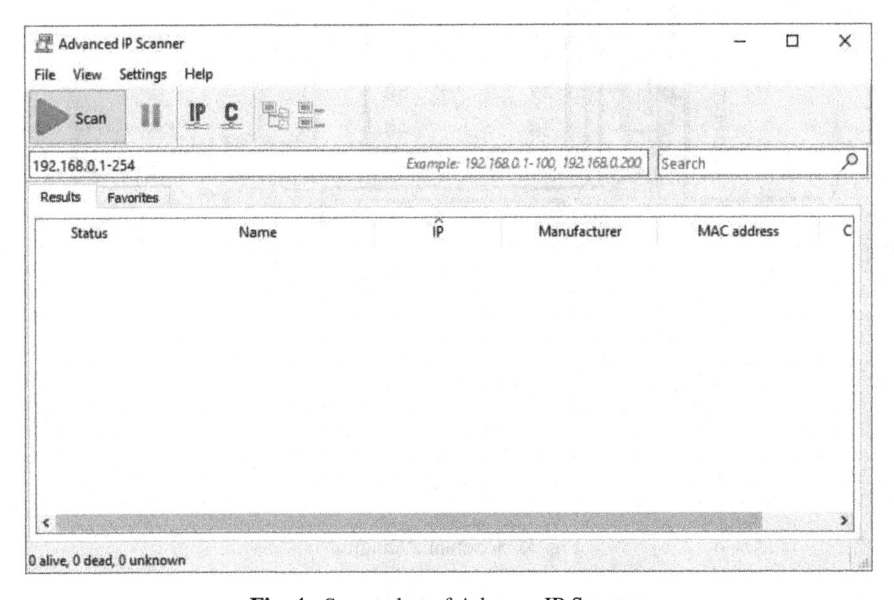

Fig. 4. Screenshot of Advance IP Scanner

4.2 VNC Viewer

VNC stands for Virtual Network Computing. It is a cross-platform screen-sharing utility designed for remote management of other computers. This means that a remote user can control the computer's keyboard, mouse, and screen with another device just as if they were directly in front of the computer. Control nearby PCs and mobile devices via VNC Viewer. VNC viewer software allows for remote access to and control of a device from a computer, tablet, or smartphone. VNC Viewer is available from:

https://www.realvnc.com/en/connect/download/viewer/. The screenshot of advanced IP scanner is shown in Fig. 5.

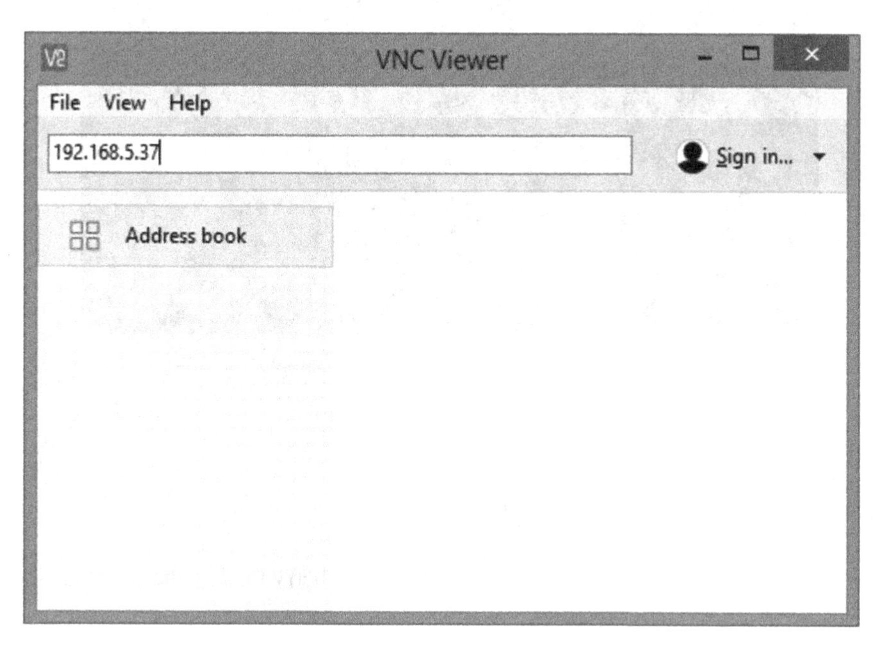

Fig. 5. Screenshot of VNC Viewer

5 Results

The system's primary goal is to prevent the spread of COVID-19 in public spaces including offices, malls, and other retail establishments. The system can monitor a person's body temperature and recognize facial masks. When there are fewer persons present than the stated limit, one may pass the IR sensor and go to the next level. When the temperature sensor measures their body temperature and determines that it is below the predetermined limit, the USB camera turns on and confirms that they are wearing the mask. The door automatically opens when the wearer's body temperature falls below the threshold and the mask is discernible; otherwise, access is not allowed. Any person whose body temperature exceeds the threshold or who is not wearing a face mask will get a warning through a buzzer or other integrated alert device. The setup of hardware is shown in Fig. 6.

The View of Advanced IP Scanner and VNC Viewer is shown in Fig. 7. The Advanced IP Scanner is used to scan the IP Address of the Raspberry Pi and VNC Viewer is used to view the Rasbian OS Screen. The steps for Advanced IP Scanner are as follows,

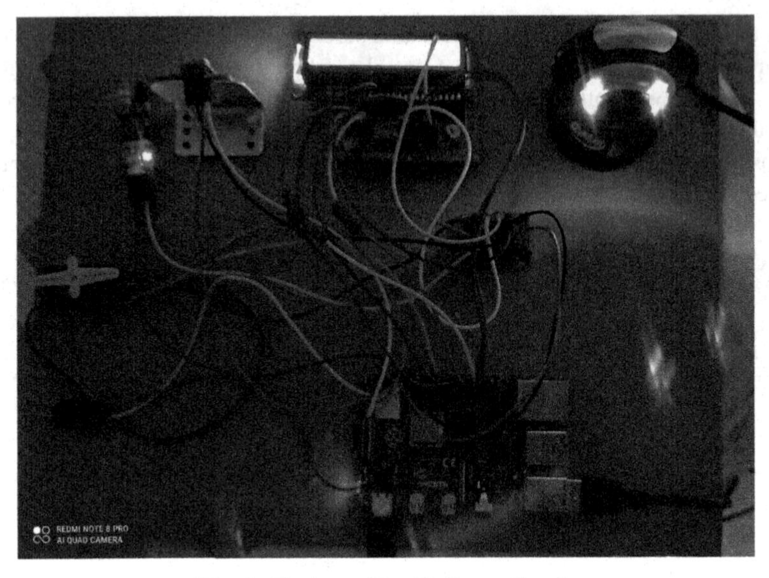

Fig. 6. Hardware kit with Power Supply

Step 1: Open Advanced IP Scanner.
Step 2: Click on Scan Button to scan the IP Address.
Step 3: Copy the Address of the Raspberry Pi.
Step 4: Open VNC Viewer.
Step 5: In VNC Connect, Paste the Address of the Raspberry Pi, then the Rasbian OS is open.

The object temperature in Fig. 8 is barely below the threshold value, but the mask status is None, meaning there is no mask. Gate is thus closed.

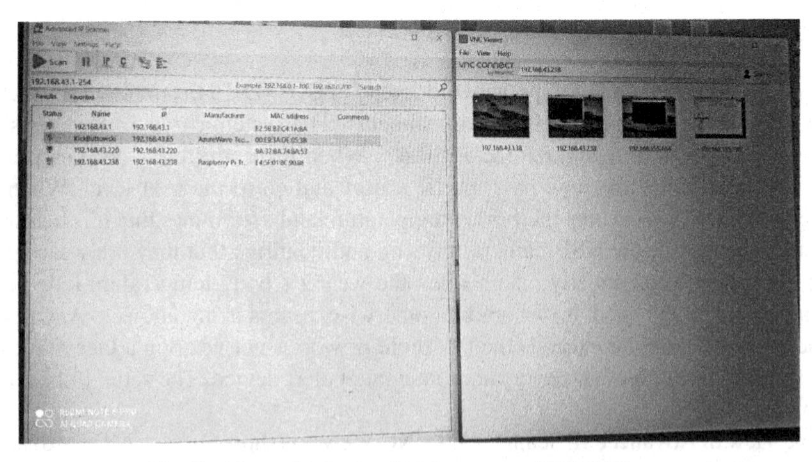

Fig. 7. View of Advanced IP Scanner and VNC Viewer

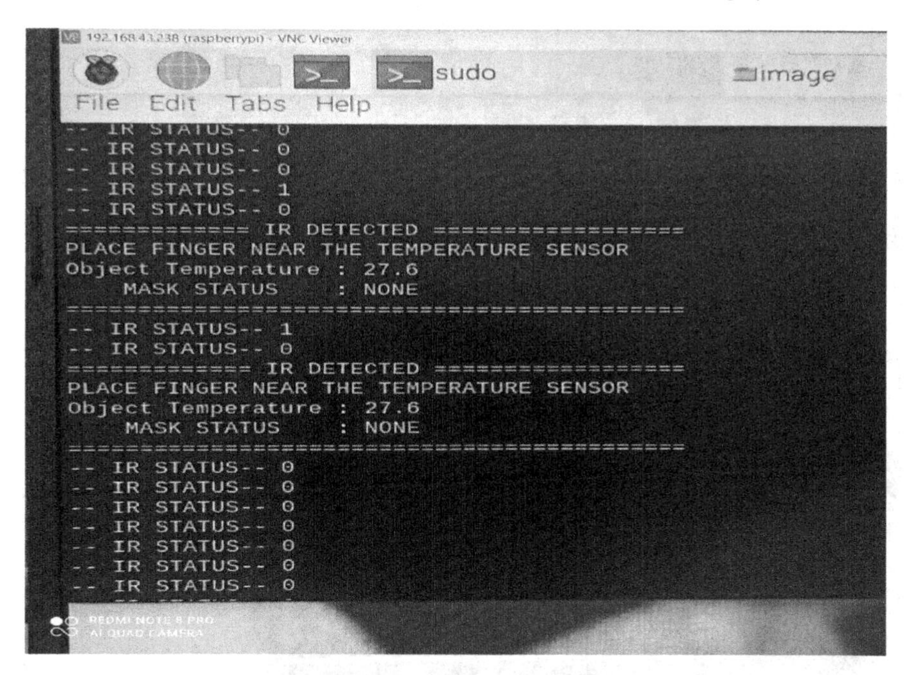

Fig. 8. Output view with No Mask

```
============= IR DETECTED ==================
PLACE FINGER NEAR THE TEMPERATURE SENSOR
Object Temperature : 28.5
   MASK STATUS     : YES
 ++++++++   OPERATING GATE  ++++++
=============================================
REDMI NOTE 8 PRO
```

Fig. 9. Output view with Mask

In Fig. 9, the mask status is Yes, and the object temperature is below the threshold value. Operation Gate is now open. Figure 10 shows the face of the person with mask.

Table 1 shows the validation results of the experiments. The performance parameters like accuracy, precision, recall, f1-score are calculated.

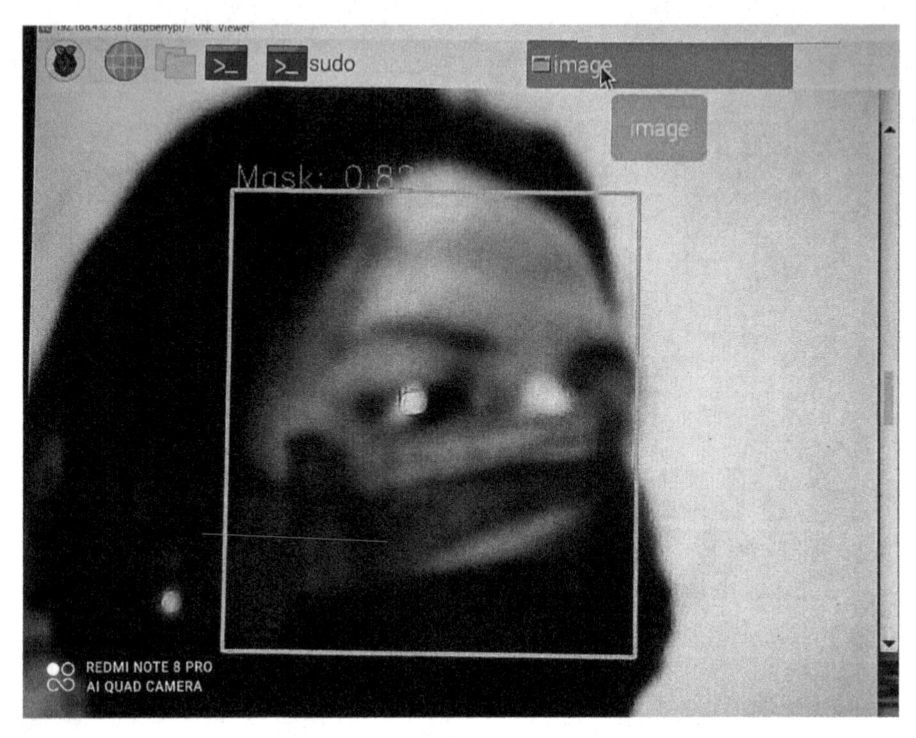

Fig. 10. Mask View

Table 1. Validation Results.

Parameters	Precision	Recall	F1 score	Accuracy
With 'mask'	0.97	0.98	0.98	0.99
With 'no mask'	0.99	0.99	1.00	0.99
Body temperature	0.98	0.97	0.99	0.99

6 Conclusion

The automatic door unlocking system monitors body temperature and looks for face masks to improve public safety. It uses infrared imaging and facial mask recognition technologies. The device incorporates a Raspberry Pi 4 for face mask recognition, an MLX90614 for body temperature monitoring, and a real-time deep learning engine. The results show a high level of accuracy in distinguishing between mask wearers and non-users. The method is effective for measuring body temperature and spotting masks. The doors will open if a person is covered up and their body temperature is below the threshold; otherwise, the buzzer will sound.

Future improvements will concentrate on enhancing these stages' accuracy, performance, merging various activities, and producing a mobile app with an intuitive user interface for monitoring. As a result, the authorities may react right away in accordance with pandemic safety guidelines.

References

1. Ghosh, A., Nundy, S., Mallick, T.K.: How India is dealing with COVID-19 pandemic. Sens Int. **1**, 100021 (2020)
2. Hidayat, A, et al.: Designing IoT-based independent pulse oximetry kit as an early detection tool for Covid-19 symptoms. In: 3rd International Conference on Computer and Informatics Engineering (IC2IE), pp. 443–448. Yogyakarta, Indonesia (2020)
3. Boano, C.A., Lasagni, M., Römer, K., Lange, T.: Accurate temperature measurements for medical research using body sensor networks. In: 14th IEEE International Symposium on Object/Component/Service-Oriented Real-Time Distributed Computing Workshops, pp. 189–198. Newport Beach, CA, USA (2011)
4. Vedaei, S.S., et al.: COVID-SAFE: an IoT-based system for automated health monitoring and surveillance in post-pandemic life. IEEE Access **8**, 188538–188551 (2020). https://doi.org/10.1109/ACCESS.2020.3030194
5. Yahiya, F.H., Yusuf, Y.M., Abidin, H.Z., Rahman, R.A.: Development of a PIC-based wireless Sensor node utilizing XBEE Technology. In: 2nd IEEE International Conference on Information Management and Engineering, pp. 116–120. IEEE, Chengdu, China (2010)
6. Ammar, A., Brach, M., Trabelsi, K., Chtourou, H., Boukhris, O., Masmoudi, L.: Effects of COVID-19 home confinement on eating behaviour and physical activity: results of the ECLB-COVID19 international online survey. Nutrients **12**(6), 1–14 (2020)
7. Varshini, B., Yogesh, H.R., Pasha, S.D., Suhail, M., Madhumitha, V., Sasi, A.: IoT-Enabled smart doors for monitoring body temperature and face mask detection. Global Transitions Proc. **2**(2), 246–254 (2021)
8. Deora, G., Godhula, R., Udpikar, V.: Study of masked face detection approach in video analytics. In: Conference on Advances in Signal Processing (CASP), pp. 196–200. IEEE, Pune, India (2016)
9. Baskaran, K., Baskaran, P., Rajaram, V., Kumaratharan, N.: IoT based COVID preventive system for work environment. In: 2020 Fourth International Conference on I-SMAC (IoT in Social, Mobile, Analytics and Cloud) (I-SMAC), pp. 65–71. IEEE, Palladam, India (2020)
10. Lim, M.G., Chuah, J.H.: Durian types recognition using deep learning techniques. In: 9th IEEE Control and System Graduate Research Colloquium (ICSGRC), pp.183–187. IEEE, Shah Alam, Malaysia (2018)
11. Kioumars, A.H., Tang, L.: Wireless network for health monitoring: heart rate and temperature sensor. In: Fifth International Conference on Sensing Technology, pp. 362–369. IEEE, Palmerston North, New Zealand (2011)
12. Vishwesh, M.S., Nikhil, D.K., Savita, S.: Social distancing and face mask detection from CCTV camera. Int. J. Eng. Res. Technol. (IJERT) **10**(8), 285–287 (2021)

Deep Learning Based Face Recognition System for Automated Identification

Prashant Ahlawat[1], Navpreet Kaur[1], Charnpreet Kaur[1], Santosh Kumar[1], and Hitesh Kumar Sharma[2](✉)

[1] Computer Science Engineering, University Institute of Engineering, Chandigarh University, Gharuan, Mohali, Punjab, India

[2] School of Computer Science, University of Petroleum and Energy Studies (UPES), Energy Acres, Bidholi, Dehradun, Uttarakhand 248007, India

hkshitesh@gmail.com

Abstract. Face recognition is a fundamental task in computer vision with numerous applications in various domains, including surveillance, security, access control, and human-computer interaction. In recent years, deep learning has revolutionized the field of face recognition by significantly improving accuracy and robustness. This abstract presents an overview of a deep learning-based face recognition system developed for automated identification purposes. The proposed system leverages convolutional neural networks (CNNs) to extract discriminative features from facial images and a classification model to match and identify individuals. The system comprises three main stages: face detection, feature extraction, and identification. In the face detection stage, a pre-trained CNN model is employed to locate and localize faces within input images or video frames accurately. The detected faces are then aligned and normalized to account for variations in pose, scale, and illumination. Next, a deep CNN-based feature extraction network is utilized to capture high-level representations from the aligned face regions. This network learns hierarchical features that are robust to variations in facial appearance, such as expression, occlusion, and aging. The extracted features are typically represented as a compact and discriminative embedding vector, facilitating efficient and accurate face matching. To enhance the system's performance, various techniques can be employed, such as data augmentation, model fine-tuning, and face verification to handle challenging scenarios, including pose variations, illumination changes, and partial occlusions. The proposed deep learning-based face recognition system has shown remarkable accuracy and robustness in automated identification tasks. It has the potential to be deployed in real-world applications, including surveillance systems, access control in secure facilities, and personalized user experiences in human-computer interaction.

Keywords: Face Recognition · Machine Learning · Attendance System · BPH

1 Introduction

A sheet of paper is typically used to record attendance, along with any other comments that may be required, and on which the student's name and any other pertinent information is written. Each student will receive a copy of this piece of paper, which will serve

R. K. Challa et al. (Eds.): ICAIoT 2023, CCIS 1930, pp. 60–72, 2024.
https://doi.org/10.1007/978-3-031-48781-1_6

as their registration. As an illustration, the lecturer might fill out the date on the registration form before passing the paper around to each student in the group. The lecturer occasionally gave out papers before the session started, to make the most of everyone's time and the most efficient use of the speaker's time. Fig. 1 shows the face recognition system.

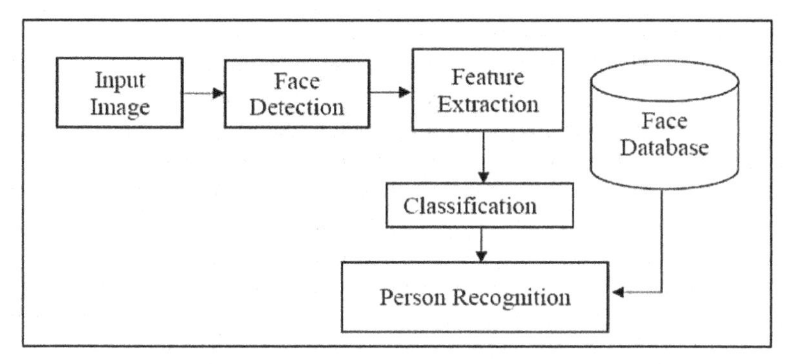

Fig. 1. Face recognition system

The two methods that have been thus far discussed have both been used on a sizable scale for a sizable amount of time. On the other hand, problems frequently arise when the typical or traditional technique of recording attendance is applied. They could take a long time, particularly when pupils are expected to respond when the speaker calls their names. It will be difficult to keep the attendance report up to date until it is time to write the report since it is conceivable that the registration form or the attendance sheet have been misplaced. While using the conventional approach, lecturers or teachers must accurately record the attendance of each student by marking their presence on the registration sheet or attendance form.

2 Literature Review

Authentication is a computer-based type of communication that serves a very important role in preserving system control [1]. Face recognition is currently utilised in a wide range of applications and has grown to be a crucial part of biometric authentication. Examples of these applications include network security, human-computer interfaces, and video monitoring systems. [2] The requirement that students attend a physical educational facility is the element of the virtual platform that presents the greatest challenge for pupils. The technique takes a lot of time and effort to complete by hand. [3] One of the most useful applications of image processing is the recognition of faces, which is crucial in the technology world. Recognizing human faces is a current issue that needs to be addressed for the sake of authentication, particularly in the context of taking student attendance. [4] Educational institutions are concerned about the consistency of their students' performance in the modern environment. One element influencing the general fall in pupils' academic performance is the low attendance rate. Your attendance may be recorded using a variety of techniques, the two most common being signing in and having pupils raise their hands. The process took longer and presented several limitations and challenges.

3 Methodology

The rising desire for higher levels of security as well as the quick uptake of mobile devices may be to blame for the spike in interest in facial recognition research over the past several years. Whatever the case, field researchers have warmly welcomed this excitement. In any case, the unexpected increase in interest has been a really fortunate development. Face recognition technology has applications in many fields, including but not limited to access control, identity verification, surveillance, security, and social media networks. In the computer vision subfield of face detection, the OpenCV approach is a well-liked option. The AdaBoost approach is then employed as the face detector after having previously been used to extract the feature photos into a sizable sample set by first extracting the face Haar features included within the image [5]. This happens following its use to extract the feature images from a sizable sample set. Here, the goal is to teach a deep neural network how to recognize people by their faces and output their faces. In order for the neural network to automatically recognise the many elements of a face and produce numbers based on those features, it appears that the neural network must be trained. If you feed in multiple photographs of the same person, the neural network's output will be very similar or close, however if you pass in multiple images of different persons, the output will be considerably different. The neural network's output can be thought of as an identifier for a certain person's face. The output of the neural network can be compared to a face-specific identification number. The Local Binary Patterns Histogram method was put out as a potential fix for the issue in 2006. It is constructed on top of the framework offered by the neighbourhood binary operator. Due to how easily it can be calculated and how powerfully it can select, it is used extensively in the field of face recognition. The following actions must be taken in order to reach this objective: (i) the creation of datasets; (ii) the collecting of faces; (iii) the recognition of distinctive traits; and (iv) the classification of the faces.

Figs. 2 and 3 show the algorithm and data flow process.

A dialogue window that asks for the user id and name, in that order, appears on the screen during the initial run of the application. The following step is selecting the "Take Images" option by clicking the pertinent button after entering the necessary data in the "name" and "id" text boxes. The working computer's camera will open when you select the Take Images option from the drop-down menu, and it will begin capturing pictures of the subject. The file with this ID and Name's name is Student Details.csv, and both of them are kept in a folder called Student Details. The Documents area of the navigation bar contains the Student Details folder.

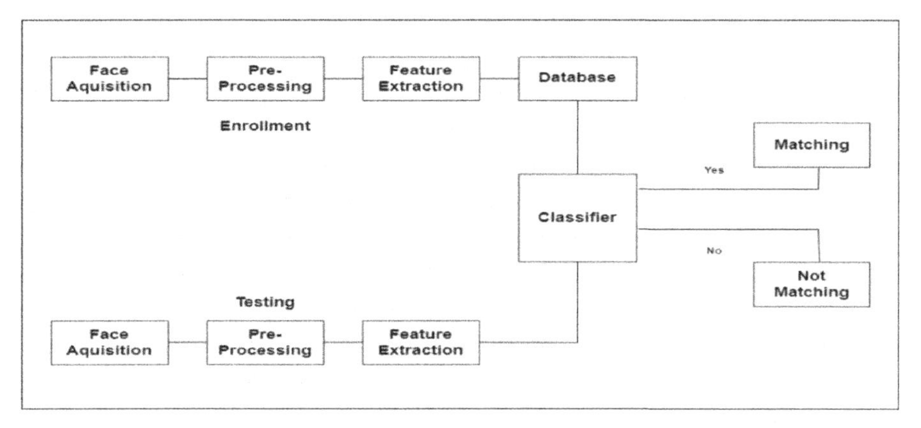

Fig. 2. Algorithm and Data flow process.

These sixty photographs are kept in a folder called Training Image and are used as an example. After the procedure has been properly finished, a notification that the photographs have been saved will be obtained once the process is complete. Once several picture samples are collected, the Train Image button is selected in order to train the images that have been collected. Now it only takes a few seconds to teach the computer to recognise the images, which also results in the creation of a Trainer.yml file and the storage of the images in the Training Image Label folder. Each of the initial parameters has now been fully implemented as of this writing. The following step is to choose the Track images after choosing Take pictures and Train pictures.

4 Dataset Description

For this work, the Facial Images Dataset available at [https://www.kaggle.com/datasets/apollo2506/facial-recognition-dataset] is used. It consists of 3950 images with different facial expressions. The dataset consists of basically 5 categories of facial expression [10]. The dataset taken for this research work is publicly available facial expression dataset.

The description of the image dataset and number of images in each class has given in Table 1 (Fig. 4).

4.1 Dataset Standardization

An image generator is created to standardize the input images. It will be used to adjust the images in such a way that the mean of pixel intensity will be zero and standard deviation will become 1. The old pixel value of an image will be replaced by new values calculated using following formula. In this formula each value will subtract mean and divide the result with standard deviation [6].

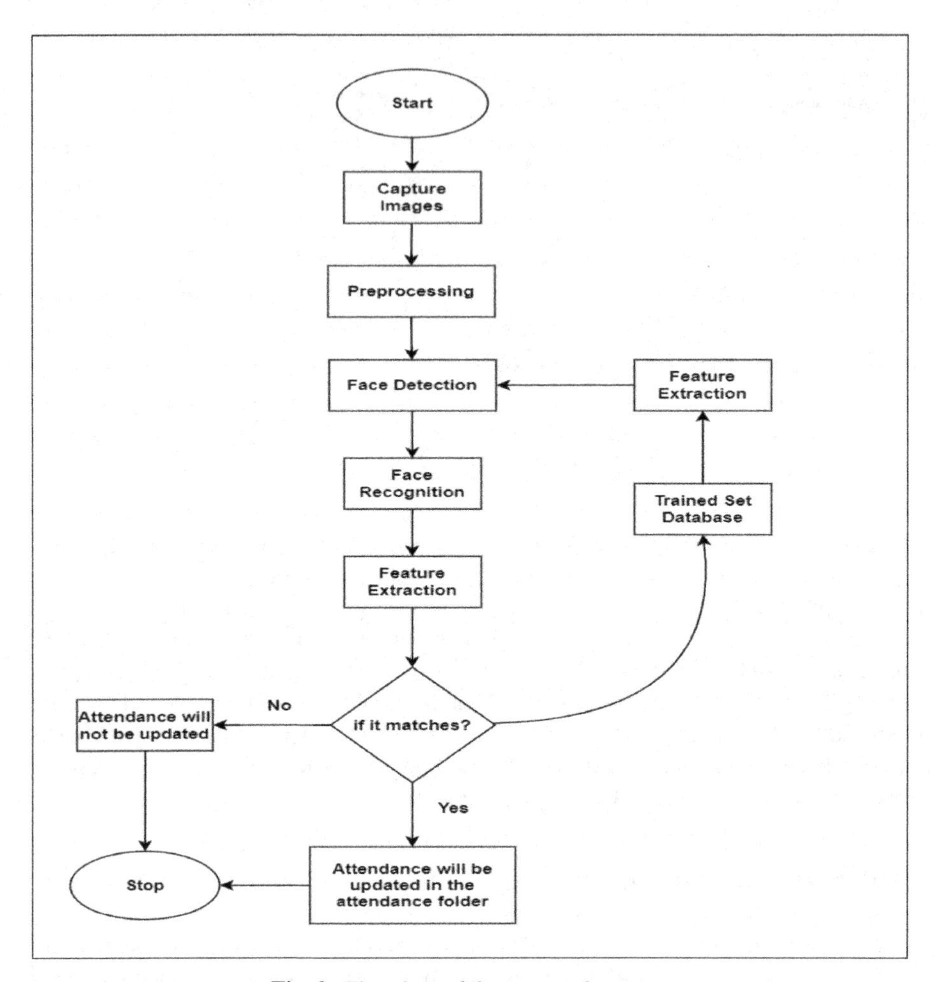

Fig. 3. Flowchart of the proposed system.

Table 1. Facial Image Dataset Description [https://www.kaggle.com/datasets/apollo2506/facial-recognition-dataset]

Facial Expression Class	Images
Angry	540
Sad	890
Happy	770
Neutral	760
Surprise	990

Fig. 4. Dataset sample images.

4.2 Class Imbalance in Image Dataset M

The major challenge in plant disease or medical image dataset is class imbalance problem. There are not equal numbers of images in all classes of these datasets. In such case, biasness will be for the class which has a greater number of images [7]. The frequency of each class or label has been plotted in Fig. 5.

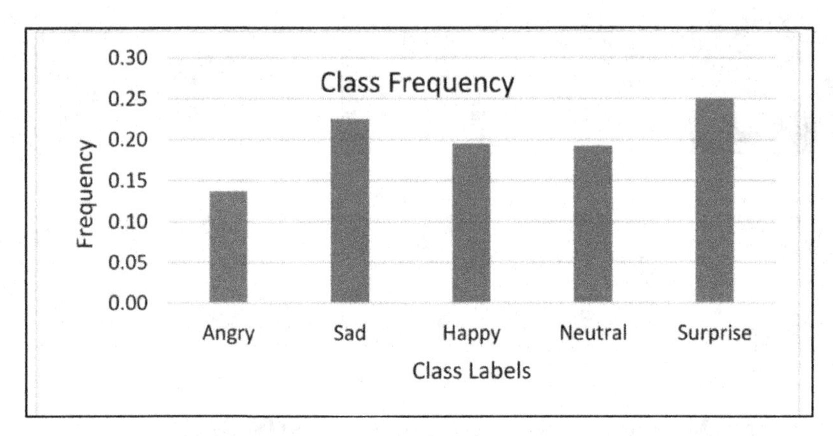

Fig. 5. Class frequency plot

It can be seen in Fig. 5 that, there is a large variation in number of images in each class. This data imbalance generates a biasness in the model and reduce the performance of model. To overcome this data imbalance challenge the following mathematical formulation is applied which is explained in detail in following subsection [11, 13].

4.3 Impact of Class Imbalance on Loss Function

The cross-entropy loss formula is given in Eq. (1).

$$L_{cross\ entropy}(X_i) = -(Y_i Log(f(X_i)) + (1 - Y_i)Log(1 - f(X_i))) \tag{1}$$

X_i and Y_i denotes the input feature and corresponding label, f(X_i) represents the output of the model. Which is probability that output is positive.This formula is written for overall average cross entropy loss for complete dataset D of size N and it is given below in Eq. (2).

$$L_{cross\ entropy}(D) = -\frac{1}{N}(\sum_{positive} Log(f(X_i) + \sum_{Negative} Log(1 - f(X_i))) \tag{2}$$

This formulation shows clearly that the loss will be dominated by negative labels if there is large imbalance in dataset with very few frequencies of positive labels. Fig. 6 shows the positive and negative class frequency of the five given classes before applying weight factor [12].

$$Freq_{positive} = \frac{No.\ of\ Positive\ Samples}{N} \tag{3}$$

$$Freq_{negative} = \frac{No.\ of\ Negative\ Samples}{N} \quad (4)$$

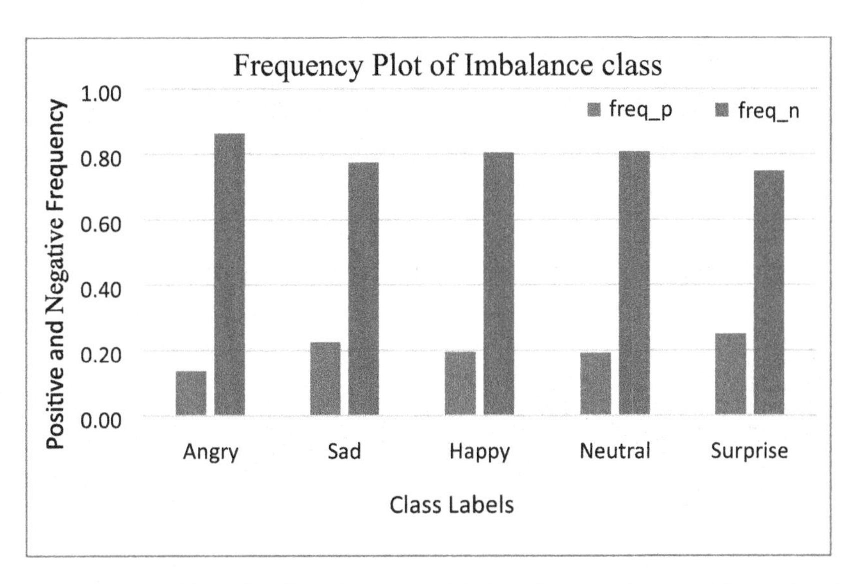

Fig. 6. Positive and Negative Class Frequency of 5 given Classes before Applying Weight Factor.

As shown in Fig. 6, contribution of positive labels is less than the negative labels. However, for accurate results this contribution should be equal. The one possible way to make equal contribution is to multiply each class frequency with a class specific weight factor $W_{positive}$ and $W_{negative}$, so each class will contribute equally in classification model [8, 9].

So, the formulation will be represented as,

$$W_{positive} \times Freq_{positive} = W_{negative} \times Freq_{negative} \quad (5)$$

which can be done simply by taking,

$$W_{positive} = Freq_{negative}$$
$$W_{negative} = Freq_{positive}$$

Using the above formulation, the class imbalance problem is dealt with. To verify the formulation, the frequency graph is plotted and it shows the expected chart. It can be seen in Fig. 5 that both frequencies are balanced. (Table 2)

As shown in Fig. 7, after applying weight factor to positive and negative class the contribution to loss function is equally distributed. The calculation of loss function is given in Eq. (6).

$$L^{W}_{cross\ entropy}(X) = -(W_{postive} Y Log(f(X)) + W_{negative}(1 - Y)Log(1 - f(X))) \quad (6)$$

This adjustment of loss function will eliminate class imbalance problem and biasness problem with imbalance dataset [14, 15].

Table 2. Computational table for data imbalance.

Class Labels	Total Images	freq_p	freq_n	w_p	w_n	freq_p	freq_n
Angry	540	0.14	0.86	0.86	0.14	0.12	0.12
Sad	890	0.23	0.77	0.77	0.23	0.17	0.17
Happy	770	0.19	0.81	0.81	0.19	0.16	0.16
Neutral	760	0.19	0.81	0.81	0.19	0.16	0.16
Surprise	990	0.25	0.75	0.75	0.25	0.19	0.19

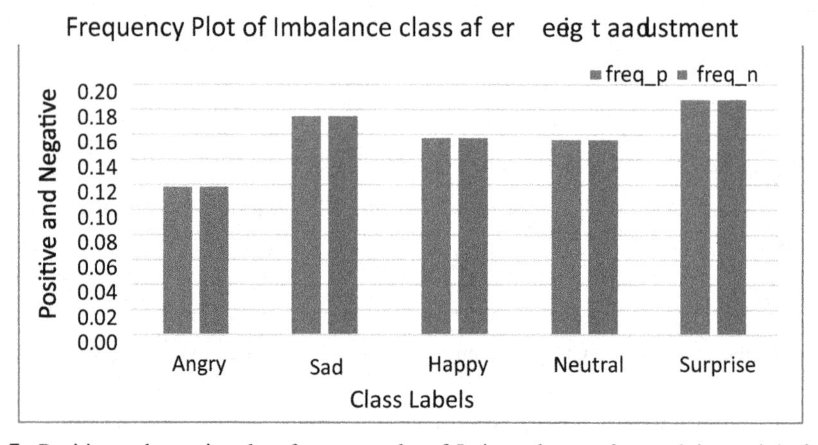

Fig. 7. Positive and negative class frequency plot of 5 given classes after applying weight factor.

5 Model Methodology

The proposed face recognition framework is carried out by building up a convolutional neural organization which is a productive model for picture characterization. CNN models are prepared by first taking care of a bunch of face pictures alongside their marks. These pictures are then gone through a pile of layers including convolutional, ReLU, pooling, and completely associated layers. These pictures are taken as bunches. In the proposed framework, a group size of 64 was given. The model was prepared to utilize 50 epochs. At first, the model concentrates little highlights, and as the preparation interaction advances more definite highlights will be separated. The greater part of the pre-processing is done consequently which is one of the significant benefits of CNN. Notwithstanding that information pictures were resized. Expansion is additionally applied which builds the size of the dataset by applying activities like turn, shear, and so forth. During the preparation interaction, the model finds highlights and designs and learns them. This information is then used later to discover the expression of face when another face expression picture is given as info. Unmitigated cross-entropy is utilized as misfortune work [16]. At first, the misfortune esteems would be extremely high however as the interaction progresses the misfortune work is diminished by changing the weight esteems.

The layer configuration of CNN model is shown in Table 3.

Table 3. Layer specifications of our CNN model

Proposed CNN Model specification	
Layers	Layer specifications
Optimizer	Adam
Loss	categorical_crossentropy
Metrics	['accuracy']
Conv2D Layer (1)	64 filters, 3×3 Filter size and ReLU Activation
Max-Pooling2d Layer	2×2 kernel size
Dropout	20%
Conv2D Layer (2)	128 filters, 3×3 Filter Size and ReLU Activation
Max-Pooling2d Layer	2×2 kernel
Dropout	30%
Conv2D Layer (3)	128 filters, 3×3 Filter Size and ReLU Activation
Max-Pooling2d Layer	2×2 kernel
Dropout	30%
Conv2D Layer (4)	256 filters, 3×3 Filter size and ReLU Activation
Max-Pooling2d Layer	2×2 kernel size
Dropout	20%
Conv2D Layer (5)	512 filters, 3×3 Filter size and ReLU Activation
Max-Pooling2d Layer	2×2 kernel size
Dropout	30%
Flatten layer	4608 Neurons
Dense Layer	1024 Neurons
Batch Normalization	Relu Activation
Dropout	50%
Dense Layer	512 Neurons
Batch Normalization	Relu Activation
Dropout	25%
Output layer	Softmax 5 classes

The proposed model contains

Total params: 10,101,765
Trainable params: 10,097,541
Non-trainable params: 4,224

6 Experimental Results

The heat map of confusion matrix is given in Fig. 8.

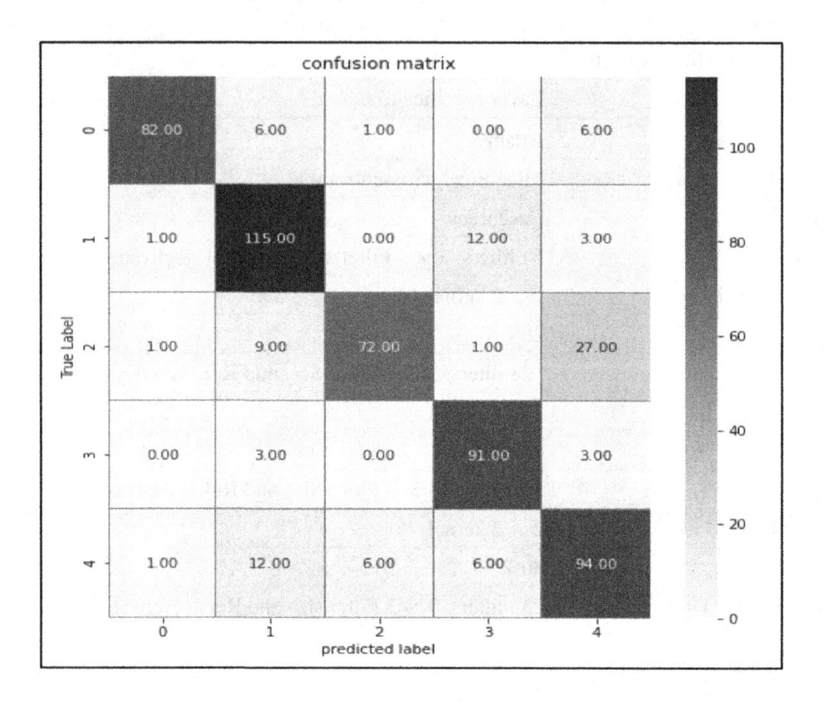

Fig. 8. Heat map of Confusion matrix for our proposed model

The results of training and validation accuracy are shown in Fig. 9(a) and (b). Accuracy rate exceeded 80% limit while in the training phase. Similarly, validation accuracy is around 80%. For this data set the result is appreciable and it can easily exceed more in more epochs.

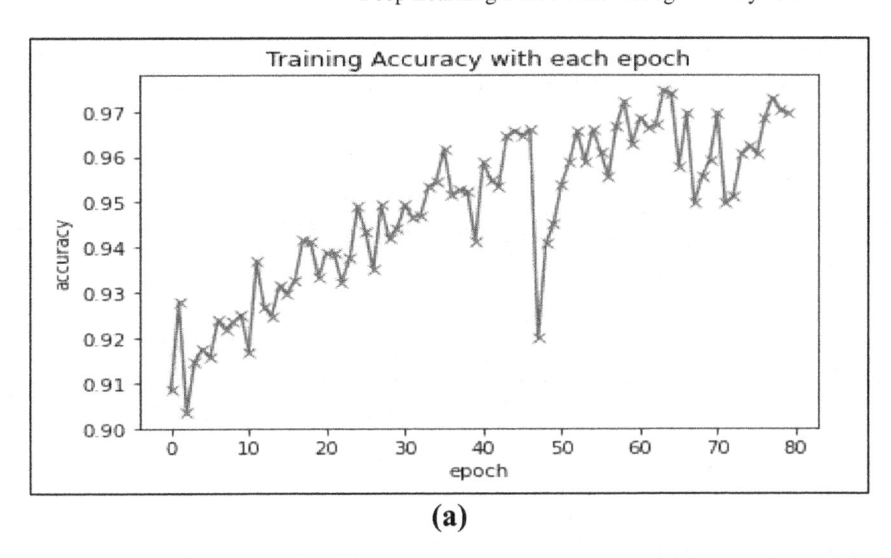

(a)

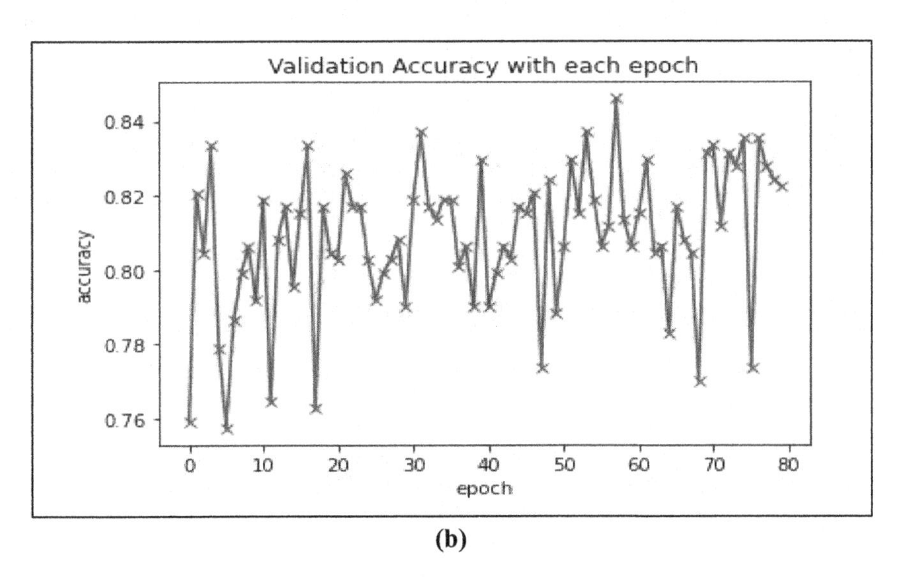

(b)

Fig. 9. Training and Validation Accuracy of Model.

7 Conclusion

Hence, in order to track the children's whereabouts, this work created a procedure for the management and administration of attendance. Particularly in situations where a sizable portion of students have previously recorded their attendance, it helps to reduce the amount of time and effort needed. Python is the programming language that is utilised to implement the system from start to finish. Face recognition techniques were employed to accurately record the children's attendance. This list of students' attendance may also be used for other purposes, such as to identify who is and is not present for exams in order to resolve examination-related issues.

References

1. Chen, Y., Jiang, H., Li, C., Jia, X.: Profound component extraction and characterization of Hyperspectral pictures dependent on Convolutional Neural Network (CNN). IEEE Trans. Geosci. Remote Sens. **54**(10), 62326251 (2016)
2. Chuang, M.-C., Hwang, J.-N., Williams, K.: A feature learning and object recognition framework for underwater fish images. IEEE Trans. Image Process. **25**(4), 1862–1872 (2016)
3. Chuang, M.-C., Hwang, J.-N., Williams, K.: Regulated and unsupervised highlight extraction methods for underwater fish species recognition. In: IEEE Conference Distributions, pp. 33–40 (2014)
4. Kim, H., et al.: Image-based monitoring of jellyfish using deep learning architecture. IEEE Sensors J. **16**(8), 2215–2216 (2016)
5. Silva, C., Welfer, D., Gioda, F.P., Dornelles, C.: Cattle brand recognition using convolutional neural network and support vector machines. IEEE Latin Am. Trans. **15**(2), 310–316 (2017)
6. Sharma, H.K.: E-COCOMO: the extended cost constructive model for cleanroom software engineering. Database Syst. J. **4**(4), 3–11 (2013)
7. Sharma, H.K., Kumar, S., Dubey, S., Gupta, P.: Auto-selection and management of dynamic SGA parameters in RDBMS. In: 2015 2nd International Conference on Computing for Sustainable Global Development (INDIACom), pp. 1763–1768. IEEE (2015)
8. Khanchi, I., Ahmed, E., Sharma, H.K.: Automated framework for Realtime sentiment analysis. In 5th International Conference on Next Generation Computing Technologies (NGCT-2019) (2020)
9. Agarwal, A., Gupta, S., Choudhury, T.: Continuous and Integrated Software Development using DevOps. In: 2018 International Conference on Advances in Computing and Communication Engineering (ICACCE), pp. 290–293 (2018). https://doi.org/10.1109/ICACCE.2018.8458052
10. Chen, C., Zhong, Z., Yang, C., Cheng, Y., Deng, W.: Deep learning for face recognition: a comprehensive survey. IEEE Trans. Pattern Anal. Mach. Intell. (TPAMI) (2022). https://doi.org/10.1109/TPAMI.2022.3056791
11. Duan, Z., Zhang, X., Li, M.: FSA-net: a light-weight feature shrinking and aggregating network for real-time face recognition. Pattern Recogn. Lett. **154**, 129–136 (2022)
12. Jin, Y., Xu, Y., Dong, S., Yan, J.: DSFD: dual shot face detector. IEEE Trans. Pattern Anal, Mach. Intell. (2022). https://doi.org/10.1109/TPAMI.2021.3082544
13. Cai, Y., Guo, H., Zhang, Z., Wang, L.: Deep dual attention networks for face recognition. Pattern Recogn. **125**, 108308 (2022)
14. Cheng, Y., Chen, C., Deng, W.: Learning spatial-aware and occlusion-robust representations for occluded face recognition. IEEE Trans. Pattern Anal. Mach. Intell. (2021). https://doi.org/10.1109/TPAMI.2021.3081622
15. Li, X., Liu, Z., Chen, J., Lu, H.: Stacked dense network with attentive aggregation for robust face recognition. IEEE Trans. Cybern. 1–14 (2021)
16. Zhang, W., Liu, H., Zhu, S., Zhang, S., Zhang, Y.: DPSG-Net: deep pose-sensitive graph convolutional network for face alignment. Pattern Recogn. Lett. **153**, 52–59 (2021)

Desirability Function Analysis Based Optimization of the On-Machine Diameter Measurement Using Machine Vision Under RGB Light

Rohit Zende(✉) and Raju Pawade

Department of Mechanical Engineering, Dr. Babasaheb Ambedkar Technological University, Lonere, Maharashtra 402103, India
zenderohit@gmail.com

Abstract. IoT-based machine vision system for dimensional inspection pays more attention to dimensional measurements of the components in a manufacturing system with the aim of a swift, reliable, error-proof measuring system with minimum human intervention. The use of an IoT system integrated with a machine vision system has a great advantage which improves the overall performance of the system. Such a system can work under a complex system environment. This research focuses on developing an IoT-based machine vision system for diameter measurement. Diameter measurements are performed under RGB light sources to investigate the errors in measurements. The approach of desirability function analysis is applied in order to reduce measurement errors which are further confirmed by a Taguchi method. Experimental results show that red light gives the lowest percentage error in diameter measurement with a value of ±0.2014% while blue light shows the highest percentage error in diameter measurement as ±0.6320%. However, the green light measurement method has a moderate percentage error in diameter measurement with a percentage error of ±0.5528%.

Keywords: IoT · Machine vision · Diameter measurement · Desirability Function Analysis

1 Introduction

Nowadays, most industries are planning to adopt the Industry 4.0 revolution in their day-to-day operations. Industry 4.0 involves the digitalization of processes, machines, and systems. Although the systems are automated in Industry 4.0, systems are communicating with each other by using the Internet. There are various technologies used in Industry 4.0, the majority among them being "cloud computing", "augmented reality", "big data", "virtual reality", the "Cyber-Physical System" (CPS), the "Internet of Things" (IoT), "Additive Manufacturing", and "Artificial Intelligence" (AI) [1, 2]. Intelligent manufacturing is another concept involved in Industry 4.0 where manufacturing systems are connected and are capable to make smart decisions by using AI technology

R. K. Challa et al. (Eds.): ICAIoT 2023, CCIS 1930, pp. 73–85, 2024.
https://doi.org/10.1007/978-3-031-48781-1_7

with dedicated CPS [3]. IoT involves the use of an internet connection system to communicate between the measuring environment and the decision-making environment. Several sensors that are connected to local servers are used in the measurement environment. The system's output performance is measured by the sensors and the information is collected at the server station known as cloud storage. The collected information is further used for data analysis to produce the final desired results. The IoT life cycle involves various stages from the construction to the update of the system [4]. The physical layer, data exchange layer, information integration layer, and application service layer form the four layers that combine to form the overall architecture of the Internet of Things system. [5]. The physical layer involves various sensors, microcontrollers, actuators, and different physical devices. In the case of the first layer the communication between the sensors and the measuring environment is performed. The data exchange layer mainly involves the use of Internet connections to connect the sensors, microcontrollers, and actuators with the Internet server. The information integration layer includes the storing of data at a "cloud server", "cloud computing", and "big data" processing of the stored data. In the case of the application service layer, the processed data is communicated to the end user. This is the last layer where an end user is going to control and monitor the system.

In a manufacturing system, the component is machined using various cutting tools and to check the conformity of the dimensions within a specified tolerance; the measurement of the geometrical features of the machined component is needed. Thus, from a quality control point of view measurement of the machined components are equally important. This measurement of the geometrical features is carried out using various sensors and machine vision systems. The machine vision system involves the use of the camera to capture the image which is further used for image processing and analysis. Thus, in IoT based measurement system, the machine vision system acts as a bridge between the measuring environment and the cloud server. In the case of the turning process, cylindrical components are machined, therefore to inspect the machined part geometrical features like diameter, run out, roundness, and cylindricity of the cylindrical component needs to be measured by using machine vision and IoT systems. Thus, more flexibility and efficiency in the production system could be the advantage of a machine vision system that measures the geometrical features on-machine and in-process.

Ray [6] developed an IoT-based thermal comfort device to measure the temperature and humidity of the system. The experimental results show the usefulness and capability of online monitoring using an IoT measurement system. Marques et al. [7] used an IoT system for monitoring the particles present in the air inside the building. The experiments result shows the promising advantage of the IoT system for performance monitoring of the air particles for healthcare application. Wei and Tan [8] employed cameras to measure the diameter of the rods. The on-machine camera setup was developed for the measurement and it was found that the flexibility of the machine vision system is advantageous for diameter measurement. Jianming et al. [9] developed an algorithm to compensate for the error in diameter measurement using a "machine vision system". The results show that the proposed "machine vision system" can achieve high precision in diameter measurement. Tan et al. [10] devised a modeling-based diameter measurement system to get a desired ellipse to fit the geometry. The accuracy of the model was checked with a checkerboard calibration plate on a lathe machine which proves the usefulness of the system for on-machine diameter measurement.

Sait et al. [11] experimented with applying the desirability function analysis (DFA) technique "to optimize the machining parameter of turning glass-fiber reinforced plastic (GFRP) pipes". Experimental results show the usefulness of the DFA method in multi-objective optimization. Singaravel and Selvaraj [12] optimized "the machining parameters" during the turning of EN25 steel by using a DFA method. The experimental results show that the DFA method can be compared with other optimization methods to find better results. A similar method by Jenarthanan and Jeyapaul [13] was employed to optimize the machining parameters in the milling of "glass-fiber-reinforced plastic (GFRP)". For the design of the experiment, the Taguchi method was used, and the DFA method was used to optimize output responses like surface roughness, machining force, and delamination factor. Galgali et al. [14] applied the Taguchi-based DFA method to find the optimal size of the distributed generation method in power generation under different load conditions. The research work shows the wide applicability of the DFA method to optimize the system parameters.

From the literature review; IoT is found to be one of the important technologies for process monitoring as the machine vision system measures the various parameters of the objects during the machining process. However, the use of IoT-based machine vision technology still requires more attention to develop a reliable measurement system. DFA can be used to optimize the parameters of "the machine vision system". This research work explains the development of IoT based machine vision system for the measurement of the diameter of the component in the turning process. The on-machine measurement is performed using this system. The diameters are measured under RGB light sources and errors in diameters are further optimized using a DFA method.

2 Experimental Setup

The experimental setup for the machine vision system is shown in Fig. 1. As the internet and software technologies are available at ease; smartphones are playing a significant role in manufacturing and quality control. In this work, a smartphone with a 50-megapixel camera was used for image acquisition. The smartphone is further communicated to the computer system using a wireless local area network having a network speed of 100 Mbps. As a result, the captured image was easily transferred to the storage location of the computer system. In image processing, usually, the digital image is described in terms of RGB colored space where the value of each color ranges from 0 to 255 [15, 16]. RGB has a monochromatic wavelength which reduces the effect of noise because of illumination. Therefore; the effect of external RGB light sources is considered for the image measurement analysis. To measure the acquired image of the workpiece system; calibration is essential. Hence, a 40 mm slip gauge was used for image calibration. Taguchi "design of experiment" (DOE) is a powerful methodology that describes the relationship between input factors and their corresponding responses by creating an orthogonal array with a minimal number of experiments. [17–19]. In this research, external cylindrical turning operation with three parameters namely "cutting speed (m/min), feed rate (mm/rev), and depth of cut (mm)" were considered as input parameters and each one having three levels.

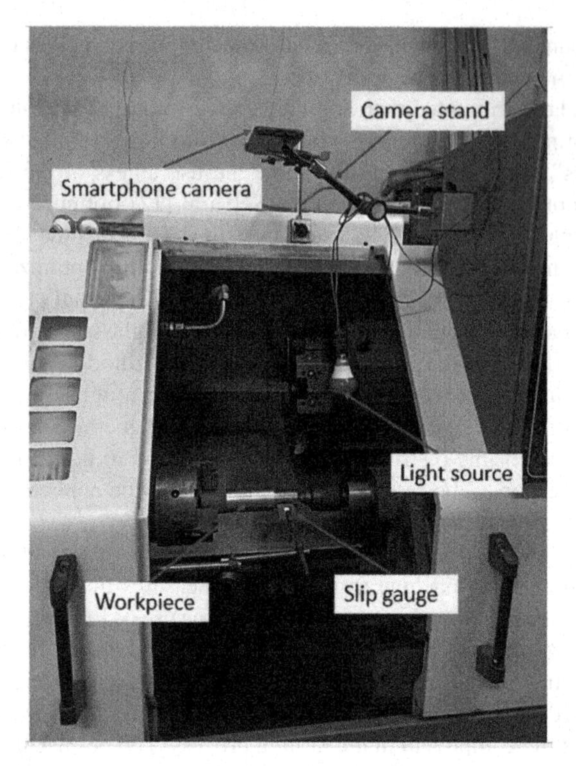

Fig. 1. Experimental setup of the machine vision system

Table 1. Taguchi DOE with nine experimental runs

Expt. Run	CS (m/min)	FR (mm/rev)	DC (mm)
1	80	0.15	0.3
2	80	0.25	0.6
3	80	0.35	0.9
4	100	0.15	0.6
5	100	0.25	0.9
6	100	0.35	0.3
7	120	0.15	0.9
8	120	0.25	0.3
9	120	0.35	0.6

where, CS = Cutting speed, FR = Feed rate, and DC = Depth of cut

Taguchi DoE array was constructed for conducting the experimental runs (see Table 1). To machine the workpiece's diameter, a total of nine experimental runs were carried out. These nine experimental runs are repeated two times and the average output reading was used for the measurement and analysis. EN8, medium carbon steel was used as a workpiece for machining while cutting tool inserts having specification as CNMG 120408-HM-PC9030 was chosen. The image acquisition was performed under individual red, green, and blue light sources separately. Each light source having a 3-W power rating was used for the image acquisition. Before capturing the image, during the experiment, the slip gauge edge was made parallel to the workpiece to reduce errors in pixel measurements of diameter values. Computer numerical control (CNC) turning center from Ace Designers, Jobber XL was used to perform the machining operation.

3 Methodology

Figure 2 shows the methodology for the proposed machine vision system. After performing each experimental run, the image is acquired by using a smartphone camera under RGB light sources individually. The greatest distance possible between any two points present on the periphery of the circle is considered as the diameter value for measurement. To measure the diameter, three random rotational positions of the workpiece were used for image acquisition. During image acquisition, the camera's intrinsic parameters were kept constant. Thus, under red light, three images were captured. The value corresponding to an average of three images is then considered. Likewise, nine experimental runs were performed two times called replicas. Therefore, the final pixel value is the average of two replicas for each experimental run.

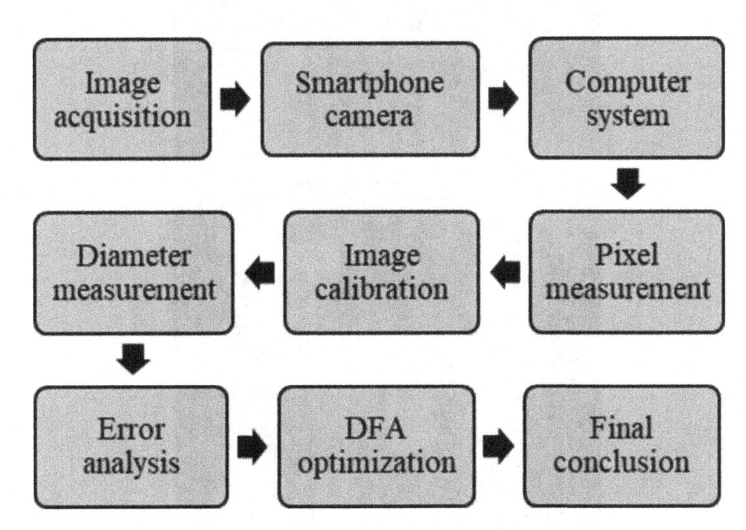

Fig. 2. Machine vision methodology for the diameter measurement

DFA is found to be one of the best techniques used in multi-objective optimization. In the DFA optimization method, rank is assigned to the maximum value of the composite desirability. An experimental run having rank # 1 is considered as an optimized experimental run. Steps to calculate the composite desirability are explained in [20]. Based on the Taguchi DoE and explained methodology; measurement of the diameter values in pixel lengths for nine experimental runs was performed shown in Table 2.

Table 2. Diameter values in pixels under RGB lights

	Red light		Green light		Blue light	
Expt. run	S (pixel)	D (pixel)	S (pixel)	D (pixel)	S (pixel)	D (pixel)
1	404.45	490.37	408.19	496.64	410.39	499.71
2	407.10	464.85	406.59	465.96	413.97	469.08
3	408.00	416.34	407.68	418.54	412.64	424.84
4	408.44	302.22	410.13	304.27	414.57	306.7
5	409.27	363.93	411.56	364.46	415.81	374.25
6	409.97	334.48	412.70	334.03	417.01	341.76
7	410.97	445.10	413.90	444.61	418.22	449.68
8	412.00	406.68	417.73	403.36	420.08	404.52
9	413.38	486.83	415.31	490.76	424.11	494.42

where, S = Slip gauge, and D = Diameter of workpiece

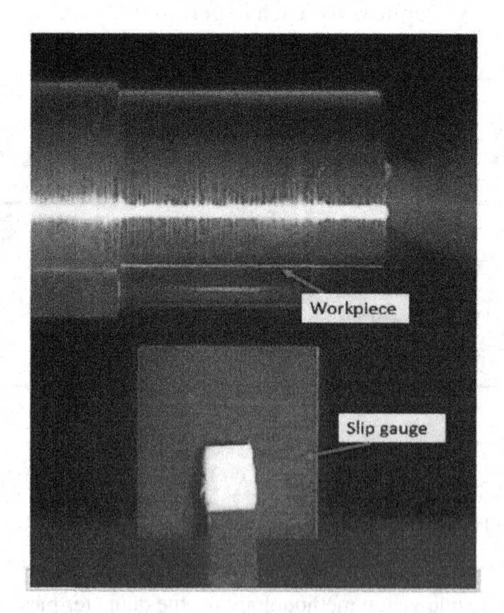

Fig. 3. Image captured under red light

An image of the slip gauge is used for calibration purpose. Pixel measurements of the captured images were performed using Coslab software (version 3.7.11) during

Fig. 4. Pixel measurement in Coslab software

measurement. The captured images are then converted into a grayscale image. The image captured using a smartphone camera under red light is shown in Fig. 3 whereas, the pixel measurement using Coslab software is shown in Fig. 4. Slip gauge having 40 mm length was used in experimentation. Pixel value was calibrated on the slip gauge image and based on the pixel calibration, the pixel values of the diameter are converted into the millimeter length. The pixel calibration was performed for each experimental run by using Eq. (1). The converted pixel values in millimeter length are shown in Table 3 in which the desired values are the final required values that are to be achieved after the machining of the workpiece.

$$\text{Pixel calibration} = 40/(\text{Pixel length}) \tag{1}$$

Table 3. Diameter values in a millimeter length

Diameter (mm) under different light sources				
Expt. run	Desired diameter values (mm)	Red light	Green light	Blue light
1	48.4	48.4975	48.6675	48.7059
2	45.4	45.6743	45.8408	45.325
3	40.6	40.8176	41.0655	41.1826
4	29.2	29.5975	29.6755	29.5921
5	35.2	35.5687	35.4223	36.002
6	32.2	32.6346	32.3751	32.7819
7	43.6	43.3219	42.9679	43.0089
8	39.4	39.4835	38.624	38.5184
9	47.8	47.1073	47.2669	46.6313

4 Results and Discussion

As discussed earlier, slip gauges are used for pixel calibration. Figure 5 shows the histogram plot for pixel values of slip gauges under RGB lights.

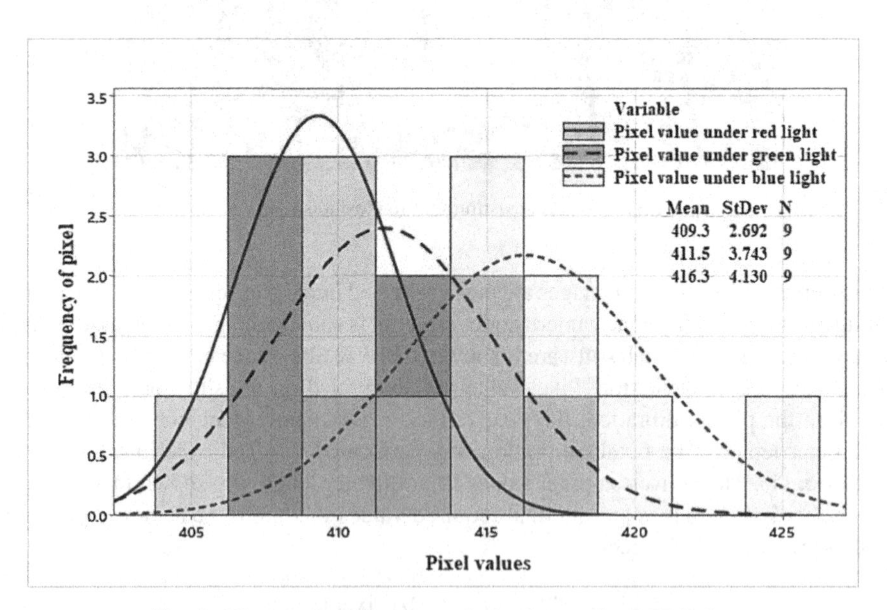

Fig. 5. Histogram of slip gauge pixel values under RGB lights

From Fig. 5 it is observed that the pixel values of the slip gauge under red light falls approximately between 405 to 415 with an approximate mean value of 409.3. The pixel values of the slip gauge under green light falls approximately between 405 to 420 with a mean value of 411.5 while the pixel values of the slip gauge under blue light falls approximately between 410 to 425 with a mean value of 416.3. The histogram analysis shows that red light has a minimum pixel value range while blue light has the highest pixel value range. The green light has a moderate pixel value range which lies in between red and blue light. The reason to change the pixel values under different light sources is due to the wavelength property of the light which affects the edge determination of the objects. The image preprocessing steps are avoided here purposefully. This research work aims to decide the best light source for robust measurement with minimum error. The percentage error in diameter measurements is calculated by using Eq. (2).

$$\text{Percentage error} = \{(\text{Measured diameter value} - \text{Desired diameter value}) \div \text{Desired diameter value}\} \times 100 \quad (2)$$

Using Eq. (2); percentage errors are calculated and are shown in Table 4.

Table 4. Percentage errors for the diameter measurements under RGB lights

		Red light		Green light		Blue light	
Expt. Run	De (mm)	D (mm)	PE (±)	D (mm)	PE (±)	D (mm)	PE (±)
1	48.4	48.4975	0.2014	48.6675	0.5528	48.7059	0.6320
2	45.4	45.6743	0.6041	45.8408	0.9709	45.3250	0.1651
3	40.6	40.8176	0.5361	41.0655	1.1467	41.1826	1.4350
4	29.2	29.5975	1.3613	29.6755	1.6283	29.5921	1.3428
5	35.2	35.5687	1.0474	35.4223	0.6315	36.0020	2.2785
6	32.2	32.6346	1.3496	32.3751	0.5438	32.7819	1.8073
7	43.6	43.3219	0.6379	42.9679	1.4498	43.0089	1.3556
8	39.4	39.4835	0.2119	38.6240	1.9696	38.5184	2.2376
9	47.8	47.1073	1.4492	47.2669	1.1154	46.6313	2.4450

where, De = Desired diameter value, D = Measured diameter value, and PE = Percentage error

Percentage error can be positive or negative, but their magnitude reflects the deviation from the required value. As a result, absolute values of percentage errors were utilized for the analysis which is shown in Table 4. To assign the ranks to the percentage errors; the DFA method was applied to the values in Table 4 with desired minimum percentage error. The output of the DFA method is shown in Table 5. While calculating the desirability, an equal amount of importance was given to measurements under RGB light sources. Therefore, a weight of 0.33 was considered to calculate the individual desirability for diameters under RGB lights. From Table 5 it is observed that experimental run # 1 has the highest composite desirability having an assigned rank of # 1.

Table 5. DFA method on percentage errors in diameter measurements

The percentage error in diameter measurements (±)					
Expt. run	Red light	Green light	Blue light	Composite desirability	Rank
1	0.2014	0.5528	0.6320	0.974430	1
2	0.6041	0.9709	0.1651	0.921245	2
3	0.5361	1.1467	1.4350	0.831625	3
4	1.3613	1.6283	1.3428	0.589168	7
5	1.0474	0.6315	2.2785	0.657381	6
6	1.3496	0.5438	1.8073	0.658218	5
7	0.6379	1.4498	1.3556	0.786954	4
8	0.2119	1.9696	2.2376	0.000000	8
9	1.4492	1.1154	2.4450	0.000000	8

Therefore, experimental run #1 is considered an initial experimental run that has a cutting speed of 80 m/min, a depth of cut at 0.3 mm, and a feed rate of 0.15 mm/rev. Using Minitab software, Taguchi analysis of composite desirability was performed. Figure 6 shows the main effects plot for the composite desirability.

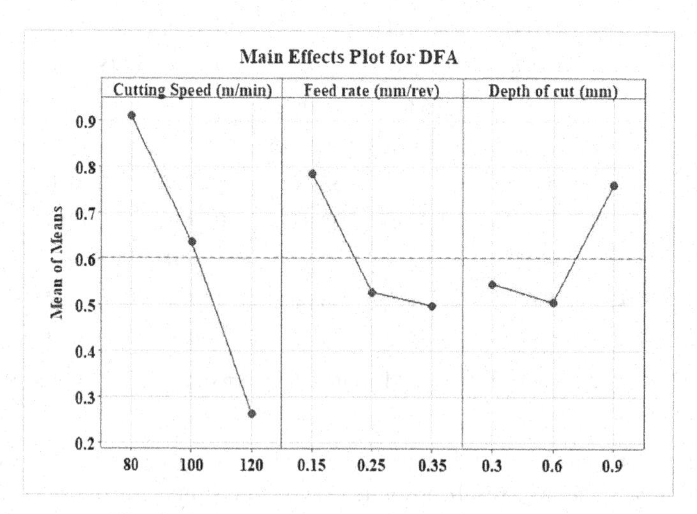

Fig. 6. Main effects plot for composite desirability

It is evident in Fig. 6 that the cutting speed at 80 m/min, feed rate at 0.15 mm/rev, and depth of cut at 0.9 mm give a higher composite desirability value. Therefore, these conditions were considered optimal machining conditions for which a confirmation experiment was performed to optimize the percentage error. This research work aims to study the errors in the measurement of diameters under RGB light sources. Therefore, percentage errors were calculated using Eq. (2) for RGB light sources. Table 6 shows the outcomes of the confirmation experiments.

Table 6. Confirmation experimental outcomes for percentage errors

		Percentage error (±)			
Sr. No	Light source	IC	OC	Im	Remarks
1	Red	0.2014	0.4028	0.2014	Non-beneficial
2	Green	0.5528	0.7619	0.2091	Non-beneficial
3	Blue	0.6320	0.3986	-0.2334	Beneficial
	Overall Percentage error (±)			**0.1771**	**Non-beneficial**

where IC = Initial conditions using the DFA method, OC = Optimal conditions using Taguchi analysis, and Im = Improvements in percentage error

Table 6 indicates an increase in percentage errors for red and green lights, while a decrease in percentage errors was observed for blue light under optimal conditions. Therefore, the minimum percentage errors given by the DFA method are for a red light and green light with a value ±0.2014% and ±0.5528% respectively. While minimum percentage error obtained by the Taguchi analysis is for blue light with a value of ±0.3986%. Under optimal conditions, the overall improvement in percentage error was found to be 0.1771%, which was determined to be non-beneficial. Consequently, the final values were selected from the values provided by the DFA method. The final percentage error values for the RGB lights were determined as ±0.2014%, ±0.5528%, and ±0.6320%. The confirmation experiment results indicate that the red light source provides the minimum percentage error in measurement, while the blue light gives the maximum percentage error in measurement.

5 Conclusions

The purpose of this research work is to devise a machine vision system that can measure diameter with a minimum amount of error. The following conclusions are made under the experimental findings:

- The IoT-based machine vision system is found to be useful and can monitor the diameter during in-process machining. Machine vision system measurement comes under non–contact type measurement therefore using the development of intelligent algorithm system precision can be further increased.
- The DFA method is found to be effective in the optimization of the dimensional measurements of the machined components. Taguchi analysis was used to confirm the values given by the DFA method which is found to be effective.
- The confirmation results show that the red light source gives a minimum percentage error of ±0.2014% while blue light gives a maximum percentage error of ±0.6320% in measurement.

- Future work should be carried out to increase the accuracy and robustness of the system by incorporating image processing techniques and the use of different machine learning algorithms to enhance the performance of the machine vision system.

Acknowledgments. This study was supported by the TEQIP III program of the Indian Ministry of Human Resource Development (MHRD), and the authors express gratitude to MHRD India for the funding as part of the program. The experiment was carried out at the Mechanical Engineering Department's facility of the Center for Advanced Machining Technology, Dr. Babasaheb Ambedkar Technological University, Lonere, Maharashtra, India (402103).

References

1. da Silva, V.L., Kovaleski, J.L., Pagani, R.N., Silva, J.D.M., Corsi, A.: Implementation of Industry 4.0 concept in companies: empirical evidences. Int. J. Comput. Integr. Manuf. **33**, 325–342 (2020). https://doi.org/10.1080/0951192X.2019.1699258
2. da Silva, N.A., Abreu, J.L., Orsolin Klingenberg, C., Antunes Junior, J.A.V., Lacerda, D.P.: Industry 4.0 and micro and small enterprises: systematic literature review and analysis. Prod. Manuf. Res. **10**, 696–726 (2022)
3. Zhou, J., Li, P., Zhou, Y., Wang, B., Zang, J., Meng, L.: Toward New-Generation Intelligent Manufacturing. Engineering **4**, 11–20 (2018). https://doi.org/10.1016/j.eng.2018.01.002
4. Rahman, L.F., Ozcelebi, T., Lukkien, J.: Understanding IoT systems: a life cycle approach. Procedia Comput. Sci. **130**, 1057–1062 (2018). https://doi.org/10.1016/j.procs.2018.04.148
5. Balestrieri, E., Vito, L.D., Lamonaca, F., Picariello, F., Rapuano, S., Tudosa, I.: Research challenges in measurements for Internet of Things systems. Acta Imecko **7**, 82–94 (2018). https://doi.org/10.21014/acta_imeko.v7i4.675
6. Ray, P.P.: An Internet of Things based approach to thermal comfort measurement and monitoring. In: 2016 3rd International Conference on Advanced Computing and Communication Systems (ICACCS), pp. 1–7. IEEE, Coimbatore, India (2016). https://doi.org/10.1109/ICACCS.2016.7586398
7. Marques, G., Roque Ferreira, C., Pitarma, R.: A system based on the Internet of Things for real-time particle monitoring in buildings. Int. J. Environ. Res. Public Health **15**, 821 (2018). https://doi.org/10.3390/ijerph15040821
8. Wei, G., Tan, Q.: Measurement of shaft diameters by machine vision. Appl. Opt. **50**, 3246–3253 (2011). https://doi.org/10.1364/AO.50.003246
9. Wang, J., Gao, B., Zhang, X., Duan, X., Li, X.: Error correction for high-precision measurement of cylindrical objects diameter based on machine vision. In: 2015 12th IEEE International Conference on Electronic Measurement & Instruments (ICEMI), pp. 1113–1117. IEEE, Qingdao, China (2015). https://doi.org/10.1109/ICEMI.2015.7494414
10. Tan, Q., Kou, Y., Miao, J., Liu, S., Chai, B.: A model of diameter measurement based on the machine vision. Symmetry. **13**, 187 (2021). https://doi.org/10.3390/sym13020187
11. Naveen Sait, A., Aravindan, S., Noorul Haq, A.: Optimisation of machining parameters of glass-fibre-reinforced plastic (GFRP) pipes by desirability function analysis using Taguchi technique. Int. J. Adv. Manuf. Technol. **43**, 581–589 (2009). https://doi.org/10.1007/s00170-008-1731-y
12. Singaravel, B., Selvaraj, T.: Application of desirability function analysis and utility concept for selection of optimum cutting parameters in turning operation. J. Adv. Manuf. Syst. **15**, 1–11 (2016). https://doi.org/10.1142/S0219686716500013

13. Jenarthanan, M.P., Jeyapaul, R.: Optimisation of machining parameters on milling of GFRP composites by desirability function analysis using Taguchi method. Int. J. Eng. Sci. Technol. **5**, 22–36 (2018). https://doi.org/10.4314/ijest.v5i4.3
14. Galgali, V.S., Ramachandran, M., Vaidya, G.A.: Multi-objective optimal sizing of distributed generation by application of Taguchi desirability function analysis. SN Appl. Sci. **1**, 742 (2019). https://doi.org/10.1007/s42452-019-0738-3
15. Kodagali, J.A., Balaji, S.: Computer vision and image analysis based techniques for automatic characterization of fruits a review. Int. J. Comput. Appl. **50**, 6–12 (2012). https://doi.org/10.5120/7773-0856
16. Prats-Montalbán, J.M., de Juan, A., Ferrer, A.: Multivariate image analysis: a review with applications. Chemom. Intell. Lab. Syst. **107**, 1–23 (2011). https://doi.org/10.1016/j.chemolab.2011.03.002
17. Chalisgaonkar, R., Kumar, J.: Optimization of WEDM process of pure titanium with multiple performance characteristics using Taguchi's DOE approach and utility concept. Front. Mech. Eng. **8**, 201–214 (2013). https://doi.org/10.1007/s11465-013-0256-8
18. Yünlü, L., Çolak, O., Kurbanoğlu, C.: Taguchi DOE analysis of surface integrity for high pressure jet assisted machining of Inconel 718. Procedia CIRP. **13**, 333–338 (2014). https://doi.org/10.1016/j.procir.2014.04.056
19. Tambolkar, P., Ponkshe, A., Mulay, V., Bewoor, A.: Use of Taguchi DOE for CFD simultion to maximize the reusability of working fluids of centrifugal filter. Procedia Manuf. **46**, 608–614 (2020). https://doi.org/10.1016/j.promfg.2020.03.087
20. Ramanujam, R., Maiyar, L.M., Venkatesan, K., Vasan, M.: Multi-response optimization using ANOVA and desirability function analysis: a case study in end milling of Inconel alloy. ARPN J. Eng Appl. Sci. **9**, 457–463 (2014)

Analysis of Detection and Recognition of Human Face Using Support Vector Machine

Shaikh Abdul Hannan[1], Pushparaj[2(✉)], Mohammed Waseem Ashfaque[3], Anil Lamba[4], and Anil Kumar[5]

[1] Computer Science and Information Technology, AlBaha University, AlBaha, Kingdom of Saudi Arabia
[2] Electronics and Communication Engineering, National Institute of Technical Teachers Training and Research, Chandigarh, India
pushprajpal@gmail.com
[3] Computer Science and I.T., Dr. B.A.M. University, Aurangabad, India
[4] Cyber Security, Technology Risk and Resilience Oversight American Express, New York, USA
[5] Government Millennium Polytechnic, Chamba, Himachal Pradesh, India

Abstract. Recently, Support Vector Machines (SVMs) have been suggested as an additional technique for design identification. In this study, a face detection framework with demeanor recognition utilizing Support Vector Machines (SVMs) with a double tree identification process is used to tackle the face recognition issue. It plays a crucial role in applications including face delineation data set management, human PC interaction, and video examination. Through two steps, the face recognition with appearance appraisal framework accomplishes look recognition. The captured image is initially processed to identify the face, and then the perceived appearance. The first two stages of the framework handle facial recognition and face trimming using picture handling, while the third stage handles converting the altered image's colors from RGB to grayscale and applying the appropriate smoothing channel. We employ a simple method to create our own SVM classifier with a Gaussian component that can recognize eyes in grayscale photos as the first step towards a part-based face identifier. The architecture of an iterative bootstrapping method is discussed in detail, and we show how choosing boundary values will usually result in the best outcomes. The findings of our study are in line with earlier studies, and the challenges encountered are typical for anybody developing an article identification system. Large support vector SVM classifiers are sluggish, and accuracy is highly reliant on the caliber and diversity of training data.

Keywords: Recognition · Face Detection · Support Vector Machine · Edge detection · Image processing

1 Introduction

In light of the learning hypothesis, support vector machines are the controlled learning computations for both arrangement and example recognition. A hyperplane is constructed as a choice plane to isolate the data of interest between two classes according

R. K. Challa et al. (Eds.): ICAIoT 2023, CCIS 1930, pp. 86–98, 2024.
https://doi.org/10.1007/978-3-031-48781-1_8

to the fundamental standard of a support vector machine (SVM). Support vectors are information foci that are somewhat off from being exactly identical to the hyperplane. SVM is perhaps one of the best methods for order problems [1].

Face recognition technology has a wide range of applications, including personality verification, access control, and spying. Face recognition research activities and interests have fundamentally increased over the past few years. Different alterations to face images should be manageable by a face recognition framework. Facial recognition is an amazing test due to two reasons. A facial image is susceptible to shifts in viewpoint, illumination, and attitude. Possessing the option to control perspective alterations is essential for a convincing depiction. The second involves using the selected representation to group a different facial image [2].

Facial features including the eyes, nose, mouth, and jawline are recognized using mathematical component-based techniques as face descriptors, areas, distances, and locations between the highlights are used as properties and relations. This class of approaches heavily relies on the extraction and estimate of face highlights, despite being conservative and effective in achieving information reduction and indifferent to variations in lighting and perspective. Unfortunately, the highlight extraction, estimate, and calculations that have been developed to date have not been sufficiently reliable to meet this demand [3].

It's interesting to note that layout coordination and brain techniques typically operate clearly on a picture-based representation of faces, such as a pixel power cluster. When compared to mathematical component-based algorithms, this class of solutions has proved more practical and straightforward to implement since the detection and estimate of mathematical facial parts are not required. The eigenface method, which relies on the Karhunen-Loeve change (KLT) or the Principal Component Analysis (PCA) for face representation and identification, is one of the best layout-matching techniques. The projection of each face picture to the premise in the eigenface space, which is how the data set addresses each face picture, is called a vector of loads. For facial recognition, the closest distance is often used [4]. In the interest of analyzing photographs using, image processing is sophisticated equipment that integrates multiple areas like Machine Learning, Deep Learning, AI, and Big Data. Compare the various criteria (rotating, the intensity of brightening change, etc.). According to big data and neural network technologies, these have witnessed a rapid and significant transformation [16]. Image processing, that has earlier been accepted for its capability of identification, is currently conducting very rapid growth because of its many applications in the health sector, including the detection of skin and lung cancers, as well as lung and breast cancers [17–19]. Identifying people on video surveillance cameras is also utilized in the security sector. The system for face recognition starts with finding individuals in photographs or videotapes. The conclusion of this determination obviously influences the extent to which the face recognition system behaves. The head shape, skin tone, or facial features among some existing methods are specifically used for facial recognition, while others combine many approaches into one [20].

1.1 Objectives of the Study

- Combining current methods in an effort to develop a face detection and identification system capable of quickly and accurately classifying a large number of faces.
- To examine Face Recognition and Face Detection using SVM

2 Literature Review

While nose recognition has been employed as a reaffirmation framework near the eyes, the Haar Outpouring Classifier is being used to accomplish the distinguishing proof of the eyes. Skin edge area, morphological overseers, and support vector machines—all utilized by perceptible producers of face disclosure—are now being studied. Then, a significant element of the affirmation tool is made up of Hoard (Histogram of Situated Inclinations) qualities that were derived from a sizable collection of face photos. Once these Hoard features have been put together for a face/client, an SVM model is ready to predict faces that are enabled in the structure [5].

Using a support vector machine, a Gabor channel bank, and a head-part analysis, a face may be identified. In this method, the elements are eliminated using a Gabor channel bank, and the low-layered highlight vectors are then created by performing a bit PCA on the channel bank's output. Finally, SVM is used to create order [6].

Prakash et al., [7] proposed a face recognition framework built on SVM. To eliminate the highlights from facial images, Gabor channels are used. Prior to extracting the highlights, homomorphism separating is used for preprocessing. The highlights that are extracted contribute to the SVM for testing and preparation purposes. When compared to other strategies, this one stands out since it makes use of the Yale face data set B.

Par, S.C et al. [8] used Gabor wavelet highlights with PCA and KPCA for face recognition. This approach uses Gabor wavelet change and provides a top-to-bottom comparison of Gabor PCA (linear) and Gabor KPCA (non-direct). The ORL data set was used as the test set, and results showed that the Gabor PCA outperformed the Gabor KPCA as a face identification algorithm.

Applying support vector machines for facial recognition, Romdhani et al. [9] compared the presentation of an SVM calculation with a PCA calculation and applied the calculation to both the check and the recognized proof. SVM's error rate throughout the check was almost half that of PCA. This illustrates how SVM provides more accurate data in the face space.

3 Proposed Methodology

3.1 Support Vector Machines

Support Vector Machine (SVM) is a supervised learning algorithm used for classification and regression analysis. SVM is widely used in various applications, including image classification, text classification, and bioinformatics. It has a solid theoretical foundation and can handle high-dimensional data well. However, SVM can be sensitive to the choice of kernel function and parameters, and it may require careful tuning for optimal performance.

In the square case, where the classes are linearly separable, SVM seeks to find the hyperplane that maximizes the margin between the two classes. The margin is the distance between the hyperplane and the closest data points from each class. SVM uses the support vectors, which are the data points closest to the hyperplane, to define the hyperplane.

In the irregular case, where the classes are not linearly separable, SVM uses a technique called kernel trick to map the data to a higher-dimensional space where it becomes separable. The kernel function computes the similarity between data points in the original space and transforms it into the higher-dimensional space. SVM then finds the hyperplane that separates the data with the largest margin.

3.1.1 Square Case

Consider a set of marks (yi, yi 1) that categorize input tests into positive and negative categories and a collection of l vectors (xi, xi R n, 1 I l) that address input tests. In the uncommon scenario that the two classes are directly divisible, there is an isolating hyperplane (w, b) representing the capability, and sign (f(x)) tells which side of the hyperplane x lies at the end of the day—the class of x. The isolating hyperplane's vector w may be written as a direct combination of the loads I and xi (also known as a double representation of w):

$$f(x) = <w.x> + b, \tag{1}$$

$$w = \sum_{1 \le i \le l} a_i y_i x_i \tag{2}$$

Therefore, the decision function f(x) has the following dual representation:

$$f(x) = \sum_{1 \le i \le l} a_i y_i <x_i.x> + b. \tag{3}$$

To create a straight SVM, identify the insertion characteristics I and offset b such that the hyperplane (w, b) segregates positive data from negative samples with the maximum edge. Not all information vectors xi are suitable for the two fold representation of w; these vectors are referred as support vectors since they have weight I > 0 and structure w.

3.1.2 An Irregular Case

Problems seldom occur when it is simple to distinguish between excellent and bad examples in real life. Non-direct support vector classifiers locate loads I of the double articulation of the isolating hyperplane's vector w to address the issue of straight division in the component space. They use a guide that is often not straightforward to transform input space X into an element space F: X F, x 7 (x):

$$w = \sum_{1 \le i \le l} a_i y_i \emptyset(x_i) \tag{4}$$

While the construction of the decision function f(x) is as follows:

$$f(x) = \sum_{1 \le i \le l} a_i y_i <\emptyset(x_i), \emptyset(x)> + b. \tag{5}$$

Working directly in a highly layered space like F, where test preparation pictures are often relatively different, would be computationally costly. Anyhow, we may pick an area F that is controlled by a component K and described by a component capability K(x, y), K(x, y) = (x) • (y) >, which manages the spot item in F. It is therefore possible to register the choice capability (Eq. (5)) by just using the portion capability, and it can be demonstrated that defining the most extreme time isolating hyperplane is equivalent to addressing the following improvement problem.

3.2 Training

3.2.1 Input Data

As the eyes are one of the most distinctive features of the human face, they are considered to be the most memorable feature. In our study, we analyzed 1066 preparatory images of human faces and identified 2132 instances of eyes. Each instance was divided into 20×20 pixels and was represented by a 400-layered vector that included the darker regions of the original image. In addition to this, we generated artificially manufactured positive examples using various techniques.

In our future work, we plan to provide instances of eyes with varying splendor, distinctiveness, and orientation. To prepare for this, we collected a set of images that included faces as well as images without faces. We randomly scaled these images and excluded the regions that contained positive examples to provide negative examples for the main cycle of preparation, which focused on the eyes.

3.2.2 SVM Settings

We modified the boundary for our SVM to a number between 800 and 1600 and decided to use the Gaussian bit. It was agreed that the limit for the I loads C would be 10. SVM classifiers that are steady with prior exploration and have values between 1200 and 1600 had the highest performance (near 5–6.3 for standardized input information) [10].

3.2.3 Bootstrapping

The primary focus of the previously described introduction material, SVM classifier capacity is ready. Next, the newly acquired capacity is applied to the face- and faceless preparation pictures, and false positives as well as unfavorable support vectors are utilized to arrange the unfavorable example set for the following cycle. The excellent example continues throughout each cycle in the same way.

The accuracy of the classifier generally improves during the initial 4–6 cycles, depending on the number of negative samples. For example, in the training set after the initial stage, a classifier with 1200 samples identifies 50–500 false positives in each image and has approximately 300 support vectors. However, after the fourth cycle, the number of support vectors increases to over 1300, while the false positives per image decrease to 0–10 [11].

3.2.4 Detection Process

To find objects in a picture, a classifier window is frequently run over it at all conceivable sizes and positions. A range of scales were selected where eyes are likely to be visible in pictures. A pyramid of scaled pictures is built to validate a photograph, with the size rising by 1.1 times for each level inside the realm of possibility. The classifier window is then run over each picture in the pyramid, with a stage of 2 pixels in each of the two heads. According to a capability evaluation, the classifier window neglects a portion of the articles with a scale step of 1.1, which should be evident when physically evaluating probable locations. Future work should concentrate on both, but for now, this step size is a decent compromise between precision and speed [12].

3.2.5 Tempo of Instruction

Due to the continuous execution of the preliminary interaction, it takes 3–5 days to prepare 4 cycles of a classifier capacity. The majority of the time is spent receiving false advantages while testing the prepared abilities on the information image set. For example, a capability with 1000 support vectors can repair a 20×20 pixel in around 7 ms, but testing takes more than a day when there are many patches in each picture and more than 1,000 images. Future work is anticipated to greatly reduce testing time since a fluid-lowered set vectors approach can cause a drop of more than 100% in the normal number of support vectors used to analyze each picture repair [13].

4 Results

A test set of 25 erroneous front-facing face images with a total of 50 eyes was employed to evaluate the SVM classifiers. In these images, eye regions were physically separated, and any window inside a 10-pixel boundary that was precisely categorized was considered a right hit. The total numbers of false positives and misleading negatives for five capabilities with different section borders are shown in Table 1.

Table 1. Rate counts for various kernel parameters

Count Parameter	False Negative	False Positive
700	13	72
800	16	53
900	18	41
1000	19	35
1200	12	62
1300	17	54
1400	20	13
1500	58	52
1600	63	17
1700	72	69

The cumulative rate counts for kernel parameters are a way of evaluating the performance of a support vector machine (SVM) classifier across a range of kernel parameter values. For each value of the kernel parameter, the cumulative rate count represents the percentage of data points that are correctly classified up to that point. The kernel parameter is a crucial parameter in SVMs that determines the shape of the decision boundary and can significantly impact the performance of the classifier. Figure 1 reflects the counts rate for various kernel parameters like false positive and false negative.

From Fig. 1 (a), by plotting the cumulative rate counts as a function of the kernel parameter value, we can identify the optimal kernel parameter value that maximizes the classifier's performance. This approach provides a more nuanced understanding of the classifier's behavior than simply looking at the overall classification accuracy. Additionally, it can help us identify cases where the classifier's performance plateaus or decreases as the kernel parameter value increases, indicating over fitting or other issues. Overall, the cumulative rate counts provide a useful tool for evaluating the performance of SVM classifiers and optimizing their parameters for specific applications.

From Fig. 1 (b), a lack of diversity in the preliminary positive information has to be blamed for the misrepresenting negatives. As an illustration, often-used portraits of people come from a single source, including about 20 distinct individuals, and show little variation in lighting and viewpoint points. We wish to increase the preparation set using two images from different sources and erroneous tests produced by modifying the images. Additionally, because the picture pyramid erred on the side of underestimating the smallest size at which eyes could be recognized, several eyeballs in the evaluation set was missed by the identifier. Future research will address this issue as well [14].

Given that a classifier examines more than 50,000 patches for each image and only marks 0 to 10 of them as eyes, the seemingly high numbers of false positives are less of a concern. Some of the misleading benefits in the robotized testing actually do suggest an eye but vanish from the 10-pixel edge region, while others frequently only show up in a few spots in the image Fig. 2.

Since hiding up-sides will be blended, the penalty for doing so when evaluating potential part combinations to create a face shouldn't be severe, and we anticipate that it will be very difficult to spot a face that contains a fake part (see Fig. 3) [15]. Changes in the evaluation set's false positive rate and support vector count for the classifier can be seen in Table 2.

Table 2 shows the changes in the evaluation set's false positive rate and support vector count for the classifier with a Count of 2000. The false positive rate refers to the percentage of negative instances that are wrongly classified as positive. It is a measure of the classifier's accuracy in predicting negative instances.

Count and Support Vector Classifier (SVC) are two different concepts in machine learning. Count refers to the number of times a particular feature appears in a dataset. On the other hand, the Support Vector Classifier is a type of supervised learning algorithm used for classification tasks. It is a form of discriminative classifier that seeks to find the hyperplane that best separates the classes in the input data.

Looking at the table, we can see that as the number of support vectors increases, the false positive rate also increases. This indicates that increasing the number of support

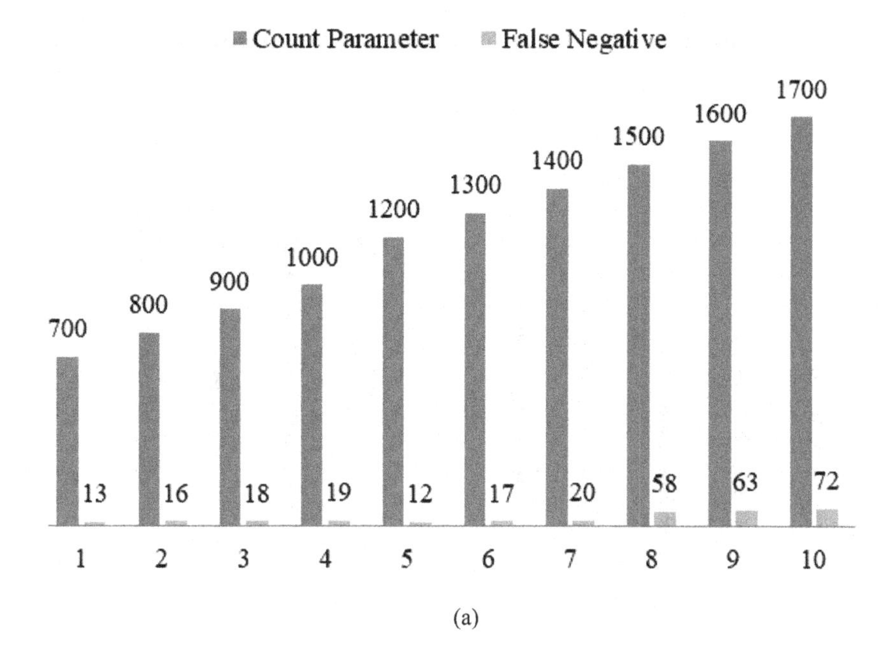

(a)

Rate counts for false positive parameters

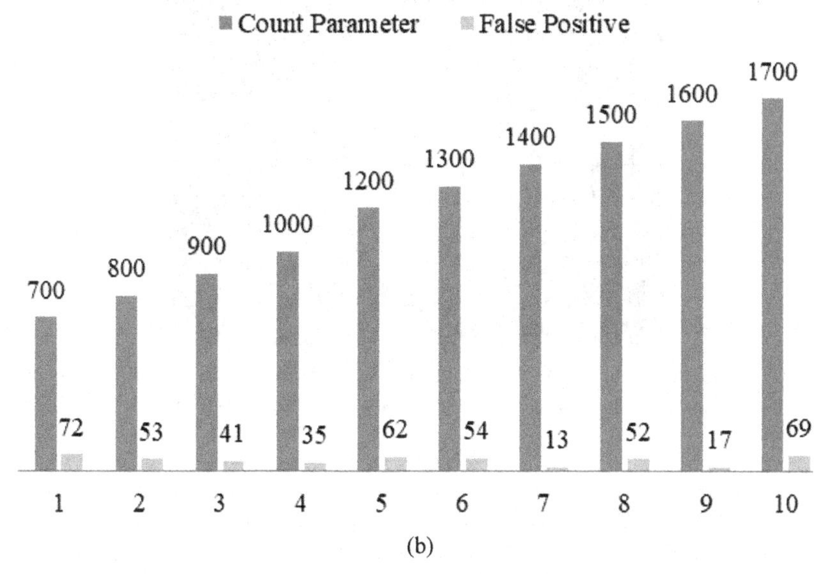

(b)

Fig. 1. Counts rate for various kernel parameters like false positive and false negative (a) Rate counts for various kernel parameters (b) Rate counts for false positive parameters (c) Cumulative Rate counts for kernel parameters.

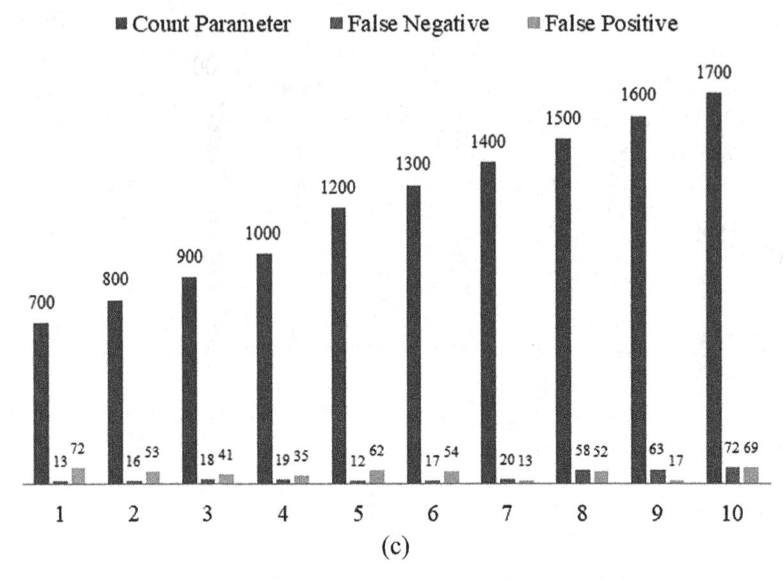

(c)

Fig. 1. *(continued)*

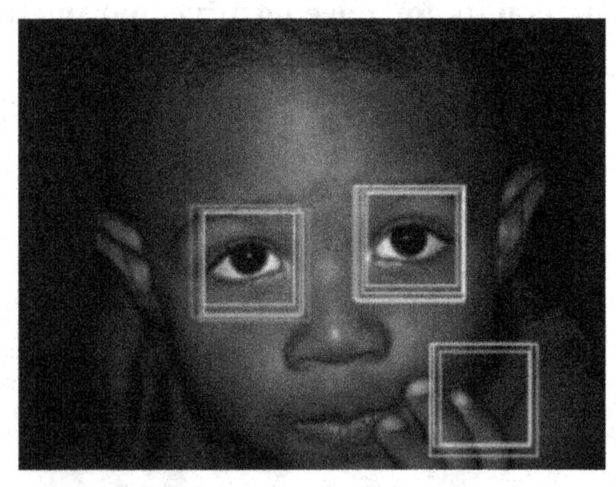

Fig. 2. Eye finder in the real world. After the third cycle of preparation, the classifier really presents a few false positives on the evaluation image (area with plenty of information surrounding the child's fingers).

vectors in the classifier may lead to overfitting, causing the classifier to become less accurate on unseen data.

However, we can also see that the number of support vectors does not always have a direct relationship with the false positive rate. For example, when the count is 0 or 2, the number of support vectors is 500 and 2000, respectively, but the false positive rate

Table 2. Changes in the evaluation set's false positive rate and support vector count for the classifier = 2000

Counts	Support Vectors	False Positive
0	500	1500
1	2500	1900
2	2000	1500
3	1000	2000
4	1500	2500
5	2000	2100

is the same at 1500. This could be due to differences in the data distribution and the complexity of the classification problem.

A confusion matrix is not applicable in the case of Fig. 3, which shows the relationship between the false positive rate and the number of support vectors for a specific classifier and count value. The Fig. 3 shows the changes in the evaluation set's false positive rate and support vector count for the classifier with a Count of 2000. It provides a visual representation of the relationship between these variables, which can be helpful in understanding the behavior of the classifier:

(a) **Count vs. Support Vector Classifier:** This plot shows the relationship between the number of support vectors and the false positive rate for different counts. As the number of support vectors increases, the false positive rate also tends to increase. However, there is variability in this relationship across different counts, indicating that the optimal number of support vectors may depend on the specific characteristics of the data.

(b) **Count vs. Count Rate:** This plot shows the count rate, which is the ratio of the number of support vectors to the count, as a function of count. We can see that the count rate tends to decrease as the count increases, indicating that the classifier may require more support vectors to achieve the same level of performance on larger datasets.

(c) **Combined Representation of Count, Classifier, and Count Rate:** This plot combines the information from (a) and (b) to show how the count, support vector classifier, and count rate are related to the false positive rate. We can see that the false positive rate generally increases as the count rate decreases, which is consistent with the observation in (b). Overall, the figure provides valuable insights into the behavior of the classifier and how its performance is affected by the count, support vector classifier, and count rate. It can be helpful in guiding the selection of parameters for the classifier and optimizing its performance on different datasets.

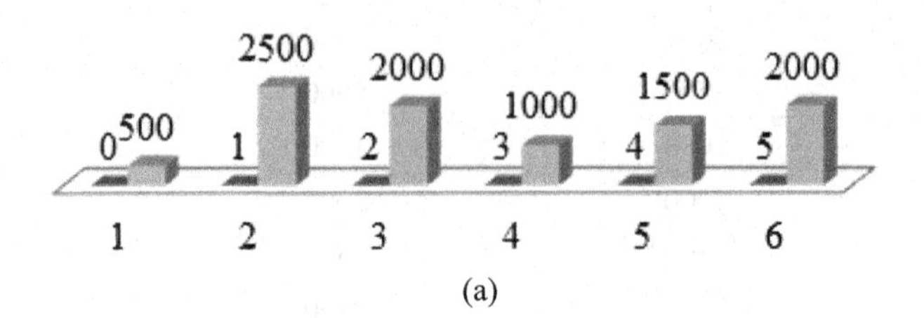

(a)

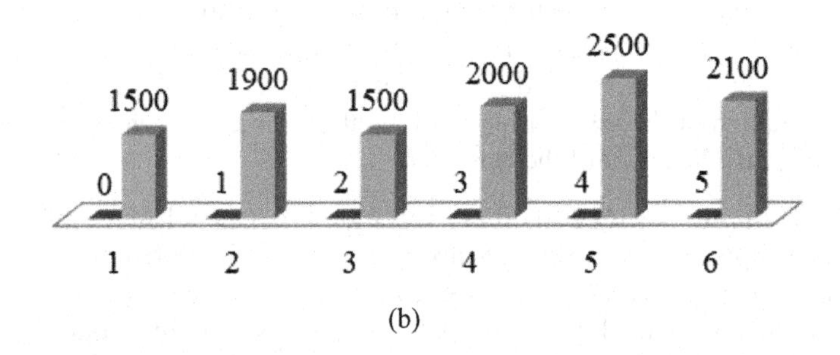

(b)

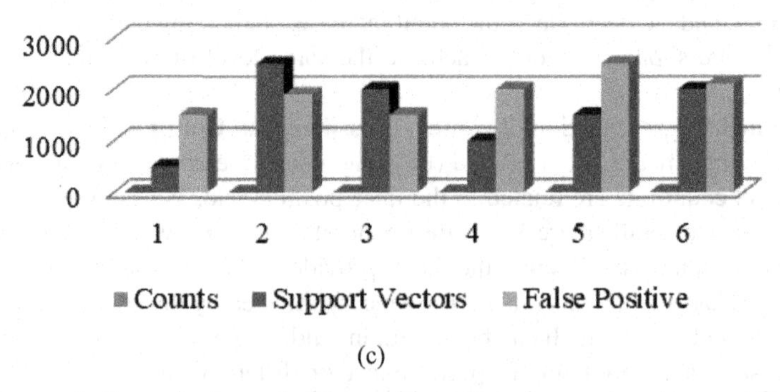

(c)

Fig. 3. Changes in the evaluation set's false positive rate and support vector count for the classifier with = 2000 (a) Count Vs Support vector classifier (b) Count Vs count rate (c) Combined representation of count, classifier and count rate.

5 Conclusion and Future Scope

The process of computing a face-recognizability proof using the skin edge identification, facial component extraction, Haar course plan, and possible Histogram of organizer has been illustrated. Following pre-dealing with phase estimator groups, the image is classified into face- and non-face-region using the Support Vector Machine (SVM) concept. After the previous picture highlights have been stored in a document, a fresh photo highlight is taken to test with all the face highlights. We have presented the face recognition tests that make use of straight SVMs and a two-tree grouping scheme. The connection between various methodologies shows that the SVMs may be effectively trained for face recognition. The trial findings demonstrate that for face recognition, SVMs are a favored computation over the closest focus method.

We directly implemented a support vector machine and discovered unanticipated drawbacks. The type of preparation information is crucial to the accuracy of an SVM classifier; specifically, sufficient variability in test preparation is anticipated to account for a variety of possible locations, lighting conditions, and halfway obstructions in the images to be recognized. Additionally, using a classifier with a small number of large support vectors is inappropriate for usage in web-based frameworks. As a first step towards speeding up the classifier, we want to use a flowing decreased set vectors technique. We next plan to expand the preparation cycle by including fake varieties in input data to address accuracy. Picking diverse facial features and training classifiers to detect them calls for further efforts to create a face-discriminating framework.

References

1. Fasel, B., Luettin, J.: Automatic facial expression analysis: a survey. Pattern Recogn. **36**(1), 259–275 (2003)
2. Heisele, B., Pontil, M.: Face detection in still gray images. Massachusetts Inst of Tech Cambridge Ma Center For Biological and Computational Learning (2000). https://doi.org/10.21236/ada459705
3. Heisele, B., Serre, T., Poggio, T.: A component-based framework for face detection and identification. Int. J. Comput. Vis. **74**, 167–181 (2006)
4. Gray, G.A., Kolda, T.G.: APPSPACK 4.0: Asynchronous Parallel Pattern Search for Derivative-Free Optimization, Sandia Report SAND2004-6391, Sandia National Laboratories, Livermore, CA (2004)
5. Kayaick, G.A.: Multi-view face detection using Gabor filters and support vector machine. Technical report IDE0852 (2008)
6. Cristianini, N., Shawe-Taylor, J.: An Introduction to Support Vector Machines and Other Kernel-based Learning Methods. Cambridge University Press, Cambridge, UK (2000)
7. Prakash, N., Singh, Y.: Support vector for face recognition. Int. Res. J. Eng. Technol. **2**(8) (2015)
8. Park, S.C.: Face Detection and Facial Expression Recognition using Geometric Features and Micro Texture Distribution. MS thesis, Dankook Univ. (2010)
9. Romdhani, S., Torr, P., Schölkopf, B.: Efficient face detection by a cascaded support-vector machine expansion. In: Royal Society of London Proceedings Series A 460, 3283–3297 (2004)

10. Meshginitober, S., Aghagolzadeh, A.: Face recognition using Gabor filter bank, Kernel principal component analysis and support vector machine. Int. J. Comput. Theor. Eng. **4**(5) (2015)
11. Eitrich, T., Lang, B.: Parallel tuning of support vector machine learning parameters for large and unbalanced data sets, Preprint BUW-SC 2005/7, University of Wuppertal (2005).
12. Barbu, T.: Gabor filter based face recognition technique. In: Proceedings of the Romanian Academy, Series A, vol. 11(3), pp. 277–283 (2010)
13. Cheon, Y., Kim, D.: A natural facial expression recognition using differential-AAM and k-NNS. Pattern Recogn. **42**(7), 1340–1350 (2008)
14. Ying, Z., Xieyan, F.: Combining LBP and adaboost for facial expression recognition. In: ICSP2008, Oct. 26–29, pp. 1461–1464. Beijing, China (2008)
15. Hassouneh, A., Mutawa, A.M., Murugappan, M.: Development of a real-time emotion recognition system using facial expressions and EEG based on machine learning and deep neural network methods. Inform. Med. Unlocked **20**, 100372 (2020)
16. Chaudhary, A., Singh, S.S.: Lung cancer detection on CT images by using image processing. In: 2012 International Conference on Computing Sciences, pp. 142–146. Phagwara, India (2012)
17. Al-Tarawneh, M.S.: Lung cancer detection using image processing techniques. Int. J. Biomed. Imaging 749024 (2012)
18. Ansari, U.B., Sarode, T.: Skin cancer detection using image processing. Int. J. Eng. Sci. Technol. **4**(4), 1190–1196 (2012)
19. Deepika, C.L., Alagappan, M., Kandaswamy, A., Wassim Ferose, H., Arun, R.: Automatic, Robust Face Detection and Recognition System for Surveillance and Security Using LabVIEW (sCUBE)
20. Nagamalai, D., Renault, E., Dhanuskodi, M.: Advances in digital image processing and information technology. pp. 146-155. Springer, Berlin, Heidelberg (2011). https://doi.org/10.1007/978-3-642-24055-3_15
21. Yi-bo, L., Jun-jun, L.: Harris corner detection algorithm based on improved contourlet transform. Procedia Eng. **15**, 2239–2243 (2011)

An Enhanced PSO-Based Energy Efficient Clustering Routing Algorithm for Wireless Sensor Network

Abin John Joseph(✉), R. Asaletha, V. J. Manoj, and R. Nishanth

Cochin University of Science and Technology, Cochin, India
{abinjoj,asalethar,manojmvj,nishanthr}@cusat.ac.in

Abstract. Wireless Sensor Network (WSN) comprises of a lot of sensor nodes which are implanted in an application system to track hardware elements in a specific area, such as monitoring of temperature, ocean level tracking, pressure monitoring, and health care, in addition to various military applications. The network is made up of a sensor network that could detect, compute, and transfer data. One of the most difficult challenges in WSN is efficiency. Sensor networks have insufficient energy and are located in remote locations. As a result, restoring the battery cells in WSN is challenging. Hence suitable cluster-based methods and cluster head (CH) methodologies has to be incorporated to enhance the lifespan of the network. In this paper, a clustering-based routing strategy that considers energy in WSN is developed utilizing the Modified Particle Swarm Optimization (MPSO) algorithm, which could cluster the network and choose the appropriate cluster heads. The Modified PSO-based Energy-Efficient Clustering Routing Protocol (PSO-EECRP) of a network is an essential criterion for extending the network's life span. Numerous factors are taken into account for the proposed approach's energy efficiency, including the intra-cluster distance and the sensor nodes' residual energy. Simulations were conducted on evaluation factors such as energy consumption, Packets delivery ratio, Network Lifetime and efficiency. To validate the decrease in energy usage, the performance measures are analyzed and the findings are evaluated by comparing them with an existing algorithm.

Keywords: Wireless Sensor Network · Energy Efficient · Clustering

1 Introduction

The sensor network is a wireless communication system of mobile devices that offers a number of services, including the protection of the environment and wildlife [1]. It is predicted to bring about drastic advancements in future innovations with several application fields in WSN including health care, virtual environments, home automation, deep learning, military applications, intelligent systems, etc [2]. WSN have been a popular preference between many scientists and scholars due to the simplicity of deployment and operation. Battery-powered sensors have been commonly implemented in hostile environments where the power supply is challenging to charge up. Moreover, there is

R. K. Challa et al. (Eds.): ICAIoT 2023, CCIS 1930, pp. 99–112, 2024.
https://doi.org/10.1007/978-3-031-48781-1_9

a number of useful energy-saving methods available in the survey, that are also likely to prolong the life of the network. The advancement of clustering and routing protocols within the system serves as one of the most essential methods for preserving the energy of nodes. As a result, the simplest methods to conserve node batteries is to build an energy-aware routing protocol.

Clustering seems to be the grouping of a network's nodes based on specific properties. Each group has also been defined as a cluster, and that every cluster does have a Cluster Head (CH) who is responsible for charge of the nodes in a cluster and gathers data from other cluster nodes before sending it to a stationary or Mobile Sink node (MS). An MS could be a Base Station (BS) or a device with good capabilities. Clustering is extremely efficient in maintaining the WSN energy usage balance. In this regard, CH save energy by attempting to prevent those nodes from getting involved in the transmission of data, because cluster nodes gather information from the actual area [3].

Static can be functioned in the network based on the application requirements of WSN. For example, static sinks are adequate for essential and not commercial application fields such as sensing forest fires, mud slides, and volcanism. Utilizing MS, on the other hand, is much more beneficial for application areas that contain sensitive and private data and are time-sensitive, such as army security and monitoring. Moreover, although the use of MS in the system offers advantages such as high speed data acquiring, low energy usage, transmission of packets, and safety, it tends to make monitoring and/or managing the system more difficult than with a static sink due to node topological changes as well as the configuration of the appropriate protocols [4, 5]. The WSN network is shown in Fig. 1.

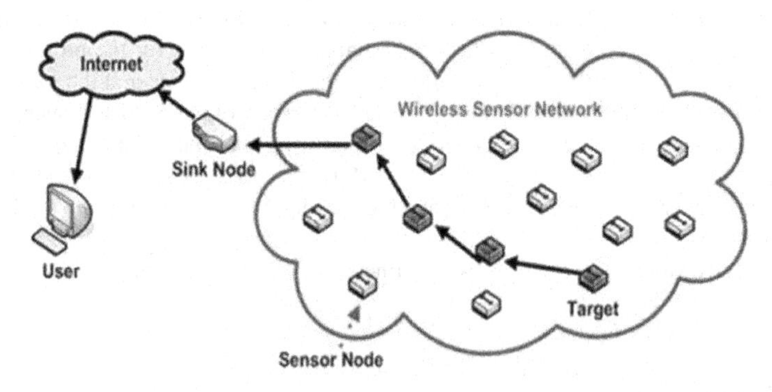

Fig. 1. WSN network

The following is the structure of the paper: The introduction is explained in Sect. 1. Section 2 provides a brief overview of similar works on methodologies for extending the lifetime of the network. Section 3 briefly describes the proposed methodology. Section 4 explains the proposed model performance and the outcomes are compared to the existing method. Section 5 concludes the article.

2 Recent Works

Energy Soaring-based Routing Algorithm (ESRA) is employed for IoT implementation in Software- Defined WSN, particularly for observing areas. The proposed ESRA system carefully selects network CH to solve the control system scheduling problems, in an effort to improve internet backbone stability and dependability along with the network's lifetime [6]. Low-Energy Adaptive Clustering Hierarchy, LEACH [7] is the most popular protocol for wireless networks. It became the first to organize using a hierarchical system. Sensor systems into group head sensors chosen at random are all in charge of collecting and aggregating data from the other sensors (nearby). Each CH sends information to the BS, to its sensors, as well as its Time Division Multiple Access (TDMA) schedule, to organize intra-cluster communication. LEACH has drawbacks, such as head selection being random, with really no regard for sensor residual energy or other factors criteria for adaptation. It is assumed that all heads will be able to reach the bottom station. When a head stops functioning, the network suffers and becomes unworkable till a new leader is elected.

Three distinct measurements such as neighboring repetition, bisection indexing, and algebraic connections, have been used for a tree-based and CH based structure for tightly packed IoT systems of WSN. This research describes Fuzzy based Reinforcement Learning for Densely Distributed Grid computing (FRLDG) model that includes all information gathering nodes in such a densely distributed WSN [8].

The Greedy algorithm and fuzzy clustering method are incorporated for effective WSN routing. The Greedy algorithm was utilized in this investigation, and a fuzzy clustering method has been designed for WSN routing efficiency. Multiple networks contain nodes. The suggested scheme tends to make processing such massive amounts of statistics, particularly in natural form, challenging for closely packed positioned heterogeneous WSN CH or BS to operate. Moreover, the WSN transfer of data to the BS took a long time [9].

The WSN architecture provided an efficient fuzzy based constant node improvement method. The major responsibility is to establish a highly secured route across various layers for WSN. The goal would be to keep track of the sensor node as well as analyze its behavior. The fuzzy-based continuous node refinement approach was created to assess its behavior. The algorithm detects and eliminates nodes that aren't appropriate for information exchange. The action among all intermediary nodes within the routing path is controlled by the sending node [10]. Table 1 compares a few recent WSN clustering and routing analyses.

Table 1. Comparison of WSN Clustering and Routing Studies.

Year	Method	Performance attribute	Major Findings	Drawbacks
2019 [11]	The artificial bee colony algorithm was utilized to collect data	Energy balanced clustering	Data aggregation optimization, mobile route Length reduction	Metrics such as both delays and packet arrival rates are ignored
2020 [12]	PoDNTs (Prioritized Requested Non parametric Dependent Trees) for WSNs	Creating a structure with a dependent endless tree-like framework and a hierarchy	Make a link between nodes	The methodology takes a long time
2020 [13]	PSO-ECSM technique	Data gathering using a multi-hop approach and MS inside the framework	Increases life of the network,	Metrics such as packet delivery rate and latency are not taken into account
2022 [3]	TEO-MCRP	In WSNs, efficient CH selection and MS route determination are done	Increasing lifetime of a network	The impact of sink mobility on network quality is not investigated

3 Proposed Method

The proposed approach is based on the properties of two algorithms, one for CH selection and the other for network routing. The proposed MPSO-EECRP technique provides an effective hybrid algorithmic architecture. When used in conjunction with a routing protocol, this strategy yields the best results.

3.1 Network Model

The suggested work in this research simulates a WSN comprising of an "n" number of sensors that are employed in a rectangular network. There are some assumptions made concerning the sensor nodes deployment [14].

- The networked area is rectangular in shape.
- The network's nodes are stationary.
- The nodes in the networked area are scattered at random.
- The initial energy is the same for all nodes.
- Sensors are assigned a unique identifier (ID) and initial energy.
- Based on the transmission distance, a node's transmission power level may be changed.

- The nodes are all homogeneous. Each node has detecting, transmitting, and transferring capabilities that are activated and although required by the routing algorithm. The nodes seem to be similar to customers, as well as consumers are not required to establish if the nodes seem to be relay nodes.

3.2 Energy Model

The energy model is based on the radio model as in [15]. This transmitter uses power required to run the sensor node's broadcasting circuits and power amplifier. The energy utilized by a node is established by two factors: The quantity of information transferred as well as the distance actually travelled. A sensor network's energy usage is directly proportionate to d^2 whenever the propagation range d is shorter than the threshold distance d_0, else it is proportional to d^4. The amount of energy used by each sensor node to transmit the data in the packet is provided by Eq. (1).

$$E_T(l, d) = \begin{cases} l \times E_{ele} + \varepsilon_{fsm} \times l \times d^2 \ if \ d < d_0 \\ l \times E_{ele} + \varepsilon_{mpm} \times l \times l \times d^4 \ if \ d \geq d_0 \end{cases} \quad (1)$$

E_{ele} is the energy expended for every bit to function the transceiver circuit, as well as the amplified energy required for the space prototype. ε_{fsm} and ε_{mpm} for the multipath model (depends on the transmitter amplifier). d_0 is the maximum transmission range.

$$d_0 = \sqrt{\frac{\varepsilon_{fsm}}{\varepsilon_{mpm}}} \quad (2)$$

Accordingly, Eq. (3) gives the resources utilized by the receiver for receiving l bit of data

$$E_R(l) = l \times E_{ele} \quad (3)$$

3.3 Cluster Head Selection Utilizing Modified Particle Swarm Optimization

Consider S as the swarm space (S ϵ R2) and f: S Gϵ R to be the cost function or fitness function. The fitness function is employed to determine where CH should be placed. The primary goal of a function would be to improve the cumulative effect of the mean distance from sensor devices in a cluster, remaining energy, and headcount (the number of occasions when a node performed as the CH).

Let $\sum = (x_1, x_2, x_3, \ldots .x_n)$ be for the collection of particles within swarm S assumed about experimentation. There are N particles exist, each of which includes a velocity vector and V_i position vector x_i

$$x_i = (x_{i1}, x_{i2}, x_{i3}, \ldots .x_{iN})^T, v_i = (v_{i1}, v_{i2}, v_{i3}, \ldots v_{iN})^T$$

where I = 1,2,3,...,N and M is the dimension. To keep track of the best global positioning, maintains the local (p_i) and global best (p_g) position between the particles in collection $\Pi = (p_1, p_2, \ldots P_M)$ which provides best particle locations ever visited.

$$p_i = (p_{i1}, p_{i2}, \ldots p_{iN})^T, p_i(t) = \arg\min f_i(t), p_g(t) = \arg\min f(p_i(t)) \quad (4)$$

Finally, in the following expression, function f(x(t)) for the ith particle:
$F(x_i(t)) = \text{optimize}(\acute{\alpha}_1 x_1 + \acute{\alpha}_2 x_2 + (1 - \acute{\alpha}_1 - \acute{\alpha}_2) x_3)$

$$x_1 = \sum_{n_j \in c_k} \left\{ \frac{\| n_j, x_i \|}{|C_k|} \right. \quad (5)$$

$$x_2 = \frac{\sum_{i=1}^{N} E(p_i)}{\sum_{j=1}^{|C_k|} E(n_j)}, E_{min} \le E(n_j) \le E_{max} \quad (6)$$

$$x_3 = \frac{1}{H(p_i)}, H(p_i) \ge 1 \quad (7)$$

$0 < \acute{\alpha}_1, \acute{\alpha}_2 < 1$ and $\acute{\alpha}_1 \le \acute{\alpha}_2$. $\acute{\alpha}_1, \acute{\alpha}_2, \acute{\alpha}_3$ are the parameters for weightage. However, $E(p_i)$ and $H(p_i)$ constitute consumption of energy and headcount related with p_i. n_j is the jth node of cluster k. $|C_k|$ – total number of cluster nodes. The Euclidean distance among the node n_j and particle p_i is denoted as $\| n_j, x_i \|$

3.4 Algorithm

The algorithm specifies how such packets are transferred from the initial node to the receiving node (see Fig. 2).

```
Begin
for i=1 :N :All nodes
   Selecting cluster head using Particle Swarm Optimization Algorithm
end
for i=1 :N :All nodes
   if node i is CH
   Transmit message to CH of sink()
   For j=1: All nodes
   Transfer message between cluster member and CH
   end
  end
end
   while (all nodes)
   calculate energy usage();
   if(node is non elected and energy>0)
   compute distance between i and BS
   join to the closest one
send data to the closest one
end
   end
end

ClusterHeadSelection()
begin
for i-1:All nodes
if node is not CH
choose the shortest distance among CH()
send request message;
end
end
end
```

4 Results and Simulation

This section presents the performance evaluation results from simulations to demonstrate the effectiveness of the proposed algorithm.

4.1 Simulation Environment

MATLAB was utilized to develop the proposed methodology using a computer equipped with an i7 processor, a GPU card, and 8 GB of RAM. The 100 nodes are distributed at random across a 200 × 200 m^2 area. The implementation first chooses CH and then establishes clusters. CH (x_1, y_1) computes the Euclidean distance among both the CH and BS, and also the collected information is delivered to the CH through the shortest

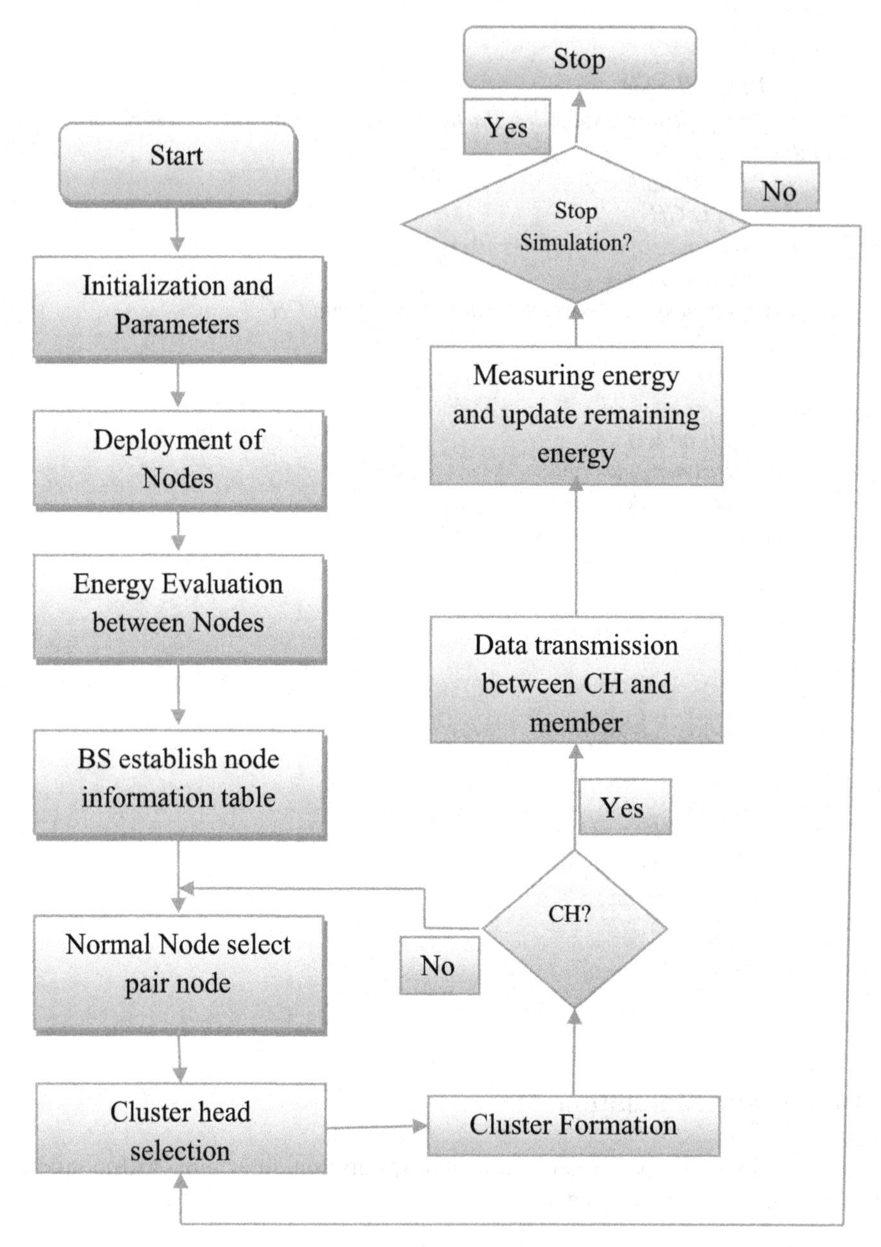

Fig. 2. Proposed Algorithm Flowchart

route from BS. A relay node is implanted inside the network's core (100, 100). Analyze a CH with location (x_1, y_1) and then a node (BS) with location (x_1, y_1) (x_2, y_2). The Euclidean partitioning between the CH and node is determined in Eq. (8):

$$\text{Distance between Nodes and BS} = (x_2 - x_1)^2 + (y_2 - y_1)^2 \tag{8}$$

4.2 Performance Evaluation

There are three kinds of sensors in a 100-node network (excluding the BS): detecting nodes, associate nodes, and CH. The experimental parameters are shown in Table 2. Several measures, such as efficiency, energy usage, packet delivery ratio, and lifetime of a network, are now being customized, which may have a considerable effect on the network. The suggested method's performance is evaluated by comparing two existing network protocols, LEACH-C and PSO-C.

Table 2. Simulation Parameters

Parameters	Values
Network dimension (m)	200 × 200
Number of sensor nodes	100
Data message	CBR traffic
CBR packet size	512 bytes
Transmission range	150m
Receiving power	1000mW
Number of Nodes	Varying from 50 to 250
Energy required to power transmitter and receiver (E_{ele})	50 nJ/bit
Amplification Energy transmission over a shorter range (E_{fsm})	12 pJ/bit/sq.m
Amplification Energy transmission over a longer range (E_{mpm})	0.0014 pJ/bit/sq.m

4.2.1 Efficiency

A WSN's efficiency is determined by its lifetime as well as the processing capacity of its sensor network. WSN node efficiency is compared to previous research. All previous techniques, including LEACH-C and PSO-C, enhance the efficiency of the suggested methodology. Table 3 compares the efficiency metrics of the proposed methods to those of previous techniques and Fig. 3 shows the efficiency comparison graph for all methods. The efficiency of the proposed PSO-EECRP method surpasses existing methods.

4.2.2 Network Lifetime

Every one of the nodes is homogeneous but also has multiple roles. This kind of distribution also directs and helps to balance the load within a cluster, it also leads to improved network management and extended network lifetime. Figure 4 depicts the life span of nodes in regards to the number of active nodes. The optimized placement of CH via swarm optimization is the reason for the proposed protocol PSO-EECRP's

Table 3. Comparison Parameters of Efficiency of the Proposed technique to currently available techniques

Methods	Efficiency (%)
(Proposed PSO-EECRP)	95
PSO-C	93
LEACH-C	89

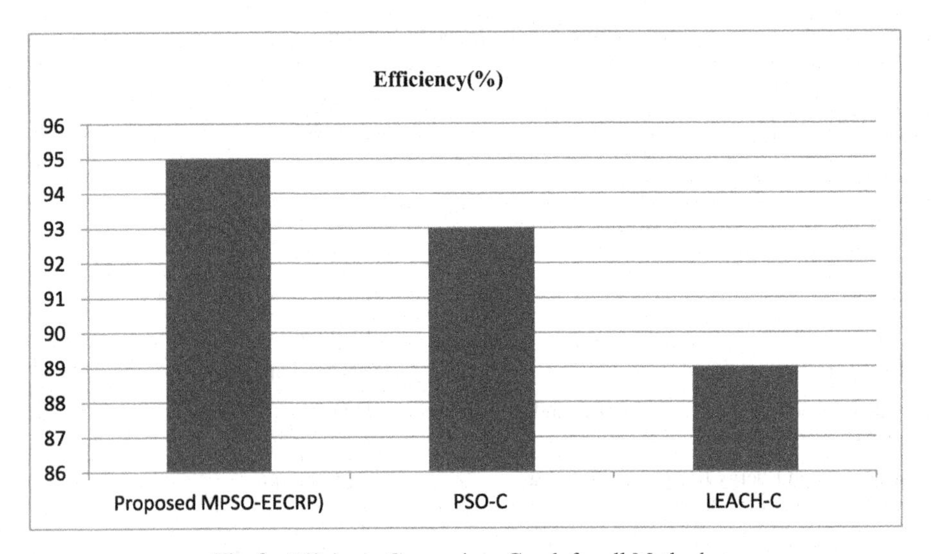

Fig. 3. Efficiency Comparison Graph for all Methods.

achievement. Using MPSO-EECRP, the cluster head is chosen depending on the distance between the cluster member node and the sink node (BT) and the remaining energy in that node. Simulation results demonstrate that the drive behind this work greatly boosts the network's life expectancy. PSO-C, on the other hand, degrades as a result that PSO functions are completely centralized at the BS. Furthermore, LEACH-C struggles significantly as a result of poor network clustering and CH selection. Table 4 shows the comparison parameters of the Maximum lifetime of the routes observed for MPSO-EECRP and Existing models. According to the findings of the simulation, the proposed protocol does a good job of balancing the network's energy consumption and extending the network's lifetime.

4.2.3 Packets Delivery Ratio (PDR)

Figure 5 depicts a significant rise in the average amount of packets received by modified MPSO-EECRP. However, the percentage of effective packet delivery remains higher than its described counterparts. Table 5 illustrates the parameters of the Packet Delivery Ratio noted for routes based on node count. As the value of PDR increases the performance of

Table 4. Comparison Parameters of Maximum Lifetime of the Routes Observed for PSO-EECRP and Existing Models.

Number of Nodes	(Proposed MPSO-EECRP)	PSO-C	LEACH-C
50	100	100	100
100	100	98	100
150	100	96	90
200	98	94	80
250	94	90	65
300	85	80	55
350	80	60	30

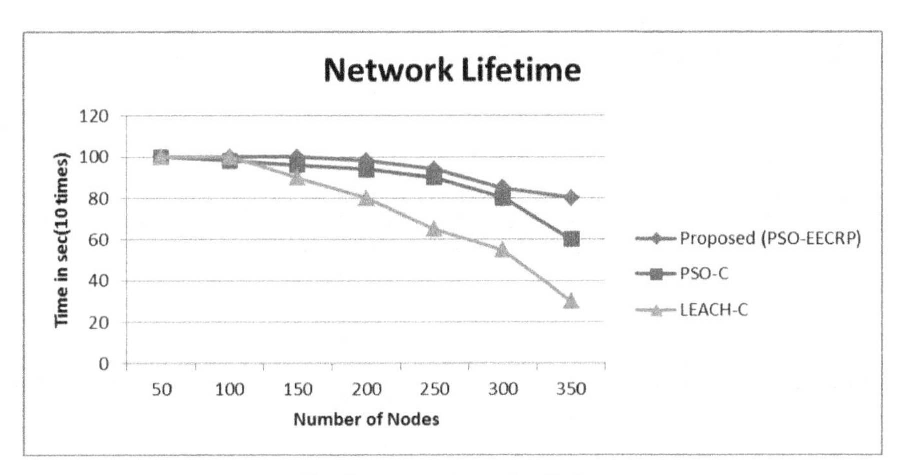

Fig. 4. Sensor Network Lifetime

the network also increases. The proposed modified MPSO-EECRP increases the number of packets delivered; it reduces communication overhead between the nodes and also it increases unnecessary overhead due to its route adaptation feature in response to topological changes in the scenario of the network.

The ratio of data packet delivered to the desired location to available by the source node is defined as PDR.

$$\text{PDR} = \frac{\text{Total number of received packets}}{\text{Total number of transmitted packets}} \quad (9)$$

where,

Total number of packets received in a destination

$$= \sum_{j=1}^{n} \text{No. of source code} \times \text{No. of packets} \times \text{Received packets}$$

$$\text{Total packet sent} = \sum_{j=1}^{n} \text{No. of nodes} \times \text{No of packets sent}$$

Table 5. Comparison Parameters of Packet Delivery Ratio Noted for Routes Based on Node Count

Number of nodes	(Proposed MPSO- EECRP)	PSO-C	LEACH-C
50	9	7	4
100	12	10	9
150	22	15	15
200	30	20	19
250	40	26	22
300	45	30	25
350	49	34	21

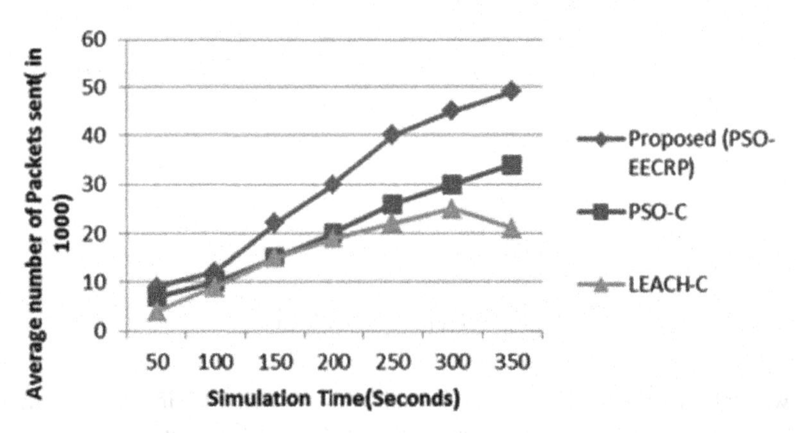

Fig. 5. Average Amount of Packets Sent During a Simulation

4.2.4 Average Energy Usage in the Network

When compared to alternative ways, the proposed method uses the least amount of energy. Additionally, both the cluster head sensors and the cluster member's sensors satisfy the energy consumption improvements. Figure 6 depicts the suggested method's effectiveness in delivering maximum sensor activities while saving energy. Table 6 depicts the average energy usage for routes based on node count.

Table 6. Comparison Parameters of Energy Usage in the Network Noted for Routes Based on Node Count

Number of nodes	(Proposed MPSO- EECRP)	PSO-C	LEACH-C
50	105	104	104
100	104	95	80
150	100	89	50
200	90	77	30
250	89	70	15
300	80	60	10
350	73	45	5

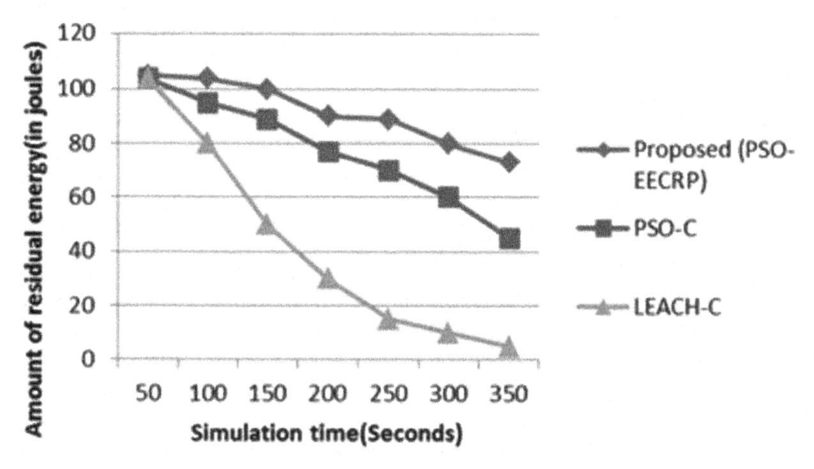

Fig. 6. Energy Usage in the Network

5 Conclusion and Future Work

On WSN, the clustering protocol allows sensor networks to decrease packets of data through data gathering. The decrease in packets decreases wireless transmission costs, and clustering routing protocols improve the network lifetime and energy usage. The proposed modified MPSO-EECRP has a 95% Efficiency. According to the simulation results, the proposed method surpasses other techniques such as PSO-C and LEACH-C. The proposed MPSO-EECRP algorithm was simulated, and various network performance measures including network lifetime, efficiency, packet delivery ratio, and total energy consumption based on node count were examined. Furthermore, Implementation of dynamic sensor nodes at larger geometrical area, as well as PSO applications in heterogeneous WSN in the future.

References

1. Singh, B., Lobiyal, D.K.: Energy-aware cluster head selection using particle swarm optimization and analysis of packet retransmissions in WSN. Procedia Technol. **4**, 171–176 (2012). https://doi.org/10.1016/j.protcy.2012.05.025
2. Yarinezhad, R., Hashemi, S.N.: Solving the load balanced clustering and routing problems in WSNs with an fpt-approximation algorithm and a grid structure. Pervasive Mob. Comput. **58**, 101033 (2019). https://doi.org/10.1016/j.pmcj.2019.101033
3. Yalçın, S., Erdem, E.: TEO-MCRP: thermal exchange optimization-based clustering routing protocol with a mobile sink for wireless sensor networks. J. King Saud Univ.-Comput. Inform. Sci. **34**, 5333–5348 (2022). https://doi.org/10.1016/j.jksuci.2022.01.007
4. Daas, M.S., Chikhi, S., Bourennane, E.: A dynamic multi-sink routing protocol for static and mobile self-organizing wireless networks. A routing protocol for Internet of Things. Ad Hoc Networks **117**, 102495 (2021). https://doi.org/10.1016/j.adhoc.2021.102495
5. Mitra, P., Mondal, S., Hassan, K.L.: Chapter 16 – Energy efficient rendezvous point-based routing in wireless sensor network with mobile sink. In: Bhattacharyya, S., Dutta, P., Samanta, D., Mukherjee, A., Pan, I. (Eds.) Recent Trends in Computational Intelligence Enabled Research, pp. 279–293. Academic Press (2021)
6. Samarji, N., Salamah, M.: ESRA: energy soaring-based routing algorithm for IoT applications in software-defined wireless sensor networks. Egypt. Inform. J. **23**(2), 215–224 (2022). https://doi.org/10.1016/j.eij.2021.12.004
7. Khan, M.A., Awan, A.A.: Intelligent on demand clustering routing protocol for wireless sensor networks. Wireless Commun. Mob. Comput. **2022**, 1–10 (2022). https://doi.org/10.1155/2022/7356733
8. Daniel, A., Baalamurugan, K.M., Vijay, R., Arjun, K.P.: Energy-aware clustering with multi-hop routing algorithm for wireless sensor networks. Intel. Autom. Soft Comput. **1**, 233–246 (2021). https://doi.org/10.32604/iasc.2021.016405
9. Pattnaik, S., Sahu, P.K.: Assimilation of fuzzy clustering approach and EHO-Greedy algorithm for efficient routing in WSN. Int. J. Commun. Syst. **33** (2020). https://doi.org/10.1002/dac.4354
10. Elavarasan, R., Chitra, K.: An efficient fuzziness based contiguous node refining scheme with cross-layer routing path in WSN. Peer-to-Peer Netw. Appl. **13**, 2099–2111 (2020). https://doi.org/10.1007/s12083-019-00825-0
11. Vijayashree, R., Dhas, C.S.G.: Energy efficient data collection with multiple mobile sink using artificial bee colony algorithm in large-scale WSN. Automatika J. Control Meas. Electron Comput. Commun. **60**, 555–563 (2019). https://doi.org/10.1080/00051144.2019.1666548
12. Yalçın, S., Erdem, E.: A mobile sink path planning for wireless sensor networks based on priority-ordered dependent nonparametric trees. Int. J. Commun. Syst. **33**, e4449 (2020). https://doi.org/10.1002/dac.4449
13. Sahoo, B.M., Amgoth, T., Pandey, H.M.: Particle swarm optimization based energy efficient clustering and sink mobility in heterogeneous wireless sensor network. Ad Hoc Netw. **106**, 102237 (2020). https://doi.org/10.1016/j.adhoc.2020.102237
14. Vimalarani, C., Subramanian, R., Sivanandam, S.N.: An enhanced PSO-based clustering energy optimization algorithm for wireless sensor network. Sci. World J. **2016**, 1–11 (2016). https://doi.org/10.1155/2016/8658760
15. Mutar, H.I., Jawad, M.M.: Enhancing energy utilization in wireless sensor networks (WSNs) by particle swarm optimization (PSO). Webology **19**, 6089–6107 (2022)

Machine Learning Based Incipient Fault Diagnosis of Induction Motor

Rahul Kumar(✉), Rajvardhan Jigyasu, Sachin Singh, and Srinivas Chikkam

Department of Electrical Engineering, National Institute of Technology, Delhi1 10036, India
212231007@nitdelhi.ac.in

Abstract. An induction motor is the most important machine used in all industries, so the health of the machine is checked regularly by the machine learning method described in this paper. The objectives of this paper is to get high accuracy in predicting the fault of a machine and reduce the training time period of the model. Prediction of electrical and mechanical faults is achieved by monitoring the acoustic signal and analyzing it to identify deviations from expected values, which can indicate the presence of faults. The AI algorithms are trained using acoustic data acquired from electrical machines. A benchmark data set available online is used to test the proposed algorithm. The data is segmented into non-overlapping epochs. Features are extracted by the Kruskal-Wallis algorithm, and then the data is trained on various models to compare their accuracy. Machine learning-based diagnostic techniques such as weighted KNN, fine KNN, wide neural networks, and coarse Gaussian SVM are used and achieve 100% accuracy. The best algorithm is weighted KNN due to its fast processing time.

Keywords: Supervised Learning · Unsupervised Learning · Acoustic Signal · Condition Monitoring · Fault Diagnosis · Feature Selection

1 Introduction

According to Industrial Revolution 4.0 and 5.0, industrial processes are getting increasingly complex as a result of the proliferation of intelligent automation. Rotating machinery is an important component of the process. Due to their durability, dependability, simplicity of maintenance, and wide range of applications, induction motors (IM) are widely used in industries. Moreover, several applications—such as the motor used to pump oxygen in hospitals—are quite crucial. The motor must operate nonstop and continuously. Any error will result in losses to one's health. Preventive maintenance and failure-oriented maintenance are the two basic types of maintenance used. In a large industry, we cannot afford a machine failure, even for a short amount of time, because there is a large loss of capital, which is not economical for a large industry.

To prevent such a loss, some preventive measures are put in place to avoid a fault condition. The application of machine learning [1, 2] is used in every field or industry to find the fault and diagnose the fault, and to do that, a large amount of data with a variety of faults is needed to improve accuracy.

R. K. Challa et al. (Eds.): ICAIoT 2023, CCIS 1930, pp. 113–127, 2024.
https://doi.org/10.1007/978-3-031-48781-1_10

To overcome the problem of unwanted shutdown and loss of lives, machine learning [3] for fault diagnosis has come into existence. Faults like inner race defect(IRD), outer race defect (ORD) and broken rotor bar are studied here. It is cheaper and safer than human resources. Because electrical machines [1, 4] are used in all industries, maintaining them is a major challenge. Predictive maintenance is to be done to find and rectify the electrical fault. Now a days, machine learning is used in every field such as in healthcare systems for predicting disease viruses, finance systems for predicting stock prices, cyber security [5], e-commerce, and transportation systems for finding the best routes [6], all of which are compared and trained to get higher accuracy.

In the context of classification, the research is going with ensemble classifier while holding the condition of 'no validation' without PCA. An automatic detection and classification of the broken rotor bar (BRB) and bearing deficiencies are studied to detect faults and achieve a maximum accuracy of 95.8% [7]. Bearing and rotor faults have been taken for the diagnosis of an electrical machine, and the result of ANN is the best with 99.18% accuracy [8]. The algorithms' performance was tested by examining their recognition rates. There were six different SVM kernels examined. Recognition rates of broken bars up to 97.9% have been attained [9]. Bearing and considering with broken rotor bar is successfully achieved 100% of accuracy [10]. By using the CNN algorithm, 99.5% accuracy is achieved in the current signature analysis regarding the faults for the category rotor faults with the consideration of flaws in bearing [11].

The publications mentioned above demonstrate the models' highly accurate design, yet many of them are unable to identify multiple defects. Also, the classifier receives irrelevant characteristics for testing and training [12]. It is typical to think of computing a high-dimensional collection of characteristics in order to acquire large information. To achieve high accuracy, only the most dominant features are used, while less important features are ignored. Dimensionality reduction strategies have been used into fault diagnosis approaches to avoid of unnecessary data that might affect diagnostic performance. Principal component analysis (PCA) is the technique that is most frequently used to reduce the number of dimensions in data sets. With a shorter processing time and higher accuracy, this work created a revolutionary defect diagnostic approach. To extract features from the fusion of four defect datasets, the obtained signal must first go through a preprocessing stage. The Kruskal- Wallis method is used to select features [13, 14], and this method is verified using the ANOVA algorithm. In both cases, we get similar results for the most dominant feature.

Before applying machine learning algorithms, information about the most common faults must be known. Today, there are many methods for condition monitoring and detecting failures in electrical equipment [15], such as

Temperature analysis [16].

Noise and vibration analysis [17, 18].

Chemical and its lubricant or coolant analysis [19] Electromagnetic field analysis [20].

Analysis of motor-current signature (MCSA) [21].

Model and using methods based on artificial intelligence [3, 22].

In this paper, models have achieved up to 100% accuracy in a resubstitution validation scheme. To make the model less complicated, only major faults in electric machines are considered, such as bearing faults, flywheel faults, piston faults, and rider belt faults, and seven main features are extracted from the data. This paper has five parts, starting at the beginning with various type of fault followed by the experiment unit and methodology. Final unit include a discussion on the results and a conclusion.

2 Types of Faults Occurring in Electrical Machines

2.1 Stator Fault

The stator winding can fail due to thermal or electrical stresses, which may result in short circuits, open circuits, or turn-to-turn faults. This fault can cause the machine to lose power, overheat, and even cause damage to the other components [23]. If the stator fault is not detected and rectified, it can lead to a complete machine break- down. The stator fault can be detected by performing various tests such as insulation resistance test, winding resistance test, and turn-to-turn resistance test. According to IEEE survey, stator fault percentage is 28% of total faults occurring on machine [24].

2.2 Rotor Fault

A rotor fault in an electrical machine refers to a problem or damage that occurs in the rotating component of the machine e.g. rotor winding damage, rotor bar defects, unbalanced rotor. This component, also known as the rotor, is responsible for generating the magnetic field that interacts with the stator to produce motion or power. A fault in the rotor can lead to a number of issues, such as reduced efficiency, increased noise or vibration, and even complete failure of the machine. According to IEEE and EPRI, rotor fault percentage's 9% and 8% respectively. In Fig. 1, the statistical studies of induction machine (IM) failure with the research work of companies like ABB and tremendous research publications on the platform of IEEE and Electric Power Research Institute [25].

2.3 Bearing Fault

A bearing fault happens when an electrical machine's bearing is damaged, worn out, or noisy. Vibrations [26] and high heat formation are results of machine failure [4, 27]. Figure 2 shows the bearing structure and its fault. There are several types of bearing faults, including [28]:

1) Inner race defect: An IRD is a defect in the inner race of the bearing. This type of fault is often caused by improper installation or contamination.
2) Outer race defect: An ORD is a defect in the outer race of the bearing. This type of fault is often caused by excessive loads or contamination.
3) Ball defect: A ball defect is a defect in one or more of the balls in the bearing. This type of fault is often caused by excessive loads or contamination.
4) Roll defect: A roll defect is a defect in one or more of the rolls in the bearing. This type of fault is often caused by improper manufacturing processes or contamination.

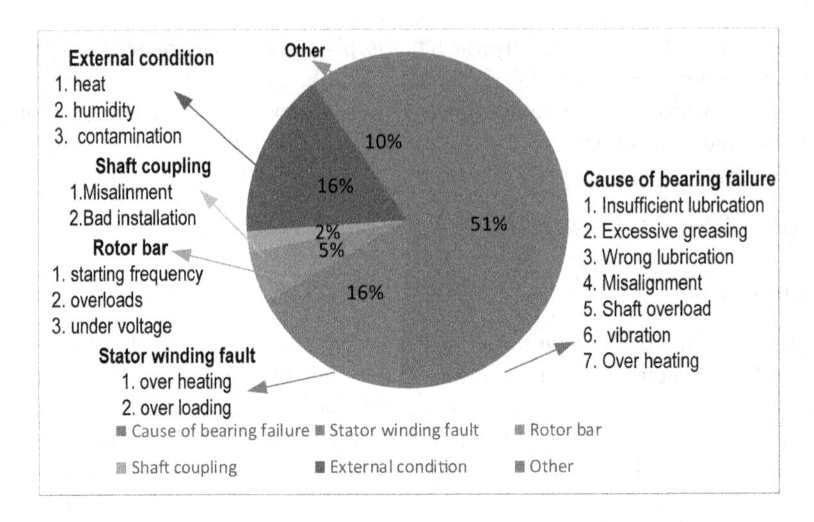

(a)

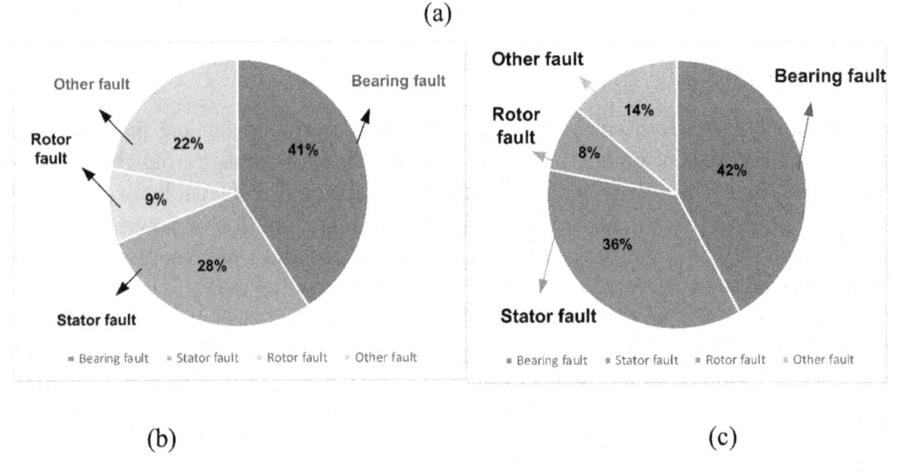

(b) (c)

Fig. 1. Study of induction motor faults (a) ABB, (b) IEEE, (c) Electric Power Research Institute

Vibration analysis [29, 30] is a common method for finding bearing faults. Testing is carried on the ground of vibration. This is analyzed according to the data synthesized from vibration levels, so that its anomalies can be identified. Ultrasonic testing is a non-destructive testing process that uses sound waves to detect bearing faults at a high frequency.

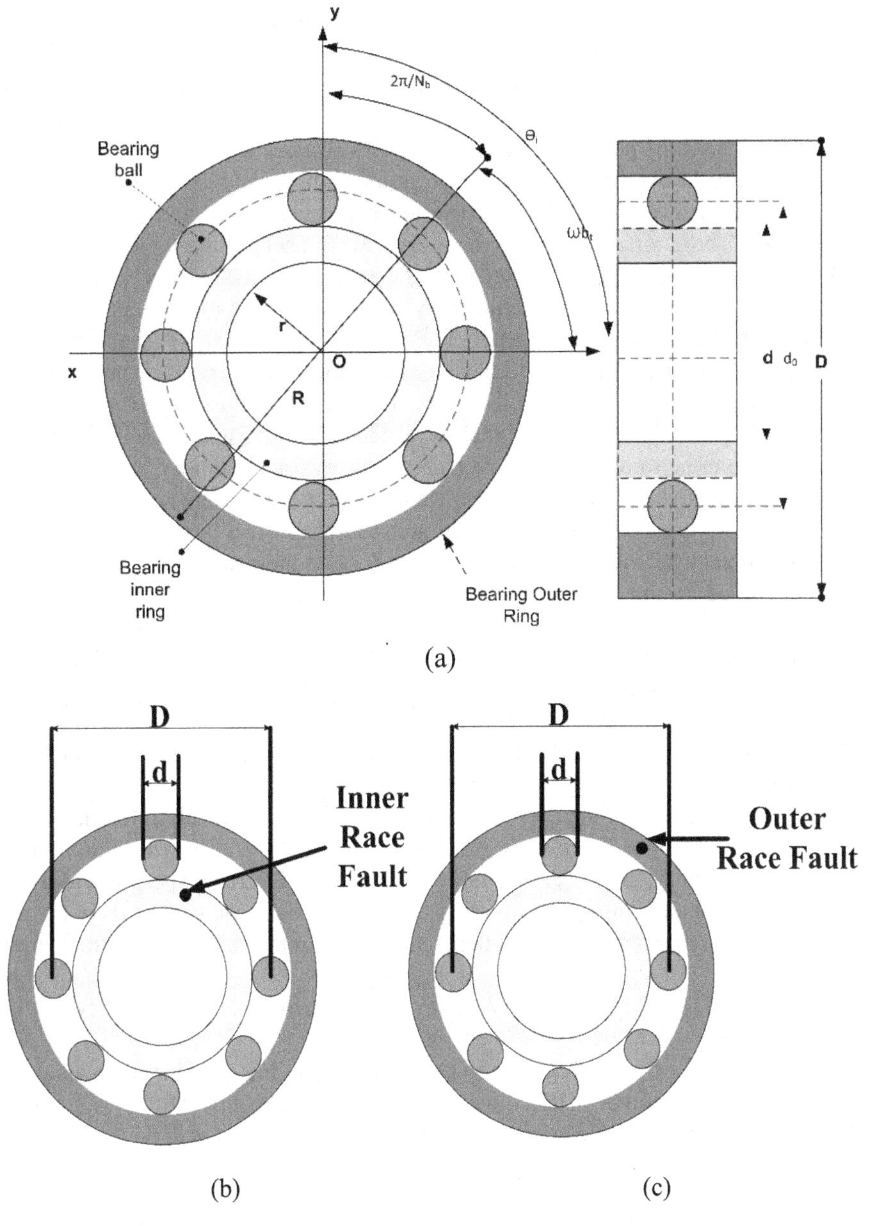

Fig. 2. (a) Ball Bearing Structure (B) Inner Race Fault (C) Outer Race Fault

3 Experimental Set-Up and Data Acquisition

The initial phase of fault diagnosis is data capture, during which machine properties are measured and recorded for later study. To collect data, a transducer is placed on the surface of the machine part. This paper focuses on acoustic signal-based fault diagnosis. Unidirectional microphones are used to gather acoustic data [31]. Only sounds that are focused in the direction of the microphone centers are picked up by unidirectional microphones. These microphones are less sensitive to background noise. A single NI 9234 data acquisition (DAQ) hardware device with many ports, an NI 9172 USB interface, and a LabVIEW-based data acquisition interface are used throughout the process to record audio through microphones. A microphone is placed at a distance of 1.5 cm from the machine. It will produce a clearer and louder sound from the machine. The output of microphones is analogue in nature. The NI 9234 is used to sample the data and transform it from analogue to digital form. Then, NI 9172 and LabVIEW interface are used to store the sampled signal on a computer. A four-channel C-series dynamic signal collection module (NI 9234) allows for the simultaneous connection of up to four microphones. As a result, if necessary, an acoustic signal can be acquired from four locations. All acoustic signals are recorded for 5 s and their sampling rate is 50 kHz, and the recorded sampled values are saved in.dat files in 24-bit PCM format. Hence, there are altogether 250,000 samples in each recording. Very few number of sets are recorded are taken from various location to determine the best spot for recording. These starting places are determined by past knowledge or human intuition as to which parts of the machine might reveal crucial details about its current state. The next step is to use Sensitive Position Analysis (SPA) to determine the best places. Twenty-four positions were considered as sensitive positions in these experiments.

In this paper, an online data set was used [31]. To make a model, different types of electrical faults are considered, such as a healthy condition, a rider belt fault, a piston fault, a flywheel fault, and a bearing fault. To get data, different types of sensors are used in experimental setups. After getting all types of data, we extract their features to differentiate the data according to their fault. Induction motor specifications are as follows:

Parameter of IM: Power: 5 HP, Voltage: 415 V, Current: 5 A, Frequency: 50 Hz, Speed: 1440 rpm

Different types of fault signals are shown below the Fig. 3. They are called raw signals, where the amplitude of the signal varies with respect to time. Multiple algorithms are used in this paper to achieve higher accuracy with acoustic signal.

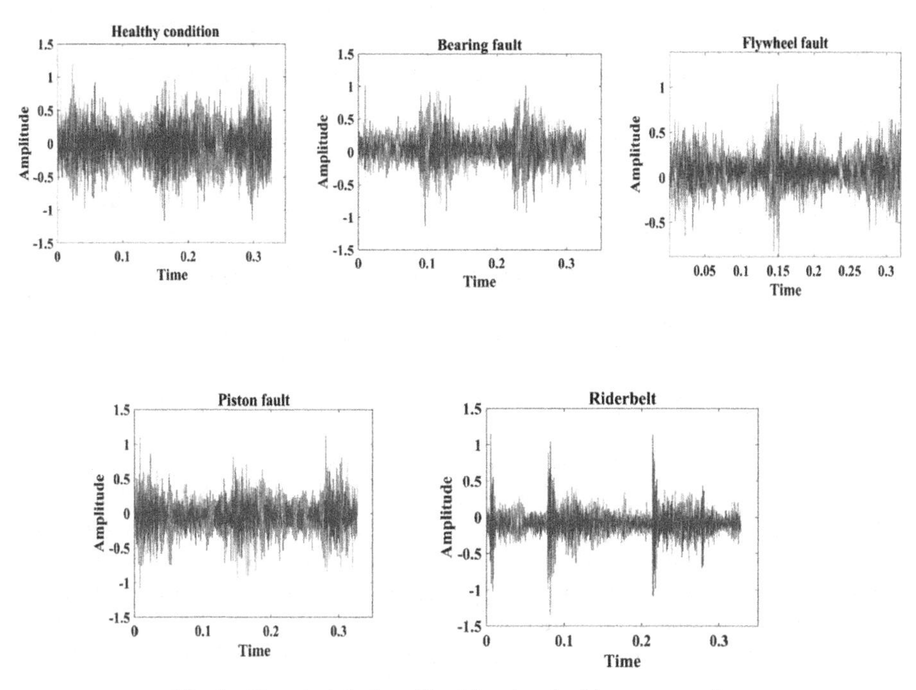

Fig. 3. Electrical fault and healthy signal with respect to time

4 Methodology

This experiment aims is to get the best result and most appropriate algorithm for the classification of many kinds of faults existing in the induction motor. The first step of the proposed methodology is to collect acoustic data from the motor under different conditions. In this work, a benchmark data set has been used in which 16385 data samples are taken for each fault and healthy condition. After that, the data is divided into segments using a non-overlapping segmentation scheme.

The time domain features such as RMS, mean, energy, variance, skewness, and standard deviation were extracted from the segmented signal. After extracting the feature, the data is normalized to get the best result. Except skewness, which was to the extent of -1 to 1, the normalization was in the range [0, 1]. Feature selection is done by the Kruskal-Wallis algorithm. After that model is trained in MATLAB using the extracted features, data has been applied to different algorithms, and accuracy has been checked. The methodology is shown in Fig. 4.

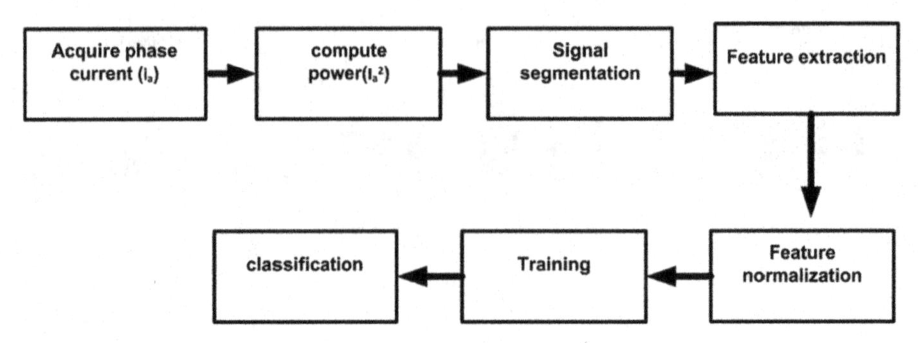

Fig. 4. Methodology of condition monitoring

4.1 Kruskal Wallis Algorithm

The Kruskal-Wallis algorithm is a non-parametric statistical test that can be used to compare three or more independent variables to determine if there are any significant differences among them. This algorithm can also be used for feature selection in machine learning and data analysis. The Kruskal-Wallis algorithm can be used to extract features by comparing the medians of feature values across different groups and identifying those features that are statistically significant in distinguishing between the groups.

4.2 Weighted KNN

Weighted K-Nearest Neighbor (KNN) is a variant of the KNN algorithm, a popular machine learning algorithm for classification and regression tasks. In KNN, the algorithm identifies the K nearest neighbors of a new data point based on a distance metric and assigns the label of the majority class among those K neighbors as the predicted label for the new data point. However, in the standard KNN, all the neighbors are given the same weight. In contrast, weighted KNN assigns weights to the K neighbors based on their distances from the new data point. The closer the neighbor is to the new data point, the higher its weight. The weight is then used to compute a weighted average of the class labels of the K neighbors, instead of just taking the majority vote.

5 Results and Discussion

In the proposed work, three types of validation schemes are used, such as resubstitution validation, holdout validation, and cross validation schemes, to test the reliability of the model. The best result is obtained from the resubstitution validation scheme. In this paper, bearing fault (b), flywheel fault (f), piston fault (p), rider belt fault (r), and healthy condition (h) data are used. Figure 5 depicts a scatter plot of features extracted from the data.

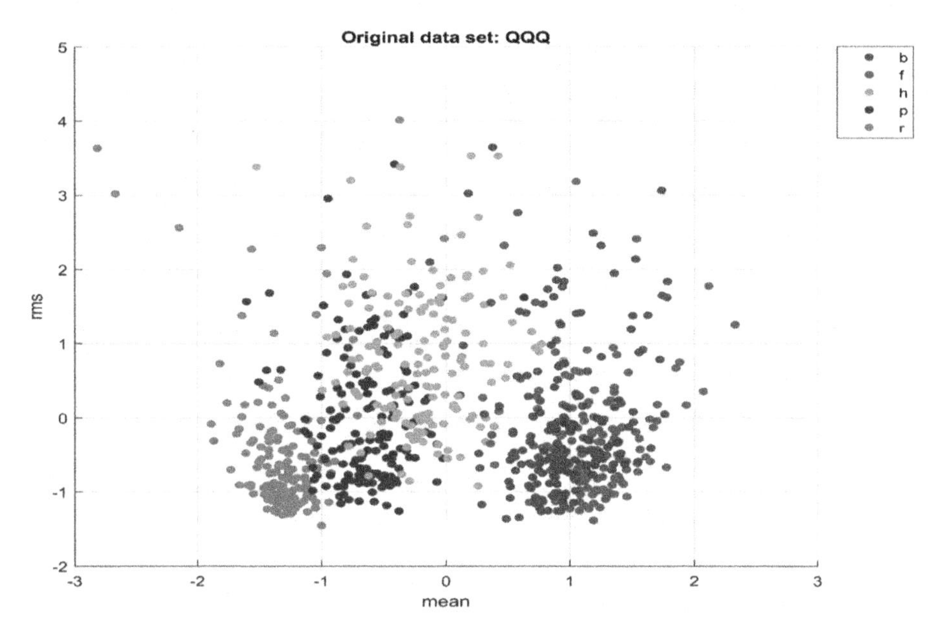

Fig. 5. Scattered plot of features for bearing, flywheel, piston, rider belt fault and healthy condition

The selection of features is one of the most crucial process, to know which feature is most essential and dominant for predicting the fault. According to the Kruskal-Wallis algorithm in Fig. 6, the "mean" feature is used as the most dominant feature; its score is 350, whereas skewness and kurtosis have a very low score in predicting the fault. By neglecting the low-score features, our model will be less complex, and the training time will decrease without compromising the accuracy of the model.

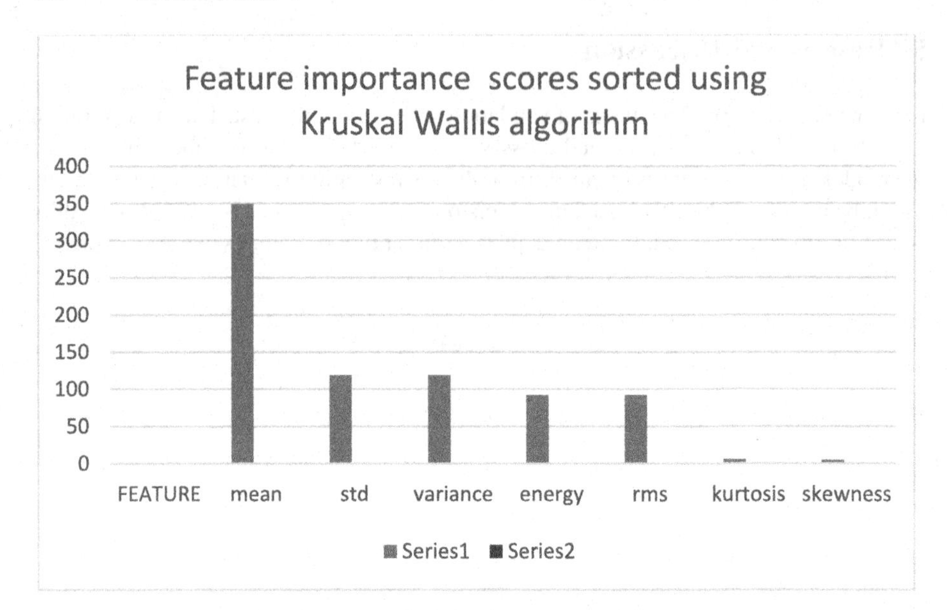

Fig. 6. Bar graph of dominant nature of features

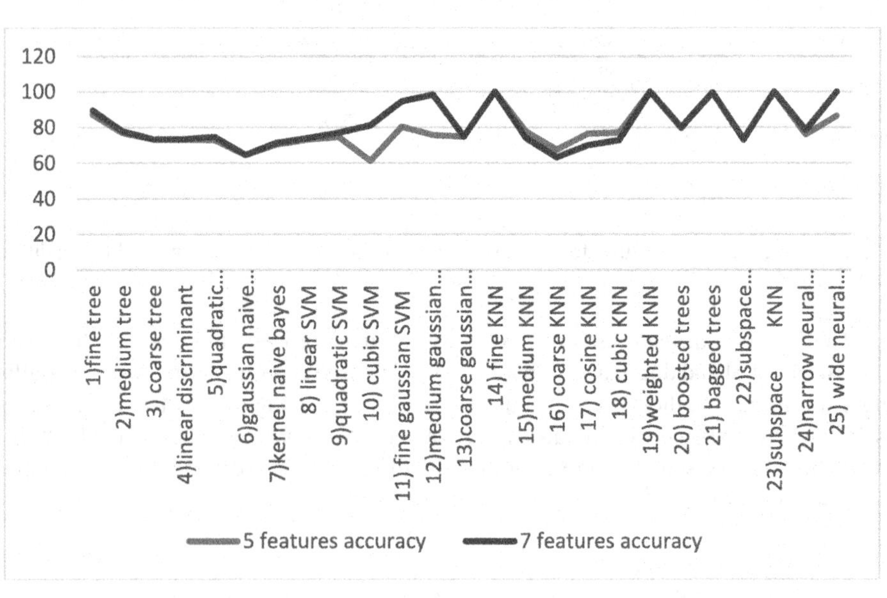

Fig. 7. Comparison of accuracy of different models with having five and seven features

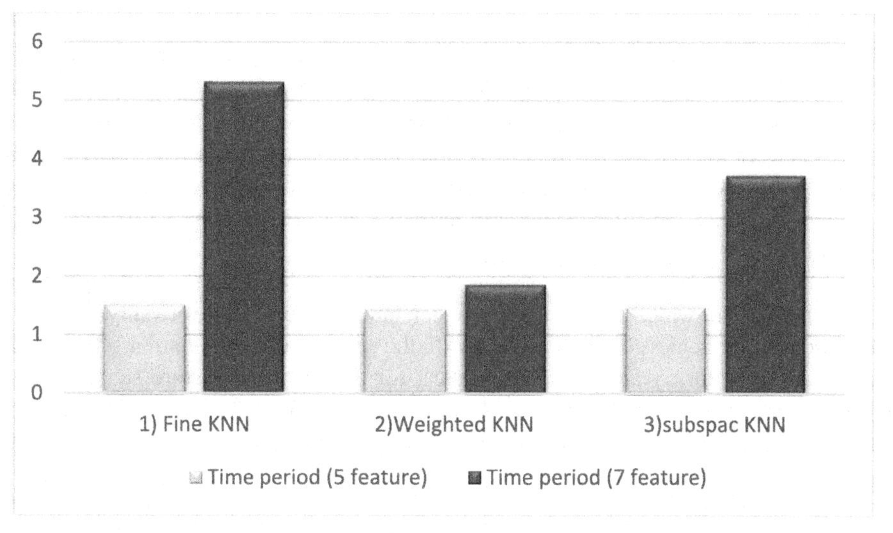

Fig. 8. Comparison of training time of model having five features and seven features

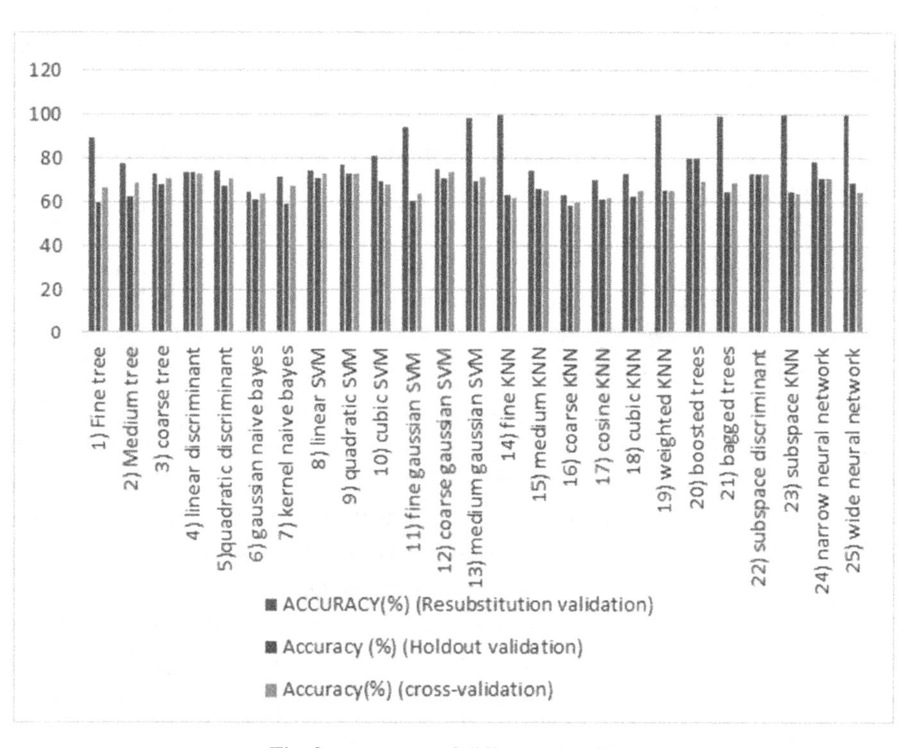

Fig. 9. Accuracy of different algorithm

The accuracy does not show much difference between the model having five features and having seven features, so kurtosis and skewness are neglected in the model. From Fig. 7, it is clear that the accuracy is nearly the same in both cases. By neglecting two features of the model, its training period becomes very short, as shown in Fig. 8.

In this work to decrease the training period of the model, only the five most essential features are selected, which are mean, standard deviation, variance, energy, and rms, as shown in Fig. 8, and the results are verified using the best three 100%-achieved algorithms.

Figure 9 gives information regarding the accuracy of different types of algorithms in bar graph form, which helps in better understanding them. Out of twenty-five algorithms, only four achieved 100% accuracy. In the resubstitution validation scheme, the percentage of accuracy achieved was highest for the extracted feature set, as shown in Fig. 9. The accuracy achieved was 100% for weighted KNN, fine KNN, subspace fine KNN, and wide neural networks; 99.5% for bagged trees; and 94.8% for the fine Gaussian algorithm.

The confusion matrix gives information about whether the number of predictive values is right or wrong. To get high accuracy, the diagonal element of the confusion matrix must have a higher percentage or be close to 100%. It displays the total positive predictive value (PPV) and false positive rate (FPR). To achieve good accuracy, our PPV must have a higher value, and our FPR must be as low as possible. As shown in Fig. 10, the confusion matrix, this model achieves 100% accuracy.

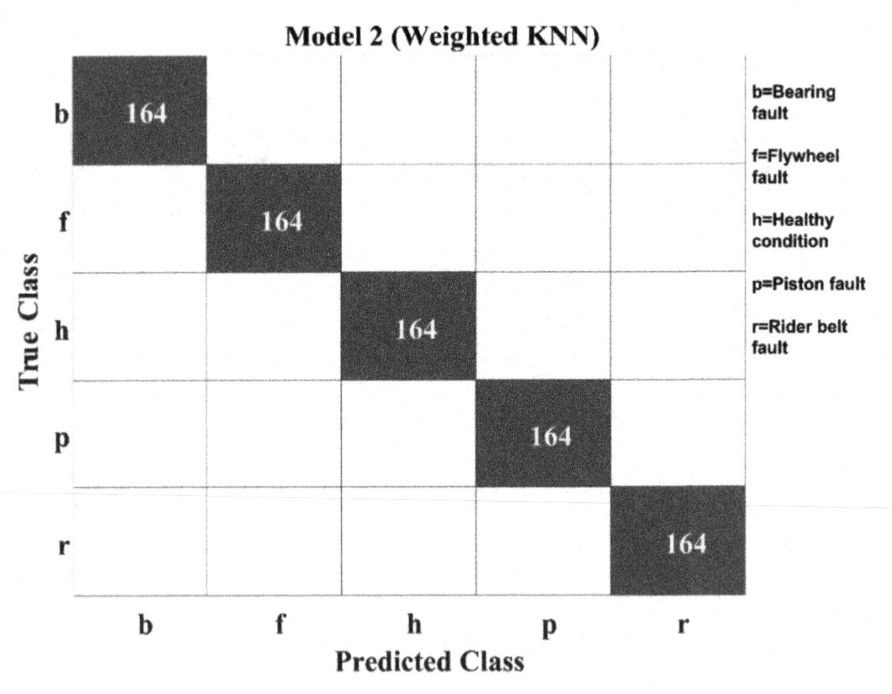

Fig. 10. Confusion matrix

6 Conclusion

The proposed method describes a model that can detect the fault with high accuracy and a short processing time. In this work, multiple AI algorithms available to detect faults are compared, and it is found that 100% accuracy is achieved for the weighted KNN, fine KNN, subspace fine KNN, and wide neural network algorithms. The acoustic signature analysis has been applied. The processing time for prediction is quite high if features are large in number; by applying the feature selection technique, the processing time has been reduced. Out of the above models with 100% accuracy, the weighted KNN algorithm performed the fastest, which suggests that weighted KNN is best for fault detection using an acoustic signature with high accuracy and less processing time. It can be stated that the classifier's accuracy relies on several deciding factors of the data sets, such as their size, duplication of data, number of null values present in a set, data quality, and, most importantly, feature selection and low data redundancy.

As future work, it can be suggested that the proposed approach be used for different monitoring schemes such as Current, Vibration and also combination of them.

References

1. Kudelina, K., et al.: Trends and challenges in intelligent condition monitoring of electrical machines using machine learning. Applied Sciences **11**(6), 2761–2761 (2021)
2. Henao, H.: Trends in fault diagnosis for electrical machines: A review of diagnostic techniques. IEEE Ind. Electron. Mag. **8**(2), 31–42 (2014)
3. Tang, S., Yuan, S., Zhu, Y.: Deep learning-based intelligent fault diagnosis methods toward rotating machinery. IEEE Access **8**, 9335–9346 (2019)
4. Zhang, J., Yi, S., Liang, G.U.O., Hongli, G.A.O., Xin, H., Hongliang, S.: A new bearing fault diagnosis method based on modified convolutional neural networks. Chin. J. Aeronaut. **33**(2), 439–447 (2020)
5. Lei, Y., Yang, B., Jiang, X., Jia, F., Li, N., Nandi, A.K.: Applications of machine learning to machine fault diagnosis: a review and roadmap. Mech. Syst. Signal Process. **138**, 106587 (2020)
6. Cui, L., Yang, S., Chen, F., Ming, Z., Lu, N., Qin, J.: A survey on application of machine learning for Internet of Things. Int. J. Mach. Learn. Cybern. **9**, 1399–1417 (2018)
7. Jigyasu, R., Shrivastava, V., Singh, S.: Smart classifier based prognostics and health management of induction motor. Mater Today Proc **43**, 355–361 (2021)
8. Jigyasu, R., Shrivastava, V., Singh, S.: Prognostics and health management of induction motor by supervised learning classifiers. IOP Conf Ser Mater Sci Eng **1168**(1), 12006 (2021)
9. Arabaci, H., Mohamed, M.A.: A knowledge-based diagnosis algorithm for broken rotor bar fault classification using FFT, principal component analysis and support vector machines. In Int. J. Intell. Eng. Informati. **8**(1), 19–37 (2020)
10. Jigyasu, R., Shrivastava, V., Singh, S.: Data fusion-based smart condition monitoring of critically applied rotating machines. In: Proceedings of Third Doctoral Symposium on Computational Intelligence: DoSCI 2022, pp. 205–218. Springer Nature Singapore, Singapore (2022)
11. Jigyasu, R., Shrivastava, V., Singh, S., Bhadoria, V.: Transfer learning based bearing and rotor fault diagnosis of induction motor. In: 2022 2nd International Conference on Advance Computing and Innovative Technologies in Engineering (ICACITE), pp. 2628–2632 (2022)

12. Gangsar, P., Tiwari, R.: Signal based condition monitoring techniques for fault detection and diagnosis of induction motors: A state-of-the-art review. Mech. Syst. Signal Process. **144**, 106908 (2020)
13. Okwuosa, C.N., Hur, J.W.: An intelligent hybrid feature selection approach for SCIM inter-turn fault classification at minor load conditions using supervised learning. In IEEE Access, https://doi.org/10.1109/ACCESS.2023.3266865
14. Li, J.: Feature selection: A data perspective. ACM computing surveys (CSUR) **50**, 1–45 (2017)
15. Fan, J., Wang, Z., Xie, Y., Yang, Z.: A theoretical analysis of deep Q-learning. Learning for Dynamics and Control **120**, 486–489 (2020)
16. Sharma, A., Jigyasu, R., Mathew, L., Chatterji, S.: Bearing fault diagnosis using frequency domain features and artificial neural networks. In: Satapathy, S., Joshi, A. (eds.) Information and Communication Technology for Intelligent Systems. Smart Innovation, Systems and Technologies, vol 107. Springer, Singapore (2019). https://doi.org/10.1007/978-981-13-1747-7_52
17. Tsypkin, M.: Induction motor condition monitoring: Vibration analysis technique-A practical implementation. In: 2011 IEEE International Electric Machines & Drives Conference (IEMDC), pp. 406–411 (2011)
18. Toh, G., Park, J.: Review of vibration-based structural health monitoring using deep learning. Applied Sciences **10**(5), 1680 (2020)
19. Yang, L., Zhang, S., Pauli, F., Charrin, C., Hameyer, K.: Material compatibility of cooling oil and winding insulation system of electrical machines. In: 2022 International Conference on Electrical Machines (ICEM), pp. 1334–1340 (2022)
20. Jigyasu, R., Mathew, L., Sharma, A.: Multiple faults diagnosis of induction motor using artificial neural network. In: Luhach, A., Singh, D., Hsiung, PA., Hawari, K., Lingras, P., Singh, P. (eds.) Advanced Informatics for Computing Research. ICAICR. Communications in Computer and Information Science, vol 955. Springer, Singapore (2019). https://doi.org/10.1007/978-981-13-3140-4_63
21. Hoang, D.T., Kang, H.J.: A motor current signal-based bearing fault diagnosis using deep learning and information fusion. IEEE Trans. Instrum. Meas. **69**(6), 3325–3333 (2019)
22. Ayas, S., Ayas, M.S.: A novel bearing fault diagnosis method using deep residual learning network. Multimed Tools Appl **81**(16), 22407–22423 (2022)
23. Glowacz, A., Glowacz, Z.: Diagnosis of stator faults of the single-phase induction motor using acoustic signals. Appl. Acoust. **117**, 20–27 (2017)
24. Verginadis, D., Antonino-Daviu, J., Karlis, A., Danikas, M.G.: Diagnosis of stator faults in synchronous generators: Short review and practical case. In: 2020 International Conference on Electrical Machines (ICEM), vol 1, pp. 1328–1334 (2020)
25. Jigyasu, R., Mathew, L., Sharma, A.: A review of condition monitoring and fault diagnosis methods for induction motor. In: Second International Conference on Intelligent Computing and Control Systems (ICICCS), pp. 1713–1721. Madurai, India (2018). https://doi.org/10.1109/ICCONS.2018.8662833
26. Ali, M.Z., Shabbir, M.N.S.K., Liang, X., Zhang, Y., Hu, T.: Machine learning-based fault diagnosis for single-and multi-faults in induction motors using measured stator currents and vibration signals. IEEE Trans. Ind. Appl. **55**(3), 2378–2391 (2019)
27. Hoang, D.T., Kang, H.J.: A survey on deep learning based bearing fault diagnosis. Neurocomputing **335**, 327–335 (2019)
28. Glowacz, A., Glowacz, W., Glowacz, Z., Kozik, J.: Early fault diagnosis of bearing and stator faults of the single-phase induction motor using acoustic signals. Measurement **113**, 1–9 (2018)
29. Zekveld, M., Hancke, G.P.: Vibration condition monitoring using machine learning. In: IECON 2018–44th Annual Conference of the IEEE Industrial Electronics Society, pp. 4742–4747 (2018)

30. Zhao, G., Sun, L., Niu, N., Wang, X., Xing, Z.: Analysis of vibration characteristics of stators of electrical machines under actual boundary. Mech. Syst. Signal Process. **185**, 109778 (2023)
31. Verma, N.K., Sevakula, R.K., Dixit, S., Salour, A.: Intelligent condition based monitoring using acoustic signals for air compressors. IEEE Trans Reliab **65**(1), 291–309 (2015). https://iitk.ac.in/idea/datasets/

Role of Federated Learning for Internet of Vehicles: A Systematic Review

P. Hiran Mani Bala and Rishu Chhabra(✉)

Chitkara University Institute of Engineering and Technology, Chitkara University, Punjab, India
rishuchh@gmail.com

Abstract. The Internet of Vehicles (IoV) is one of the most exciting and practical ways that corporations and academics are interested in, especially by employing coordinated unmanned vehicles to explore areas like the automobile industry. To provide long-term possibilities for task investigations, the IoV connects vehicles, transportation networks, and communication infrastructure. Data privacy, however, may be compromised by the coordination of information gathering from numerous sources. Federated Learning (FL) is the answer to these concerns of privacy, scalability, and high availability. A well-distributed learning framework designed for edge devices is federated learning. It makes use of large-scale processing from edge devices while allowing private data to remain locally. In this work, different categories of federated learning have been discussed. A review of various systems implementing FL for IoV has been presented followed by the applications and challenges of FL in the IoV paradigm. The paper concludes by providing future research directions for FL in the IoV.

Keywords: Federated learning · Internet of Vehicles · Machine learning · Unmanned Arial Vehicles

1 Introduction to Federated Learning

Federated Learning (FL) utilizes a centralized aggregator and provides a solution to the issues associated with many Machine Learning (ML) clients. It ensures that training data for federated learning is decentralized to protect data privacy [1]. Two key concepts of local computation and model transmission have been introduced to lower the privacy risk and cost of centralized ML systems. In FL, participants train their models by using local data and then send the model to the server for aggregation, and the server disseminates model updates. In Fig. 1, FL's high-level map process has been given [2]. In FL, local models are trained on separate vehicles before aggregating them in the cloud to enhance security, accuracy and learning efficiency [3]. In two phase mitigating scheme, an intelligent architecture with FL provides data leakage detection [4] and intelligent data transmission [5] to improve security [6]. In Unmanned Aerial Vehicles (UAV) federated deep learning applications with wireless networks, the focus is to improve the learning efficiency, learning speed significance, conscious joint data selection, resource allotment algorithm [7], and content caching method for edge computing of FL in IoV.

R. K. Challa et al. (Eds.): ICAIoT 2023, CCIS 1930, pp. 128–139, 2024.
https://doi.org/10.1007/978-3-031-48781-1_11

1.1 Categories of Federated Learning.

FL depends on five aspects which are heterogeneity, communication architecture, data partitioning, applicable machine learning models, and privacy mechanism [8]. Figure 2 depicts the classification of FL.

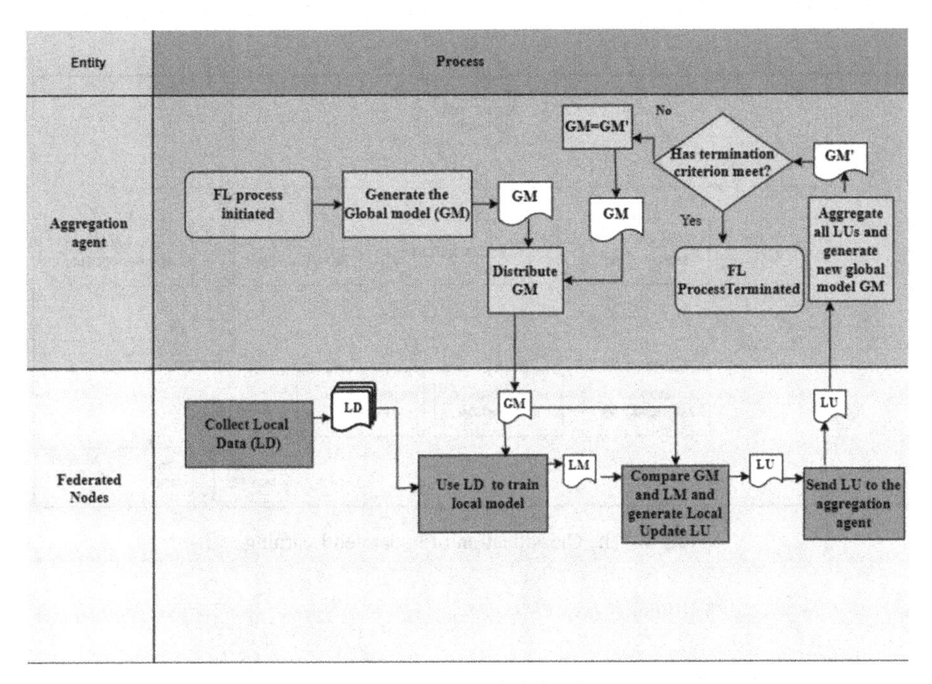

Fig. 1. Federated learning high-level process map

1.1.1 Data Partition

Based on various distributions of design, FL can be divided into the three types [9] i.e. horizontal FL, Vertical FL, and Federated Transfer Learning. Horizontal FL is employed where user attributes for two datasets overlay remarkably but user overlap is minimal. The data set is divided horizontally, with the same user attributes, but different users for training [10, 11]. In vertical FL (VFL), users are overlapped a lot and user features are overlapped a little. In VFL, data sets are divided vertically and a portion of data is considered for training wherever users are identical with different useful features [12]. In Federated Transfer learning (FTL) the users or the users' attributes are never segmented. However, it can be employed in cases when there is a lag of information or tags [13].

1.1.2 Privacy Mechanism

Using FL, clients can store data locally and transmit model information for target model training. In model aggregation, the only significant aspect of FL is model aggregation that trains their global model by integrating the model attributes from all clients thus prevent transferring the metadata throughout the training process [14]. The problems associated with calculating encrypted data have been resolved using homomorphic encryption key. In differential privacy, both ML and deep learning use gradient iteration [15], which incorporates the addition of noise to the result to implement differential privacy to safeguard user privacy [16].

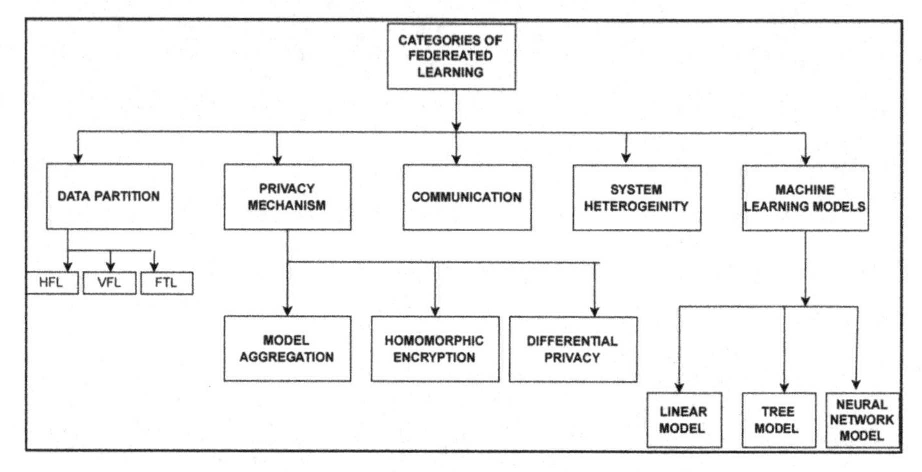

Fig. 2. The Classification of Federated Learning

1.1.3 Machine Learning Schemes

FL enhance the capability and security of the ML model. Neural network, decision tree and linear model are the three main ML models. In federated environment, the linear model of training that mitigates the security issues and attains the same accuracy as a non-private solution is proposed in [17]. The tree model in FL, random forest and gradient boost decision trees are utilized for both single and multiple decision trees [18]. The most famous model of machine learning is the neural network model that trains complex tasks. In autonomous vehicles using drones target location, trajectory planning, target recognition services plays a vital role [19]. Due to the drawback of regular connection between UAV group and base station, centralized training fails in real time but deep learning provides excellent efficiency with UAV group usually [20].

1.1.4 Communication Architecture

The problems of application scenario of FL are equipment computing and uneven distribution of user data [21]. All the participants that are under training are in touch with centralized server for the update of global model. The communication cost of FL is high for critical problems. To minimize the cost between server and local users the model data is compressed by secondary sample random rotation quantization [22].

1.1.5 Methods for Solving Heterogeneity

Different devices affect the accuracy of total training process. The diversions of heterogeneity are model heterogeneity, fault tolerance, asynchronous communication and device sampling [23]. The main factor of FL is the efficiency of unevenly distributed data from various devices. In FL, the processing of data from various devices affects the model. Asynchronous communication is the solution to many problems with dispersed devices in FL settings.

2 Related Work

This section discusses some of the related works of FL in IoV. For intelligent object detection, a two-layer FL model has been used in a 6G supported IoV environment [24, 25]. The use of a hybrid blockchain method in addition to FL by using DRL to select optimized participants which improve the learning efficiency has been proposed in [26]. Iterative model averaging is used by the deep network federated learning frame-work to train the global model by adding the local models in each round of updates [20]. For the selection of Smart Vehicles (SVs) for FL, local learners are adopted by using round robin, random scheduling using heterogeneous asynchronous FL networks [27–29]. In [30], the authors concluded that the Quality of Information (QoI) received by the SVs is dynamic and will affect the performance of FL. Therefore, to improve the QoI, the Vehicular Service Provider's (VSP's) responsibility is to select SVs of current location within important areas. With the combination of important areas and QoI, VSP can obtain beneficial on-road data [14] and trustworthy trained model updates [31] from the chosen SVs.

In a non-collaborative Stackleberg game model proposed in [32–34], the mobile devices have full information about the VSPs payment budget. For estimating the traffic with FL that more correctly captures the spatio temporal correlation of the traffic flow with the use of clustering, FedGRU approach proposed in [35] combines the GRU (Gated Recurrent Unit) to get the best overall model. A model proposed in [36] FedProx integrates the edge devices data of distributed training with the Federal Averaging (FedAvg) model maximization method that improves the reliability of the target task. In [37], Multiple Principal One Agent (MPOA) based contract optimization is being employed to maximize the revenue of VSPs in each iteration [38]. OBU in IoV can gather data and glean local knowledge. A model proposed in [39] replaces data as a service by know ledge as a service in IoV. Knowledge serves as real information and is incorporated into data intelligence. IoV cars learn about the environment and the roads in different locations using ML techniques, and they share their expertise [40]. In [41], a model has been proposed to maintain balance between the dataset computational resources and wireless resources that is affected by the combination of vehicle client selection resource and wireless resource. Table 1 gives the review of previous work done by various researchers with respect to implementation of federated learning for IoV.

3 Applications of Federated Learning

This section discusses the various applications of federated learning including its applications for IoVs.

Table 1. Review of the Federated Learning systems for IoVs

Ref no. Year	Method	Dataset/Parameters	Implementation	Outcome	Main Focus
[42] 2022	The design of federated learning collaborative authentication of protocol for shared data reduces the data leakages and propagation delay of data	Not Available	SUMO and OMNet + + platform	Authentication delay is lowest. Packet loss rate is reduced to 4.6%. Globally aggregated model is effective	Reducing the propagation delay and data leakages
[43] 2022	Hedonistic game theoretical model using horizontal based federated learning involved by fog federation	Road traffic sign dataset	SUMO and Images of traffic signs imported from Kaggle	Latency of the proposed approach is 18 ms, The. Quality of service (QoS) is improved with accuracy	Fog federation providers outperforms in terms of efficacy, scalability and stability
[44] 2022	Spike neural network which is fast training model and efficient energy with novel encoding method. Based on neuron receptive field to extract information from spatial dimension and pixel of traffic sign	Time step = T = 10, Leak rate = λ = 0.9 for Optimal settings	BelgiumTS (Belgium traffic sign) dataset	The outcome of FedSSNRFEE with accuracy approximately 95%	High accuracy and quick training for traffic sign in autonomous vehicles
[45] 2021	With the use of FL and Block chain this research develops the light weight encrypted algorithm called CPC	Not Available	TOSSIM simulator, TinyOS system	Communication cost = 13000bytes, Predicting accuracy = 84.25%, Computation time = 4000m	Predicting the road and traffic conditions

(*continued*)

Table 1. *(continued)*

Ref no. Year	Method	Dataset/Parameters	Implementation	Outcome	Main Focus
[46] 2021	The DGHV (Dijk-Gentry-Halevi-Vaikutanathan) algorithm has been enhanced	Data from Rancho Palos Verdes and San Pedro, California roads	Autonomous driving simulation in Python with the real-world Data	Reduced training loss by about 73.7%	Autonomous cars
[47] 2021	The set of intimate consensus nodes replaced neural network called GRU, differential policy, central server	Caltrans performance measurement system dataset	Consortium blockchain PySyft for FL framework	MAE = 7.96, MSE = 101.49 RMSE = 11.04	Traffic Flow Prediction
[48] 2021	Proposed three assaults, looked into the relevant defenses, and developed a privacy-preserving local model sharing algorithm using the LDP mechanism	MNIST dataset	The CNN model is adopted for model training, Python is used for implementation	Obtains the enhanced utilities, convergence strategies, QoLM (Quality of Local Model update) in federated learning process	Sharing high-quality models while protecting UAVs' privacy
[49] 2021	Promotes seamless service availability to end devices, the UAV and UGV cooperation procedure ensures constant power availability to UAVs	Data set on traffic taken from the U.S. Traffic Fatality Records	OMNET + +, Python and the scikitlearn ML library, NS-3, OverSim framework	97% of networks were covered, 89% less energy was used, and 90% of packets were delivered successfully	Energy used was 89%, less than that of other conventional methods
[50] 2021	Differential privacy (DP), Mobile Edge Computing (MEC) integrated with privacy preserving FL framework named BMFL	MNIST dataset	A cloud, 5 Mobile Edge Computing (MEC) servers and 50 devices with IoV system	Defends the backdoor attack successfully and maintains stability with an attack success rate of 9.54%	Lowering cloud communication costs and ensuring model training quality
[51] 2021	RSU infers the toxic model parameters by comparing the aggregation accuracy of various groups	KDDCup99	For deep learning Syft, Pytorch for federated learning	Epoch = 40, Accuracy = 96% for data size = 10000	To manage poisons attack

(continued)

Table 1. *(continued)*

Ref no. Year	Method	Dataset/Parameters	Implementation	Outcome	Main Focus
[36] 2021	CNN model is used as the local training model	MNIST dataset	Numerical simulation	Improves the accuracy of knowledge up to 18%	Knowledge Trading
[52] 2020	The RAOC-B rain drop optimization (RFO) algorithm is blockchain enabled	Simulation of urban mobility (SUMO) generated data	Network Simulator-2, SUMO	Highest throughput = 94%, maximum Packet delivery ratio = 0.9, End to end delay = 0.3 s for 100 nodes	Load balancing using cluster based VANET and communication by block chain to enhance privacy
[53] 2020	The supervised learning model Support vector machine (SVM) has been employed	Training data from Xiaojue Station of Shuohuang Railway to West Station of Dingzhou	SVM model based on the mixed kernel function	Federated learning accuracy is 94.21%	Intelligent control model for heavy haul trains
[54] 2020	ADMM-based algorithm	MNIST, CIFAR10 Datasets	Simulation parameters like road width, speed and road surface conditions have been considered	FL algorithms exhibits 10% increased accuracy	Sharing of knowledge in the IoV
[55] 2020	Long Short-Term Memory (LSTM) network has been used as the supervised learning model	Passenger flow data of Beijing Metro	Linux, EOS for blockchain and Node.js for test script	Optimal prediction using LSTM	Secure Railway passenger flow prediction model
[56] 2020	DRL-based node selection algorithm	MNIST dataset	Matplotlib, basemap toolkit	Improved of accuracy more than 90%	Secure Data sharing in IoV

3.1 Google Keyboard

The prediction of the next word is achieved while improving the quality of the keyboard with security and privacy [7, 57]. In building the recommended systems, building of language model is also attained.

3.2 Intelligent Medical Diagnosis Systems

In the centralized method of ML, the data gathering and processing for medical diagnosis becomes difficult because of privacy and security concerns. With FL, it is possible to use the data locally without any issues of privacy and train the model for diagnosis [57, 58]. For small and insufficient labels, federated transfer learning is the solution. To implement an integrated multi FL network on APOLLO network merges the interrelated medical system's longitudinal real world data with health outcome data for the help of doctors in forward-looking diagnosis of patients [57].

3.3 RSU Intelligence

IoV comprises of RSUs that are designed to receive data to process basic operations. Different varieties of data are received by RSUs and thus FL can be applied in various situations. One of the familiar approaches is in image processing. For autonomous vehicles, both onboard Vehicular Computing (VC) and RSU image processing is important. Collision detection and pedestrian detection also make use of image processing tasks [7].

3.4 Network Function Virtualization (NFV) Orchestration

NFV enabled network highlights the use of FL in making NFV orchestration in security/privacy services and in vehicular service delivery. Each of the networks is divided into sub-networks which may be further divided, utilizing network slicing techniques [59]. This added complexity supports the use of FL because NFV orchestrators from different network partitions can use cooperative ML training to create models capable of performing operations like VNF installation, scaling, termination, and migration. In MEC enabled orchestration of NFV, the RSU is placed near the network edge so that it can act as a network node to hold Virtual network functions (VNFs) [7].

3.5 Vehicular Intelligence

Vehicular intelligence in IoV has many applications such as forecasting the road conditions, image processing in lane detection and popular predictive maintenance [60]. The predictive maintenance uses operational data and alerts the user for maintenance of specific part by predicting the failure of the component through planned maintenance. With the use of FL, in addition to vast collection of data, predictive maintenance models have been built with greater efficiency [7].

4 Conclusions and Future Direction

Given the importance of communication in federated networks and the privacy risks associated with transferring raw data, it is mandatory to keep generated data local. In this situation, two things can be done to further minimize communication: one is to lessen the total number of iterations of communication rounds, and other is too minimize the size of message. In this paper, federated learning and its various categories have been discussed. A systematic review of different systems implementing FL in IoVs has been presented followed by applications of federated learning in different areas related to IoV. However, due to the heterogeneity of the VCs, system complexity is a major challenge. Thus, building privacy protection schemes depending on specific devices in IoV is the future direction. Resolving the tradeoff between communication cost and computational pressure is another challenge. Distributed FL is forthcoming research direction with heterogeneous data.

References

1. Zhang, C., et al.: A survey on federated learning. Knowledge-Based Syst. **216**, (Mar. 2021). https://doi.org/10.1016/J.KNOSYS.2021.106775
2. Manias, D.M., Shami, A.: Making a case for federated learning in the internet of vehicles and intelligent transportation systems. IEEE Netw. **35**(3), 88–94 (2021)
3. Tang, F., Kawamoto, Y., Kato, N., Liu, J.: Future intelligent and secure vehicular network toward 6G: Machine-learning approaches. Proc. IEEE **108**(2), 292–307 (2019)
4. Du, Z., Wu, C., Yoshinaga, T., Yau, K.-L.A., Ji, Y., Li, J.: Federated learning for vehicular internet of things: Recent advances and open issues. IEEE Open J. Comput. Soc. **1**, 45–61 (2020)
5. Brik, B., Ksentini, A., Bouaziz, M.: Federated learning for UAVs-enabled wireless networks: use cases, challenges, and open problems. IEEE Access **8**, 53841–53849 (2020)
6. Yu, Z., et al.: Mobility-aware proactive edge caching for connected vehicles using federated learning. IEEE Trans. Intell. Transp. Syst. **22**(8), 5341–5351 (2020)
7. Yang, Q., Liu, Y., Chen, T., Tong, Y.: Federated machine learning: Concept and applications. ACM Trans. Intell. Syst. Technol. **10**(2), 1–19 (2019)
8. Aono, Y., et al.: Privacy-preserving deep learning via additively homomorphic encryption. IEEE Trans. Inf. Forensics Secur. **13**(5), 1333–1345 (2017)
9. Geyer, R.C., Klein, T., Nabi, M.: Differentially private federated learning: A client level perspective. arXiv Prepr. arXiv1712.07557 (2017)
10. Wei, J.: Managed communication and consistency for fast data-parallel iterative analytics. In: Proceedings of the Sixth ACM Symposium on Cloud Computing, pp. 381–394 (2015)
11. Mansour, Y., Mohri, M., Ro, J., Suresh, A.T.: Three approaches for personalization with applications to federated learning. arXiv Prepr. arXiv2002.10619 (2020)
12. Chen, Y.-R., Rezapour, A., Tzeng, W.-G.: Privacy-preserving ridge regression on distributed data. Inf. Sci. (Ny). **451**, 34–49 (2018)
13. Kang, J., et al.: Reliable federated learning for mobile networks. IEEE Wirel. Commun. **27**(2), 72–80 (2020)
14. Wan, L., Ng, W.K., Han, S., Lee, V.C.S.: Privacy-preservation for gradient descent methods. In: Proceedings of the 13th ACM SIGKDD international conference on Knowledge discovery and data mining, pp. 775–783 (2007)

15. McMahan, B., Moore, E., Ramage, D., Hampson, S., y Arcas, B.A.: Communication-efficient learning of deep networks from decentralized data. In: Artificial intelligence and statistics, pp. 1273–1282 (2017)
16. Papernot, N., Abadi, M., Erlingsson, U., Goodfellow, I., Talwar, K.: Semi-supervised knowledge transfer for deep learning from private training data. arXiv Prepr. arXiv1610.05755 (2016)
17. Kim, H., Park, J., Bennis, M., Kim, S.-L.: Blockchained on-device federated learning. IEEE Commun. Lett. **24**(6), 1279–1283 (2019)
18. Nikolaenko, V., et al.: Privacy-preserving ridge regression on hundreds of millions of records. In: 2013 IEEE symposium on security and privacy, pp. 334–348 (2013)
19. Cheng, K.: Secureboost: a lossless federated learning framework. IEEE Intell. Syst. **36**(6), 87–98 (2021)
20. Abadi, M.: Deep learning with differential privacy. In: Proceedings of the 2016 ACM SIGSAC conference on computer and communications security, pp. 308–318 (2016)
21. Agarwal, N., Suresh, A.T., Yu, F.X.X., Kumar, S., McMahan, B.: cpSGD: Communication-efficient and differentially-private distributed SGD. Adv. Neural Inf. Process. Syst. **31** (2018)
22. Du, W., Han, Y.S., Chen, S.: Privacy-preserving multivariate statistical analysis: Linear regression and classification. In: Proceedings of the 2004 SIAM international conference on data mining, pp. 222–233 (2004)
23. Lindell, Y., Pinkas, B.: A proof of security of Yao's protocol for two-party computation. J. Cryptol. **22**(2), 161–188 (2009)
24. Saputra, Y.M., et al.: Dynamic federated learning-based economic framework for internet-of-vehicles. IEEE Trans. Mob. Comput. **1233**(c), 1–20 (2021). https://doi.org/10.1109/TMC.2021.3122436
25. Zhou, X., et al.: Two-layer federated learning with heterogeneous model aggregation for 6g supported internet of vehicles. IEEE Trans. Veh. Technol. **70**(6), 5308–5317 (2021)
26. Lu, Y., Huang, X., Zhang, K., Maharjan, S., Zhang, Y.: Blockchain empowered asynchronous federated learning for secure data sharing in internet of vehicles. IEEE Trans. Veh. Technol. **69**(4), 4298–4311 (2020)
27. Yang, H.H., Liu, Z., Quek, T.Q.S., Poor, H.V.: Scheduling policies for federated learning in wireless networks. IEEE Trans. Commun. **68**(1), 317–333 (2019)
28. Xie, C., Koyejo, S., Gupta, I.: Asynchronous federated optimization. arXiv Prepr. arXiv1903.03934 (2019)
29. Li, T., et al.: Federated optimization in heterogeneous networks. Proc. Mach. Learn. Syst. **2**, 429–450 (2020)
30. Amiri, M.M., Gündüz, D., Kulkarni, S.R., Poor, H.V.: Convergence of update aware device scheduling for federated learning at the wireless edge. IEEE Trans. Wirel. Commun. **20**(6), 3643–3658 (2021)
31. Zhan, Y., Li, P., Qu, Z., Zeng, D., Guo, S.: A learning-based incentive mechanism for federated learning. IEEE Internet Things J. **7**(7), 6360–6368 (2020)
32. Khan, L.U.: Federated learning for edge networks: Resource optimization and incentive mechanism. IEEE Commun. Mag. **58**(10), 88–93 (2020)
33. Liu, Y., Zhang, S., Zhang, C., Yu, J.J.Q.: FedGRU: privacy-preserving traffic flow prediction via federated learning. In: 2020 IEEE 23rd International Conference on Intelligent Transportation Systems (ITSC), pp. 1–6 (2020). https://doi.org/10.1109/ITSC45102.2020.9294453
34. Sarikaya, Y., Ercetin, O.: Motivating workers in federated learning: A stackelberg game perspective. IEEE Netw. Lett. **2**(1), 23–27 (2019)
35. Kang, J., Xiong, Z., Niyato, D., Xie, S., Zhang, J.: Incentive mechanism for reliable federated learning: a joint optimization approach to combining reputation and contract theory. IEEE Internet Things J. **6**(6), 10700–10714 (2019)

36. Zou, Y., Shen, F., Yan, F., Lin, J., Qiu, Y.: Reputation-based regional federated learning for knowledge trading in blockchain-enhanced IOV. In: 2021 IEEE Wireless Communications and Networking Conference (WCNC), pp. 1–6 (2021)
37. Abad, M.S.H., Ozfatura, E., Gunduz, D., Ercetin, O.: "Hierarchical federated learning across heterogeneous cellular networks. In: ICASSP 2020–2020 IEEE International Conference on Acoustics, Speech and Signal Processing (ICASSP), pp. 8866–8870 (2020)
38. Wang, S., Liu, F., Xia, H.: Content-based vehicle selection and resource allocation for federated learning in iov. In: 2021 IEEE Wireless Communications and Networking Conference Workshops (WCNCW), pp. 1–7 (2021)
39. Anand, A., Rani, S., Anand, D., Aljahdali, H.M., Kerr, D.: An efficient CNN-based deep learning model to detect malware attacks (CNN-DMA) in 5G-IoT healthcare applications. Sensors **21**(19), 6346 (2021)
40. Bonawitz, K.: Practical secure aggregation for privacy-preserving machine learning. In: Proceedings of the (2017) ACM SIGSAC Conference on Computer and Communications Security, pp. 1175–1191 (2017)
41. Price, W.N., Cohen, I.G.: Privacy in the age of medical big data. Nat. Med. **25**(1), 37–43 (2019)
42. Zhao, P., et al.: Federated learning-based collaborative authentication protocol for shared data in social IoV. IEEE Sens. J. **22**(7), 7385–7398 (2022)
43. Hammoud, A., Otrok, H., Mourad, A., Dziong, Z.: On demand fog federations for horizontal federated learning in IoV. IEEE Trans. Netw. Serv. Manag. **19**(3), 3062–3075 (2022)
44. Xie, K.: Efficient federated learning with spike neural networks for traffic sign recognition. IEEE Trans. Veh. Technol. **71**(9), 9980–9992 (2022)
45. Peng, O., et al.: Bflp: an adaptive federated learning framework for internet of vehicles. Mob. Inf. Syst. **2021**, 1–18 (2021)
46. Tao, X., Zhang, X., Liu, J., Xu, J.: Privacy-preserved federated learning for autonomous driving. IEEE Trans. Intell. Transp. Syst. **23**(7), 8423–8434 (2021)
47. Liu, Y., James, J.Q., Kang, J., Niyato, D., Zhang, S.: Privacy-preserving traffic flow prediction: a federated learning approach. IEEE Internet Things J. **7**(8), 7751–7763 (2020)
48. Wang, Y., Su, Z., Zhang, N., Benslimane, A.: Learning in the air: Secure federated learning for UAV-assisted crowdsensing. IEEE Trans. Netw. Sci. Eng. **8**(2), 1055–1069 (2020)
49. Aloqaily, M., Al Ridhawi, I., Guizani, M.: Energy-aware blockchain and federated learning-supported vehicular networks. IEEE Trans. Intell. Transp. Syst. **23**(11), 22641–22652 (2021)
50. Wang, R., Li, H., Liu, E.: Blockchain-based federated learning in mobile edge networks with application in internet of vehicles. arXiv Prepr. arXiv2103.01116 (2021)
51. Liu, H.: Blockchain and federated learning for collaborative intrusion detection in vehicular edge computing. IEEE Trans. Veh. Technol. **70**(6), 6073–6084 (2021)
52. Joshi, G.P., et al.: Toward blockchain-enabled privacy-preserving data transmission in cluster-based vehicular networks. Electronics **9**(9), 1358 (2020)
53. Hua, G., et al.: Blockchain-based federated learning for intelligent control in heavy haul railway. IEEE Access **8**, 176830–176839 (2020)
54. Chai, H., Leng, S., Chen, Y., Zhang, K.: A hierarchical blockchain-enabled federated learning algorithm for knowledge sharing in internet of vehicles. IEEE Trans. Intell. Transp. Syst. **22**(7), 3975–3986 (2020)
55. Shen, C., Zhu, L., Hua, G., Zhou, L., Zhang, L.: A blockchain based federal learning method for urban rail passenger flow prediction. In: 2020 IEEE 23rd International Conference on Intelligent Transportation Systems (ITSC), pp. 1–5 (2020)
56. Lu, Y., Huang, X., Zhang, K., Maharjan, S., Zhang, Y.: Blockchain empowered asynchronous federated learning for secure data sharing in internet of vehicles. IEEE Trans. Veh. Technol. **69**(4), 4298–4311 (2020). https://doi.org/10.1109/TVT.2020.2973651

57. Lee, J.S.H.: From discovery to practice and survivorship: building a national real-world data learning healthcare framework for military and veteran cancer patients. Clin. Pharmacol. & Ther. **106**(1), 52–57 (2019)
58. Rieke, N.: The future of digital health with federated learning. NPJ Digit. Med. **3**(1), 1–7 (2020)
59. Jain, B., Brar, G., Malhotra, J., Rani, S., Ahmed, S.H.: A cross layer protocol for traffic management in Social Internet of Vehicles. Futur. Gener. Comput. Syst. **82**, 707–714 (2018). https://doi.org/10.1016/j.future.2017.11.019
60. Seth, I., Guleria, K., Panda, S.N.: Introducing intelligence in vehicular ad hoc networks using machine learning algorithms. ECS Trans. **107**(1), 8395 (2022)

Optimal Rescheduling for Transmission Congestion Management Using Intelligent Hybrid Optimization

Ward Ul Hijaz Paul[1](✉), Anwar Shahzad Siddiqui[1], and Sheeraz Kirmani[2]

[1] Jamia Millia Islamia, New Delhi 110025, India
wardulhijazpaul@gmail.com
[2] Aligarh Muslim University, Aligarh 202002, India

Abstract. Power system congestion seems to be one of the principal issues that organizations like system operators deal with on a routine basis in the current restructured energy market. Congestion on the transmission lines creates significant issues for the efficient operational management of the deregulated power system and raises the cost of transmitting power which is an important factor of consideration. Therefore, it is crucial to examine various methods which can help in reducing this problem of congestion occurring in transmitting power in the power system network. Among the most effective methods to solve this congestion issue is rescheduling the generation side of the power system network. This research proposes a new fuzzy based hybrid optimization technique which is established on hybridization of particle swarm optimization and genetic algorithm optimization technique for minimizing this congestion cost. Utilizing the modified IEEE 57 bus system, the effectiveness of the given methodology is evaluated, and the results are discussed.

Keywords: Power Deregulation · Restructuring · Congestion Management · Optimal Rescheduling · Hybrid Optimization

1 Introduction

Power system deregulation has allowed all power purchasers and sellers equally able to access the transmission network, in these modern times. Microgrid provides a platform for integrating various distributed generations and renewables into the grid [1, 2]. The energy utilities are attempting to satisfy the necessary demand by increasing their generations, but this is being hampered by the substantial growth in populations and the rising urbanization [3]. The transmission network, which is the biggest obstacle to equitably executing the reform of the electrical sector, performs a critical role in the smooth control of a market that is competitive. The transmission lines become crowded whenever one or more of the aforementioned constraints are broken because the flow of power in a power transmission line is constrained by many factors, the stability limit, operating voltage and heat limits [4–6]. The increasing microgrid penetration to the conventional grid has added to the worries of congestion management [7, 8]. Severe outages with significant

R. K. Challa et al. (Eds.): ICAIoT 2023, CCIS 1930, pp. 140–152, 2024.
https://doi.org/10.1007/978-3-031-48781-1_12

economic and social repercussions could be the outcome of not keeping the integrity of the electrical infrastructure. Control of every electric infrastructure is very important [9].

Power system congestion prevents the desired power exchange, forcing customers to buy power from alternative sources at increased prices [10, 11]. Therefore, for the system to function successfully and reliably, managing this congestion is highly desirable and needs to be made easier. Therefore, the primary problem in the restructured power network is controlling transmission line congestion, which is the subject of much research. In this work, we propose a model based on the method of scheduling the real power of generating units that is being used.

In the publications recently published, to tackle the congestion issue, various congestion management strategies have really been explored. A number of solutions have been discussed, including physically slowing down transactions, scheduling reactive and active power and different optimization methodologies. Economic and financial aspects of this congestion in transmission lines have been discussed in [12, 13]. Coordinating the electrical transactions on the basis of priority and curtailment of load variables have been suggested in [14, 15] as a way to disperse this congestion issue. A method is described in [16] for minimizing the congestion and service costs that acknowledges real power losses. Regarding the scheduling of generated electricity and load shedding, an approach has also been proposed in [17–19] that reduces overload transmission lines in a mathematically efficient way. Stability of voltage is guaranteed while reducing the congestion issue in pool markets is discussed in [20–22] that uses thermal overload and voltage fluctuations for controlling the congestion problem based optimal power flow (OPF). Dutta et al. proposed a method for selecting the best generations based on generator sensitivity to reduce congestion problems [23]. In [24–26], a variety of optimization methods are resolved using the random search based numerical optimization methods. After taking both congestion and security cost into account, an optimization technique based on harmony search has been suggested in [27, 28] to tackle the planning of transmission expansion. Kumar and Sekhar discussed the role of the flexible alternating current transmission system contribution to the issue of these congestions [29]. Continuous constrained and unconstrained optimizations have been used tried to solve in [30] using the teaching learner-based optimizations (TLBO) approach. The effectiveness of the TLBO method is explored in [31], which demonstrates how little processing is required and how quickly convergence occurs [32]. Therefore, for dealing with these different congestion scenarios, TLBO was effectively applied upon that IEEE 57 bus test system. The findings were found to be significantly better than those of a number of other optimization methods.

This study offers a hybrid optimization approach built on the new fuzzy logic-based GA-PSO algorithm to address the issue of congestion. The proposed approach is examined on IEEE 57 bus standard network and the findings have been reviewed. This paper is discussed as follows. An introduction to congestion management is given in Sect. 1. Section 2 gives an overview of model formulation. Section 3 gives description of a new fuzzy based hybrid optimization technique. Results and discussion are presented in Sect. 4 with conclusion in Sect. 5 and the references in the end.

2 Model Formulation

The primary goal of the management of this congestion is accomplished while adhering to system limits by the scheduling method, which entails a decrease or an increase in the real power produced from the participating generation units. Each generation company requests rising or falling cost bids in order to reschedule the generation. The statement of the mathematical model is given in Eq. 1.

$$C_c = \sum\nolimits_{j \epsilon N_j} (N_j \Delta P_{G_j}^{+} = D_k \Delta P_{G_j}^{-}) \frac{dollars}{hour} \tag{1}$$

where D_k and C_k respectively are the decreasing and increasing values of the price bids by the generation companies and C_c is the cost required for the alteration of the output value participating generation units' real power.

$\Delta P_{G_j}^{-}$ and $\Delta P_{G_j}^{+}$ respectively are the deceasing and increasing values of the generating unit's real power output.

The problem goal of this research will be to minimize this cost. This minimization would be dependent on certain inequality and equality constraints which are given in Eqs. 2–10.

2.1 Equality Constraints

$$P_{Gk} = P_{Gk}^{c} + \Delta P_{Gk}^{+} - \Delta P_{Gk}^{-} \tag{2}$$

$$P_{Dj} = P_{Dj}^{c} \tag{3}$$

$$P_{Gk} - P_{Dk} = |V_j||V_k||Y_{Kj}| \cos\left(\delta_k - \delta_j - \theta_{kj}\right) \tag{4}$$

$$Q_{Gk} - Q_{Dk} = |V_j||V_k||Y_{Kj}| \sin\left(\delta_k - \delta_j - \theta_{kj}\right) \tag{5}$$

2.2 Inequality Constraints

$$P_{Gk}^{min} \le P_{Gk} \le P_{Gk}^{max} \tag{6}$$

$$Q_{Gk}^{min} \le Q_{Gk} \le Q_{Gk}^{max} \tag{7}$$

$$V_n^{min} \le V_n \le V_n^{max} \tag{8}$$

$$(P_{Gk} - P_{Gk}^{min}) = P_{Gk}^{min} \le P_{Gk} \le P_{Gk}^{max} = (P_{Gk}^{min} - P_{Gk}) \tag{9}$$

$$P_{ij} \le P_{jk}^{max} \tag{10}$$

P_{Gk} denotes the actual power generated at a certain bus k and Q_{Gk} denotes the reactive power generated at that same bus. The actual and reactive power available at bus k are denoted by P_{Dk} and Q_{Dk}, respectively. The voltages at bus j and k, meanwhile, are V_j and V_k. The voltage angles at buses j and k is denoted by δ_j and δ_k while θ_{kj} refers to the admittance angle of the line connecting the two buses. N_b, N_g and N_d represent the total number of buses, generators, and loads. Generator k provides real power P_{Gk}^{c}, and load bus j uses real power provided by P_{Dj}^{c}.

3 A New Fuzzy Based Hybrid GA-PSO Algorithm

Among the most important concerns while addressing a variety of technological issues is optimization. It seeks to establish the characteristics of processes or functions in order to achieve the best outcome under constraints. It enables cost reduction and increased productivity. Conventional analytical techniques could be used to address basic optimization problems. But many issues are multi-modal, multi-dimensional, or noisy, making them too complex for current approaches to provide an adequate resolution.

Minimization of the likelihood of early convergence in local minima could boost the effectiveness of the widely used PSO technique. This could be done by using the mutation and crossover operators found in genetic algorithm to change a few of the particles. The operators of genetic algorithm affect the optimization procedure, although, they ought to be dependent on the PSO individual's ongoing state at the time.

The suggested approach blends the benefits of GA and PSO. An initial set of particle population is created during the first stage. The system then adjusts the particles' speed and location accordingly. After then, the genetic operators of crossover and mutation are carried out $[p_e.N]$ number of times, here N denotes the quantitative measure of PSO swarm $S(t)$, and the influence factor is represented by $P_e \epsilon [0, 1]$ those controls how the GA affects the searching process. In a competition, the particles having the best possible personal solution $\Delta P(t)$ gets the selection of the particles changed by GA operators out of the swarm $S(t)$. GA operators' solution $S(t)$ is added to provisional population $CH(t)$. The paternal particle $P_i(t)$ of O_i is identified by the i value. The swarm $S(t)$ of the PSO algorithm is then combined with the provisional population. An important component of the suggested technique is the combining strategy.

The influencing factor controls how genetic operators are used. Therefore, it can have a significant effect on the algorithm's converging power. However, selecting the right value depends on the issue and may be challenging. Furthermore, it makes sense to make changes while the algorithm is running. The effect of the GA should be minimal whenever the particle swarm optimization enables the discovery of new, improved solution in successive iteration. Nevertheless, the impact of genetic operators must be enhanced whenever the PSO algorithm stalls. This might make it possible to steer the searching into fresh, potentially more fruitful territory and abandon it if the algorithm becomes stuck at the local optimum.

In this paper, a neuro fuzzy based multiple input, single output system to regulate the influence factor's value, which is later used to minimize the objective function is proposed. The information can be saved as understandable IF-THEN fuzzy system rules thanks to neuro fuzzy system. These rules' characteristics might be predetermined by a specialist or found out immediately by artificial intelligence techniques. Figure 1 shows the flowchart and Fig. 2 illustrates the schematics to proposed hybrid fuzzy system FS.

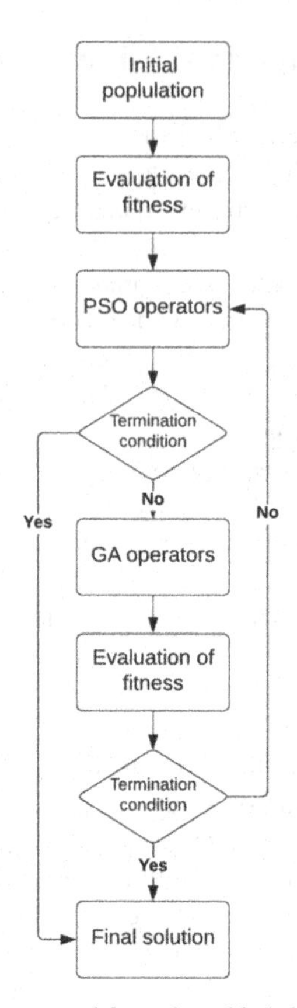

Fig. 1. Flowchart of proposed fuzzy based hybrid GA-PSO algorithm.

In the suggested hybrid algorithm, the magnitude of ΔP_e, which either enhances or decreases the influence of the GA on the process of obtaining the best possible solutions, thus, being determined by the neuro fuzzy framework. Depending upon the knowledge of the present p_e value and the normalized efficiency ΔE_{GA} of the genetic algorithm to the PSO approach, the fuzzy network decides. The large value of ΔE_{GA} suggests that GA operators can produce solution that is superior to those produced by the PSO. If this condition persists over rounds, it may indicate that the particle swarm optimization has reached a deadlock or is trapped within local minima. The Eq. 11 defines the value of ΔEN_{GA}.

$$\Delta EN_{GA} = {}^{\Delta E'_{GA}}\!/\left(\Delta E'_{GA} + \Delta E'_{PSO}\right) \tag{11}$$

where $\Delta E'_{GA}$ and $\Delta E'_{PSO}$ represent the GA and PSO efficiency calculated using Eqs. 12 and 13:

$$E'_{GA} = \sum\nolimits_{t'=t-w_0}^{t} \Delta E_{GA}(t') \Big/ \sum\nolimits_{t'=t-w_0}^{t} \left|CH(t')\right| \tag{12}$$

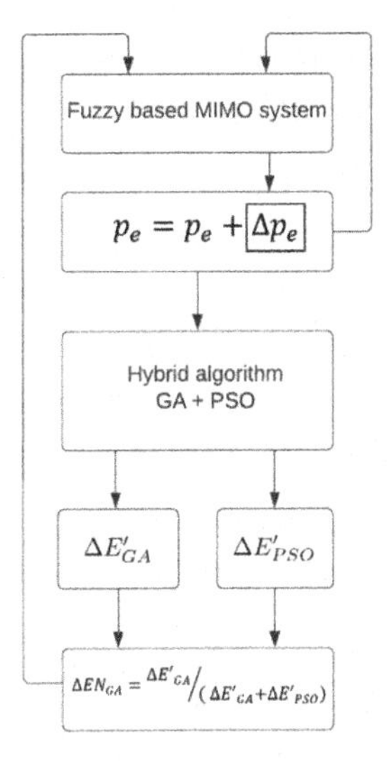

Fig. 2. Generalization of schematics of proposed fuzzy based hybrid GA-PSO algorithm.

$$E'_{PSO} = \sum\nolimits_{t'=t-w_0}^{t} \Delta E_{PSO}(t') \Big/ \sum\nolimits_{t'=t-w_0}^{t} N \tag{13}$$

In the context of a specific iteration t', $|CH(t')|$ denotes the size of the temporary population, whereas $\Delta E'_{GA}$ and $\Delta E'_{PSO}$ denote, respectively, the efficacy of GA and PSO. The best solution's overall fitness improvement or decline is used to calculate its effectiveness. They are described in Eqs. 14 and 15.

$$E_{GA(t')} = \sum\nolimits_{j=1}^{|CH(t)|} \begin{cases} f(g(t') - f(o_j)) if \;\; f(g(t')) > f(o_j) \\ 0 \quad otherwise \end{cases} \tag{14}$$

$$E_{PSO(t')} = \sum\nolimits_{i=1}^{N} \begin{cases} f(g(t') - f(x_i(t')) if \;\; f(g(t')) > f(x_i(t') \\ 0 \quad otherwise \end{cases} \tag{15}$$

In the context, o_j and $x_i(t')$ relate to the new elements produced via PSO modification and genetic operators, respectively. f and g are the mathematical functions. It's vital to remember that only solutions that help the global best solution during iteration t' have an impact on these values. In a similar way, the PSO algorithm's efficiency can be described in Eq. 16.

$$\Delta EN_{PSO} = \Delta E'_{PSO} / (\Delta E'_{GA} + \Delta E'_{PSO}) \tag{16}$$

As the ΔEN_{PSO} efficiency value is complemented by the ΔEN_{GA} measure. The precision and efficiency of the method are unaffected by the usage of such numbers alternately. The p_e component is modified accordingly subject to the value e produced by the fuzzy logic system FS using the given Eq. 17. The fuzzy logic system FS is described in the Fig. 2.

$$p_e = p_e + \Delta p_e = p_e + FS(\Delta EN_{GA}, p_e) \quad (17)$$

The findings from the observation window, or the final w_m repetitions of the loop function of the suggested technique, are taken into account for the calculations $\Delta E'_{PSO}$ and $\Delta E'_{GA}$ in order to establish computation efficacy with reliability. Additionally, the quantity of the p_e variable is changed no more frequently than every w_m iteration so that it can be observed how the modification in the variable affects the process of finding the best solution.

We are choosing the proposed hybrid optimization because it blends the advantages of both the GA and PSO techniques and helps in arriving at the best solution. This extra population serves as a safeguard against the PSO's early convergence. The usage of a fuzzy logic to regulate population size is the main component of the suggested approach. As the PSO algorithm becomes stagnant, it enables increasing the impact of the GA on the searching process.

4 Results and Discussion

The suggested fuzzy logic-based hybrid GA-PSO algorithm method is implemented using MATLAB to find an answer to this issue of congestion in the power transmission lines. By running it on the altered IEEE 57 bus system, the proposed technique is investigated. Since the decremental cost is expected to be lower than the value of marginal cost value, the incremental cost is expected to be greater. Below is a summary of the key findings of the work done.

A customized IEEE 57 bus system has been selected to evaluate the findings. The outline diagram of IEEE 57 bus test network is given in Fig. 3. This bus network consists of 50 load buses, 7 generator buses and 80 power transmission lines. Since line failures account for the majority of uncertainties in an electrical network, load fluctuations and line outage cases were included in the model. In order to encounter such outages, let us take the following case i.e., by reducing the capacity of transmission lines and overloading of transmission lines between the bus number five and six and bus number six and twelve.

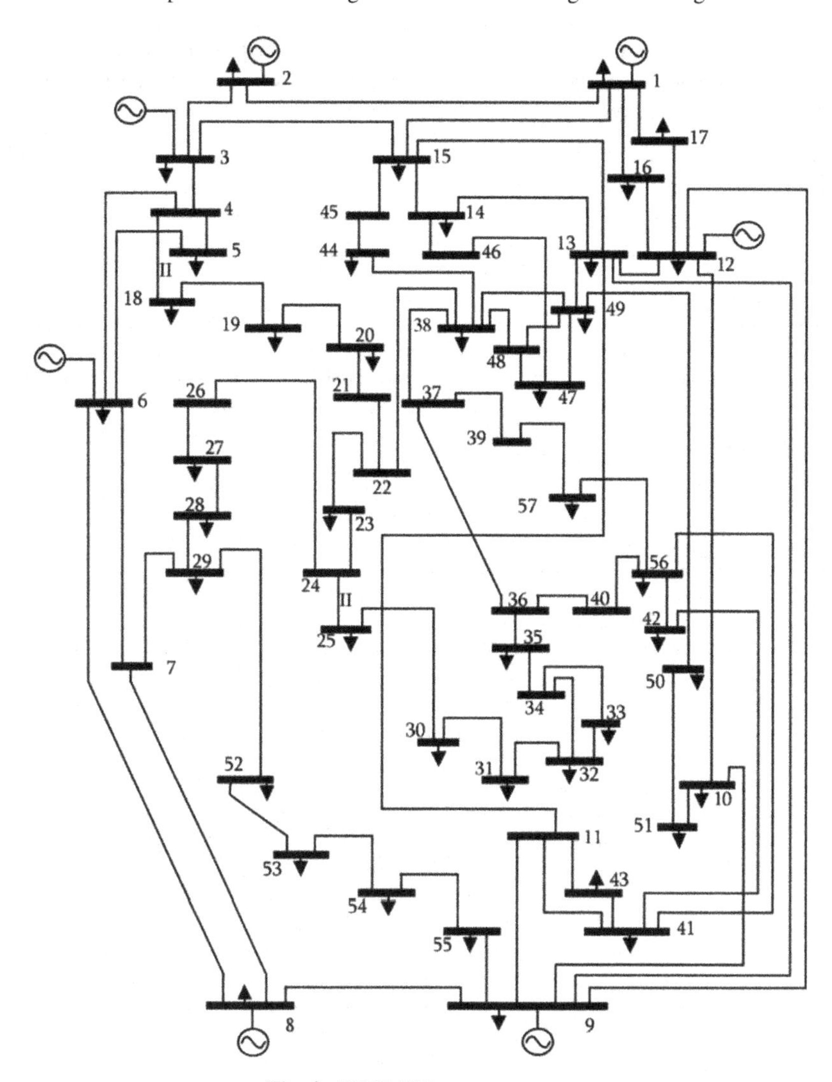

Fig. 3. IEEE 57 bus test system.

Around 184 MW of real power is running in the transmission lines between buses five and six and around 47 MW is flowing in the transmission lines between buses six and twelve. Since the transmission lines between buses six and twelve have a base load power restriction of 50 MW, and the line between buses five and six has a maximum of 200 MW. The power limit of the lines between buses five and six and the lines between buses six and twelve is considered to be 175 MW and 35 MW, respectively, for the purposes of overload simulation. Due to the lines' reduced capacity, the line between buses five and six is overcrowded by around 5.5%, while the line between buses six and twelve is overloaded by 34%, resulting in 21 MW net power violations. The suggested hybrid method allows for the most efficient rescheduling of generations to handle this 21 MW overload. Table 1 shows quantitatively this power flowing, power limits, overloading percentages, and the power violations of the different buses under consideration. Table 2 lists the rescheduling of power and the congestion costs achieved by the proposed hybrid algorithm as well as the comparative analysis with the different individual optimization techniques.

Table 1. Analysis of the overload and power violations of congested lines

Overloading	Congestion in transmission lines	Power flowing in transmission lines (MW)	Power limit of transmission lines (MW)	Overload (%) in transmission lines	Power violation (MW)
Overloading to lines 5 to 6 and 6 to 12	5 to 6	184	175	5.5	9
	6 to 12	47	35	34	12
Total power violation (MW)					21

Table 2. Comparative analysis of the proposed optimization technique against the different algorithms.

Parameter	Algorithms used		
	PSO	GA	Fuzzy hybrid algorithm
Congestion cost (Dollars/hour)	3117.7	2916.5	939.40
ΔP_{G1}	No rescheduling	−1.07	−429.31
ΔP_{G2}	No rescheduling	−24.64	+ 39.10
ΔP_{G3}	No rescheduling	+ 36.09	−4.62
ΔP_{G4}	No rescheduling	−6.23	−27.16
ΔP_{G5}	No rescheduling	−0.28	−2.05
ΔP_{G6}	No rescheduling	−1.25	+ 3.24
ΔP_{G7}	No rescheduling	−2.57	−29.71
Total rescheduled power (MWs)	76.30	72.13	536.19

Figure 4 shows the quantity of power rescheduled. According to Table 2, the proposed hybrid algorithm's congestion cost is 934.40 dollars per hour, that is the lowest of the earlier individual congestion management optimization strategies. Additionally, the overall loss in the system has also dropped from their peak value during congestion because the voltage profiles improve. Table 2 also gives a comparison of congestion costs achieved using the GA and PSO algorithms individually and also when applied as hybrid of them. Figure 5 illustrates the convergence of the fitness functions with the number of iterations using the suggested fuzzy based hybrid algorithm. The high convergence indicates how well the algorithm is performing in finding the optimal solution and provides confidence that the algorithm has found a good solution. Figure 6 shows the voltage profiles of the different buses in per unit before and after the application of the proposed fuzzy based hybrid algorithm. The voltage profiles slightly improve at almost every bus improves after the application of said algorithm. The proposed hybrid optimization improves the voltage profile of a power system by redistributing power flows to alleviate congestion and ensure that all buses in the system operate within acceptable voltage limits. Although the change in voltage values might seem small in p.u. values, the base voltage values are significant, and it aids in reducing the overall power losses in the system.

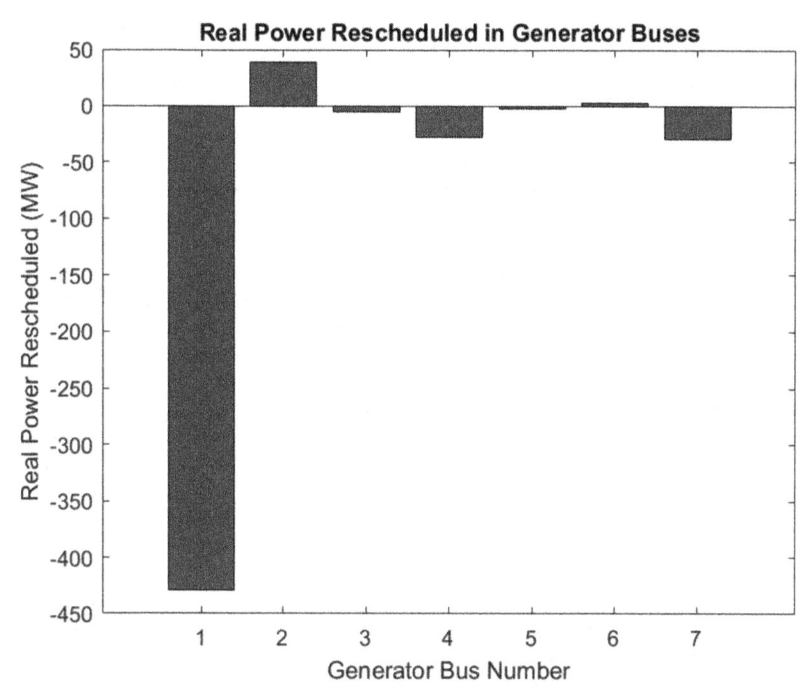

Fig. 4. Rescheduling of real power generation using proposed hybrid algorithm on IEEE 57 bus system.

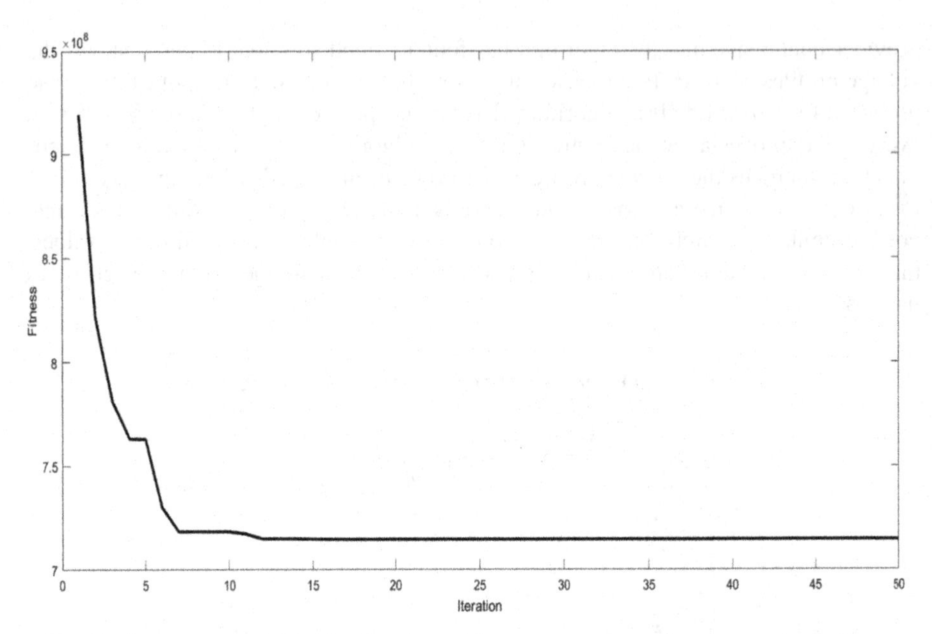

Fig. 5. Convergence of objective function subject to constraints using proposed algorithm.

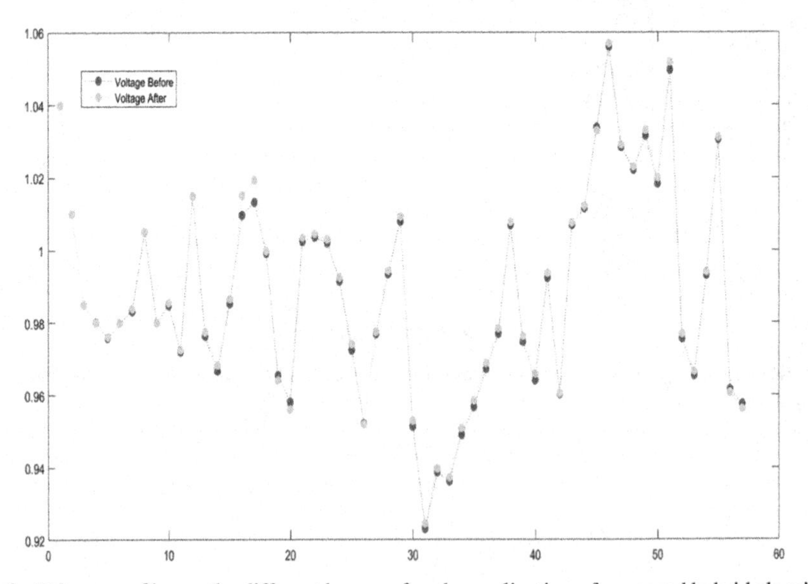

Fig. 6. Voltage profiles on the different busses after the application of proposed hybrid algorithm.

5 Conclusion

The current research illustrates a fuzzy based GA-PSO hybrid congestion management technique using the best power producing unit scheduling in the pool energy markets. Transmission line failures caused by overload and abrupt changes in load are also considered while evaluating the efficacy of this activity. The proposed intelligent technique solves the problem of congestion management by following the optimal rescheduling of the generations to handle the 21 MW overload. At the same time, it ensures the minimization of power losses along with the improvement of voltage profile. When correlated to alternative optimization methods, the proposed approach is far more cost-effective and effective at minimizing congestion. Additionally, it is discovered that there are fewer overall losses and rescheduling in the power system.

References

1. Siddiqui, O., Paul, W.U.H., Kirmani, S., Ahmad, M., Ali, D., Ali, M.S.: Voltage and frequency control in a microgrid. J. Eng. Sci. Technol. Rev. **15**, 115–124 (2022)
2. Kirmani, S., Paul, W.U.H., Bhat, M.B., Akhtar, I., Siddiqui, A.S.: Optimal Allocation of V2G Stations in a Microgrid Environment: Demand Response. In: 2023 International Conference on Power, Instrumentation, Energy and Control (PIECON), pp. 1–6. IEEE (2023)
3. Verma, S., Saha, S., Mukherjee, V.: Optimal rescheduling of real power generation for congestion management using teaching-learning-based optimization algorithm. J. Elec. Sys. Info. Technol. **5**, 889–907 (2018)
4. Sarwar, M., Siddiqui, A.S.: Congestion management in deregulated electricity market using distributed generation. In: 2015 Annual IEEE India Conference (INDICON), pp. 1–5. IEEE (2015)
5. Prashant, S.A.S., Sarwar, M., Althobaiti, A., Ghoneim, S.S.M.: Optimal location and sizing of distributed generators in power system network with power quality enhancement using fuzzy logic controlled D-STATCOM. Sustainability **14**, 3305 (2022)
6. Ali, D., Paul, W.U.H., Ali, M.S., Ahmad, M., Ashfaq, H.: Optimal placement of distribution generation sources in hybrid generation network. Smart Grid and Renewable Energy **12**, 65–80 (2021)
7. Paul, W.U.H., Siddiqui, A.S., Kirmani, S.: Demand side management strategies in residential load with renewable energy integration: a brief overview. Paripex Indian J. Res. **11**, 66–69 (2022)
8. Paul, W.U.H., Siddiqui, A.S., Kirmani, S.: Demand side management and demand response for optimal energy usage: an overview. Paripex Indian J. Res. 151–152 (2022)
9. Paul, W.U.H., Kirmani, S., Bhat, M.B., Nahvi, S.A.: Data based controller design for pmdc motor setup using system identification. Studies in Indian Place Names **40**, 11 (2020)
10. Sarwar, M., Siddiqui, A.S.: An approach to locational marginal price based zonal congestion management in deregulated electricity market. Frontiers in Energy. **10**, 240–248 (2016)
11. Ali, Md.S., Ahmad, A., Paul, W.U.H., Ali, D., Ahmad, M.: Optimal allocation of wind-based distributed generators in power distribution systems using probabilistic approach. In: Smart Energy and Advancement in Power Technologies, pp. 385–396. Springer Singapore (2023)
12. Singh, H., Papalexopoulos, A.: Transmission congestion management in competitive electricity. IEEE Trans. Power Sys. (1998)
13. Prashant, S.A.S., Sarwar, Md.: An Optimum GA-Based Solution for Economic Load Dispatch for Clean Energy. Presented at the (2022)

14. Fang, R.S., David, A.K.: Transmission congestion management in an electricity market. IEEE Trans. Power Syst. **14**, 877–883 (1999)
15. Akhtar, I., Paul, W.U.H., Kirmani, S., Asim, M.: Cost Analysis of 18 kW Solar Photovoltaic System for Smart Cities Growth in India. In: Lecture Notes in Electrical Engineering (2021)
16. Fu, J., Lamont, J.W.: A combined framework for service identification and congestion management. IEEE Trans. Power Sys. **16**, 56–61 (2001)
17. Bashir, M.U., Paul, W.U.H., Ahmad, M., Ali, D., Ali, M.: An efficient hybrid TLBO-PSO approach for congestion management employing real power generation rescheduling. Smart Grid and Renewable Energy **12**, 113–135 (2021)
18. Prashant, S.Md., Siddiqui, A.S., Ghoneim, S.S.M., Mahmoud, K., Darwish, M.M.F.: Effective transmission congestion management via optimal DG capacity using hybrid swarm optimization for contemporary power system operations. IEEE Access **10**, 71091–71106 (2022)
19. Talukdar, B.K., Sinha, A.K., Mukhopadhyay, S., Bose, A.: A computationally simple method for cost-efficient generation rescheduling and load shedding for congestion management. Int. J. Electr. Power Energy Syst. **27**, 379–388 (2005)
20. Conejo, A.J., Milano, F., Garcia-Bertrand, R.: Congestion management ensuring voltage stability. IEEE Trans. Power Syst. **21**, 357–364 (2006)
21. Ahmad, M., Ali, D., Paul, W.U.H., Ali, M.S., Ashfaq, H.: Management of energy and coordinated control of PV/HESS in islanded DC Microgrid. In: Smart Energy and Advancement in Power Technologies, pp. 325–339. Springer, Singapore (2023)
22. Capitanescu, F., Van Cutsem, T.: A unified management of congestions due to voltage instability and thermal overload. Electric Power Sys. Res. **77**, 1274–1283 (2007)
23. Dutta, S., Singh, S.P.: Optimal rescheduling of generators for congestion management based on particle swarm optimization. IEEE Trans. Power Syst. **23**, 1560–1569 (2008)
24. Jang, J.S.R., Sun, C.T., Mizutani, E.: Neuro-Fuzzy and Soft Computing: A Computational Approach to Learning and Machine Intelligence. Pearson Education, USA (1996)
25. Paul, W.U.H., Siddiqui, A.S., Kirmani, S.: Optimal positioning of distributed energy using intelligent hybrid optimization. J. Phys. Conf. Ser. (2023)
26. Paul, W.U.H., Siddiqui, A.S., Kirmani, S.: Intelligent load management system development with renewable energy for demand side management. Int. J. Adv. Eng. Manage. Res. **08**, 140–153 (2023)
27. Rastgou, A., Moshtagh, J.: Improved harmony search algorithm for transmission expansion planning with adequacy–security considerations in the deregulated power system. Int. J. Electr. Power Energy Syst. **60**, 153–164 (2014)
28. Siddiqui, A.S.: Prashant: Optimal Location and Sizing of Conglomerate DG- FACTS using an Artificial Neural Network and Heuristic Probability Distribution Methodology for Modern Power System Operations. Protection and Control of Modern Power Systems **7**, 9 (2022)
29. Kumar, A., Sekhar, C.: Congestion management with FACTS devices in deregulated electricity markets ensuring loadability limit. Int. J. Electr. Power Energy Syst. **46**, 258–273 (2013)
30. Pawar, P.J., Rao, R.V.: Erratum to: Parameter optimization of machining processes using teaching-learning-based optimization algorithm. The Int. J. Adv. Manuf. Technol. **67**, 1955 (2013)
31. Toğan, V.: Design of planar steel frames using Teaching-Learning Based Optimization. Eng. Struct. **34**, 225–232 (2012)
32. Malik, H., Iqbal, A., Joshi, P., Agrawal, S., Bakhsh, F.I. (eds.): Metaheuristic and Evolutionary Computation: Algorithms and Applications. Springer Singapore, Singapore (2021)

Predictive Load Management Using IoT and Data Analytics

Sushil Phuyal(✉), Shashwot Shrestha, Swodesh Sharma, Rachana Subedi, and Shahabuddin Khan

Department of Electrical Engineering, Institute of Engineering, Tribhuwan University, Pulchowk Campus, Lalitpur, Nepal
sushilphuyal.sp@gmail.com

Abstract. The objective of this paper is to design and implement an Internet of Things (IoT) based energy data acquisition system that incorporates a Long Short Term Memory (LSTM) model for predicting household energy demand and optimizing energy consumption with load scheduling and shifting. To achieve this goal, the system feeds 7 days' worth of energy consumption data, along with relevant features such as temperature, humidity, precipitation, and holiday information, into the LSTM model to predict the energy demand curve. The system categorizes loads into deferrable and non-deferrable loads. Based on these categories, the system applies load scheduling techniques to flatten out the demand curve and optimize energy consumption. In addition, the system shifts the deferrable loads from the grid to renewable sources during peak hours if available. This helps to reduce the burden on the main grid and promotes sustainability. Furthermore, the system takes into account the day-ahead hourly price-based tariff rate, ensuring that energy consumption is cost-effective. This is achieved with a prototype Predictive Load Management Device (PLMD). To relay information to the users, the system includes a web-based application made in Blynk that presents the information in a simple, easy-to-understand format. This reduces the complexity associated with different techniques of demand-side management (DSM) and makes it accessible to users with varying technical backgrounds. The web-based application allows users to monitor energy consumption patterns, view predictions generated by the LSTM model, and make informed decisions on energy consumption intuitively and straightforwardly.

Keywords: Internet of Things · Long Short Term Memory · Deferrable · Non-deferrable · Predictive Load Management Device · Blynk

1 Introduction

Artificial Intelligence (AI) and the Internet of Things (IoT) are getting popular and being used in many sectors. They are also being used in optimizing the management of electricity in our homes. IoT can be used for collecting large amounts of data and those data can be analyzed by AI algorithms. From those analyzed data, various predictions

R. K. Challa et al. (Eds.): ICAIoT 2023, CCIS 1930, pp. 153–168, 2024.
https://doi.org/10.1007/978-3-031-48781-1_13

and real-time insights regarding energy usage can be done. After the availability of real-time energy usage, each household can be able to use energy more wisely, resulting in less waste of energy and cheap energy cost.

AI and IoT are being increasingly incorporated by solar panels and battery storage systems. For example, the Tesla Power-wall uses AI algorithms to optimize energy storage and usage, automatically adjusting the charging and discharging of the battery based on energy consumption patterns and weather predictions [1]. With the integration of AI and IoT technology, the energy management system of every household can be revolutionized. Real-time insights and prediction are the two key features of this technique by which households can make more informed decisions about their energy consumption by using real-time data obtained.

Another advantage of AI is time series forecasting. Time series forecasting is a statistical technique used to predict future values based on past observations of a time-dependent variable. It is the process of using historical data and statistical models to predict future values of a time-based variable. This model makes an accurate prediction by analyzing key features like trends, seasonality, and patterns in the data. Because of these features, it is being used in a wide range of fields such as finance, economics, sales, weather, and more. There are different types of time series forecasting models ranging from simple moving averages to more complex machine learning algorithms such as ARIMA and LSTM [2]. One of the use cases of time series forecasting is to predict the energy demand curve of consumers. These predicted demands can be crucial data for demand-side management (DSM).

DSM refers to the measures taken to manage the demand for electricity in the power grid. It aims to optimize the use of electricity, reduce peak demand, and improve the overall efficiency of the power system [3]. DSM is required for several reasons. One of the main reasons is to reduce the need for expensive power plants and transmission infrastructure. By reducing the peak demand for electricity, it is possible to avoid the need for building additional power plants, which can be expensive and take a long time to construct. DSM can also help to reduce the strain on the existing power grid, which can improve its reliability and reduce the likelihood of blackouts. DSM can be achieved through different techniques such as load shifting, load scheduling, valley filling, etc.

The two concepts (time series forecasting and DSM) are interrelated. With the help of accurate time series forecasting, the utility can better predict the demand pattern and accordingly plan their generation and even implement more effective DSM programs. For example, by forecasting peak demand periods, utilities can offer incentives to customers to reduce their energy usage during these times, reduce the overall demand on the grid, and implement real-time variable tariff structures. There are several modules from which electricity load forecasting can be made. These include statistical models, machine learning algorithms, and engineering models. These methods can be applied at various spatial and temporal scales, depending on the need of the forecast. Neural Networks (NN) is being popular in forecasting recently. Day ahead hourly demand can be predicted by using modules of NN such as LSTM, Artificial Neural Network (ANN), Convolution Neural Network (CNN), etc. These modules can predict day-ahead hourly electricity demand with high accuracy.

An IoT smart energy management device is crucial for collecting data on energy usage patterns and presenting real-time energy charts to consumers. This device records the behaviour of consumers such as the energy consumption habits of a household or business, and by analyzing the recorded data, it can recommend and suggest the best time for shifting loads. Smart appliances and meters are two examples of IoT devices that can give users insights into their energy usage patterns and help them decide how to use minimal energy. Also, with the help of programming, these devices can automatically shift loads to off-peak hours. By providing a comprehensive view of energy usage and costs, the IoT smart energy management device empowers consumers to make informed decisions about their energy consumption and take steps to become more energy efficient. Thus, an IoT-based smart monitoring device is necessary for displaying the day-ahead electricity demand forecast. It helps the consumers to know about the load pattern of a day and involve in flattening the load curve by applying techniques of DSM. The major techniques of DSM include scheduling deferrable loads to off-peak hours and shifting lighting loads to available renewables during peak hours.

2 Problem Statement

DSM techniques can be effectively implemented only if demand can be predicted accurately, for that historical data on energy consumption is needed. To collect energy consumption data, many smart meters have been placed but they aren't utilized properly in the DSM technique, rather just to avoid reading energy bills manually. Besides these, various problems encountered while implementing DSM are shown below:

2.1 Manual Load Shifting

One of the primary challenges of DSM is the difficulty of manually shifting loads (e.g., turning off non-essential appliances during peak demand periods) in a coordinated and efficient manner. This can be labour-intensive and may not be effective in reducing overall electricity demand. No effective technology is developed for load shifting based on peak and off-peak demand.

2.2 Load Forecasting

Accurate load forecasting is essential for effective DSM, as it enables utilities to anticipate and respond to changes in demand. However, forecasting can be challenging due to the complex and dynamic nature of electricity consumption patterns, as well as the influence of external factors such as weather and economic conditions.

2.3 Integration of Distributed Energy Resources

The increasing penetration of distributed energy resources (DER), such as rooftop solar panels and small-scale wind turbines, can complicate load forecasting and make it more difficult to manage demand. This is because DER are often located at the customer's premises and may not be centrally controlled by the utility.

2.4 Lack of Communication Infrastructure

In some cases, there may be a lack of communication infrastructure (e.g., smart meters, advanced metering infrastructure) in place to enable effective DSM. This can limit the ability to monitor and control demand in real time.

2.5 Implementation of Feed-In Tariff Scheme

Feed-in tariffs (FIT) are some amount of electricity prices paid to renewable energy producers for injecting their renewable energy source into the electrical grid. Implementation of this scheme increases the continuous and stable renewable sources market development. The major problem of DSM is the flat tariffs which provide no financial incentive for consumers to alter their behaviour and shift their consumption to off-peak hours.

2.6 Problems in Time of Day/Use (ToD/ToU) Pricing Scheme

Time-of-day (ToD) or Time-of-use (ToU) pricing helps consumers to shift their energy consumption to the time of day when energy consumption is lower and it helps to avoid peak time. If all the consumers initiate to shift their consumption to the time of lower demand, then a new peak will be developed in the meantime. Thus, a different ToD/ToU scheme like real-time ToD/ToU should be proposed to different consumers to flatten the demand curve [4].

2.7 Lack of Advancement of Battery Storage Systems

Battery storage systems enhance the ability to store energy during low demand and use the stored energy during peak demand. For example, solar is an on-site generation renewable energy source, but the energy can be stored in a battery and used later. This helps to flatten the demand curve.

3 Literature Review

Neural Network (NN) has been the strongest tool for forecasting energy demand. In [5], the authors have proposed a NN model based on LSTM and CNN for one day ahead forecasting of electrical energy demand. This NN architecture consists of multiple types of input features where the inputs are processed using different types of NN components according to their specific characteristics. A data set consisting of hourly loads of about 3 years was used. With the aid of multiple input features like temperature, humidity level, holiday, and day of the week, they have made the network more efficient and feature-rich. In their model, they have used CNN to extract useful features from the additional features (e.g. temperature, humidity level, holiday, and day of the week) and RNN to model the temporal dynamics in historical load series.

In [6], the researchers have used Feed Forward ANN by adopting multiple linear regression (MLR) for load forecasting. In this model, prediction variables like temperature as the weather variable and the day of the week and hour of the day as the calendar variable are included. With the significant deployment of smart meters across end-user platforms, the dynamic visibility of energy flow among the end-users has increased significantly. The granular information of smart meters can be used to improve the load forecast accuracy and influence energy consumption patterns with DSM techniques.

Fan et al. [7] developed two Machine learning-based load prediction methods for home energy management systems (HEMS) to accurately forecast residential electricity consumption. Using an 8-week dataset of 2337 residential customers, the methods integrate the traditional sequence to sequence long short-term memory (S2S-LSTM) model with human behaviour pattern recognition. Method 1 utilizes clustering algorithms (DBSCAN, K-means, and Pearson correlation coefficient (PCC)) to recognize human behaviour patterns, with PCC outperforming S2S-LSTM. Method 2 further improves the performance of Method 1 with a modified NN architecture optimized for supervised learning in LSTM.

The study review [8] shows a selection of approaches that have used ANN, Particle Swarm Optimization (PSO), and Multi Linear Regression (MLR) to forecast electricity demand for Gokceada Island. The results obtained were analyzed using statistical error metrics such as MSE, RMSE, and MAE. The confidence interval analysis of the methods was performed. It was observed that ANN yields the highest confidence interval of 95% among the methods utilized, and the statistical error metrics have the highest correlation for ANN methods between electricity demand output and actual data.

In [9], a communication-based demand-side residential load scheduling scheme is presented. HEMS is used to manage the utility demand of the consumer. HEMS is a system that requires a metering system, sensors, and communication infrastructure. The proposed method reduces peak power consumption by implementing a few measures such as scheduling loads to off-peak hours, automatically switching off less significant loads, not letting turn on certain power-hungry appliances during peak hours, etc. Also, this model provides reserved overriding option which allows the consumer to change their decision for any specific loads to keep it switched on in a critical situation. The prime responsibilities of HEMS are to reduce electricity bills and build a bidirectional communication system between consumers and providers. With the help of HEMS, the model was able to reduce electricity bills and build a bidirectional communication system.

Zhu et al. [10] proposed a genetic algorithm that can effectively manage a large number of controllable loads in the selected area. The algorithm minimizes the cost and peak-to-average ratio by changing the load. During peak time, consumers can essentially shift or schedule their deferrable loads as the price of electricity is low. Through this technique, consumers were able to reduce their daily electricity by as much as 7.25%.

Adejumobi and Adeoti [11] modelled an optimization-based formulation of DSM techniques i.e. peak clipping and load shifting together to improve the load factor thereby reducing the maximum power demand of an industry. The outcomes from the research see some considerable reduction in energy consumption and cost without any compromise in the production or the comfort of the system.

A two-step process is presented by Javor and Raicevic [12] to apply the load-shifting technique for the demand that includes some deferrable loads. The first step involves minimizing the sum of the absolute values of the hourly demand differences, which lowers the mean demand and raises the minimum demand. While maintaining the same daily demand, a new schedule for the deferrable loads is obtained. This raises the load factor. The cost function is now minimized for buying prices in the second step, while keeping the aforementioned data as constraints, producing an optimized daily load curve with the greatest amount of savings. If deferrable loads are higher, this two-way procedure offers greater savings with a higher load factor.

Dynamic tariff is one of the important factors for achieving DSM. Stute and Kühnbach [13] compared the annual electrical cost of a city by applying three tariff structures; static electricity tariff, 3-tier ToU, and dynamic tariff based on day-ahead pricing. From these tariff rates, the dynamic tariff rate was found to have the least average cost.

In order to meet the energy gap between generation and demand, renewable sources should be utilized. In [14], Maharaja et al. proposed a bidirectional net meter that injects power from solar to the grid to meet the peak demand. They used the term bidirectional to address the energy from the distribution company to the consumer and from the consumer to the distribution company. Net energy metering displays the difference between imported and exported energy. When PV modules are connected in series, they produce a string of modules with greater voltage. When these modules are parallelly connected, they produce dc input with a higher current. It is fed to an inverter which converts dc electricity to ac electricity. This ac electricity is suitable for grid supply.

In [15], a DSM strategy for residential users based on the load-shifting technique is proposed. In the proposed model, user priority and comfort are highly considered. With the help of DSM, the cost of electricity usage is reduced and varying power consumption on an electricity supply basis is managed. A cascaded ANN [16] is utilized to construct a DSM strategy for managing peak electricity in residential buildings. A multi-layer cascaded feed-forward NN is used to implement a controller for managing peak demand using three different strategies. Using Matlab and Simulink, the suggested model was simulated and tested for an apartment building with 11 houses and varied loads. The results revealed that the proposed cascaded feed-forward NN outperformed the other methods.

4 Methodology

Higher the number of data, the more accurate the AI model will be. With this, the focus of the research is to collect the real-time energy consumption pattern of every household as much as possible. For the context of this project, the state of art Predictive Load Management Device (PLMD) was prototyped to track the energy consumption of a model house that essentially has two lighting sources and a power socket. PLMD mainly does the following as shown in Fig. 1.

The proposed system PLMD contains the following functional block diagram that contains ESP32 microcontroller, current and voltage sensors, and other circuitry required as shown in Fig. 2. In the following section, the energy measurement, passing of the data for AI processing, a compilation of the whole using Blynk IoT, and data prediction are described in more detail.

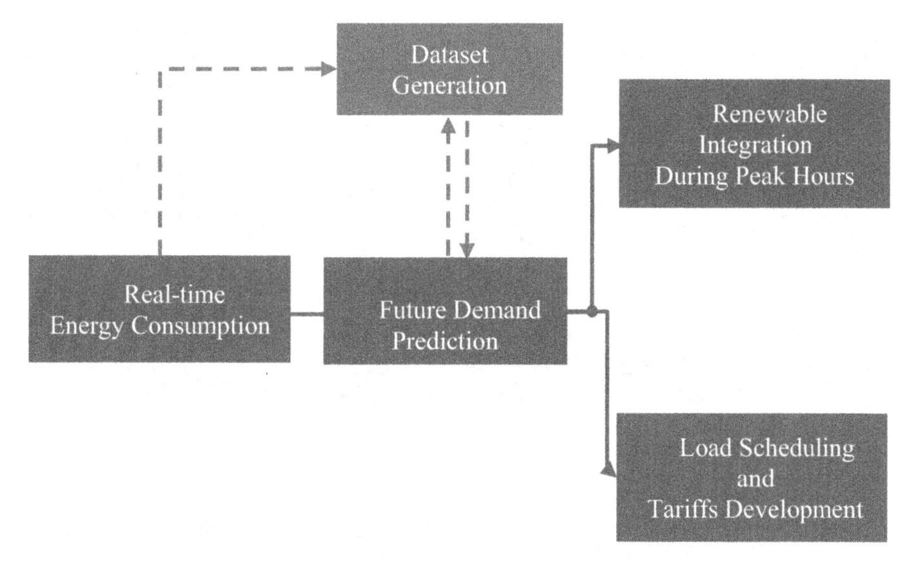

Fig. 1. Flow chart of PLMD.

4.1 Energy Measurement

ESP32 microcontroller was used for measuring the energy consumption of a house by simply connecting the current and voltage sensors in its analog to digital converter (ADC) pins to sample the voltage and current waveforms. The resulting data was then processed to calculate the instantaneous power and energy consumed by a load. The sensors used and their working are described below.

ACS712 Current Sensor. The ACS712 is a current sensor that measures AC or DC flowing through a conductor by using the Hall effect. ACS712 consists of a precise, low-offset, linear Hall circuit with a copper conduction path located near the surface of the die. Applied current flowing through this copper conduction path generates a magnetic field that the Hall IC converts into a proportional voltage [17]. The sensor can be configured for several different current ranges, including 5 A, 20 A, and 30 A (used). It works on a supply voltage of 4.5 V–5.5 V DC and has a measuring current range of −30 A to 30 A based on the sensitivity of 66 mV/A. With little tampering in the codes, the sensor was able to detect near to the accurate value of rms-current.

ZMPT101B Voltage Sensor. The ZMPT101B is a voltage sensor used for measuring the AC voltage levels up to 1000 V in most electrical project systems. It is a transformer-based sensor that works by detecting the change in the AC voltage signal and converting them into a proportional output voltage. It consists of a transformer with primary winding and secondary winding. The primary winding is parallelly connected to the AC power line of which the voltage is to be measured, and the secondary winding provides the output voltage proportional to the AC voltage being measured. It does so according to the electromagnetic induction principle. When an AC voltage is applied to the primary winding, it creates a magnetic field that induces a voltage in the secondary winding. This induced voltage is proportional to the AC voltage on the primary side. The sensitivity

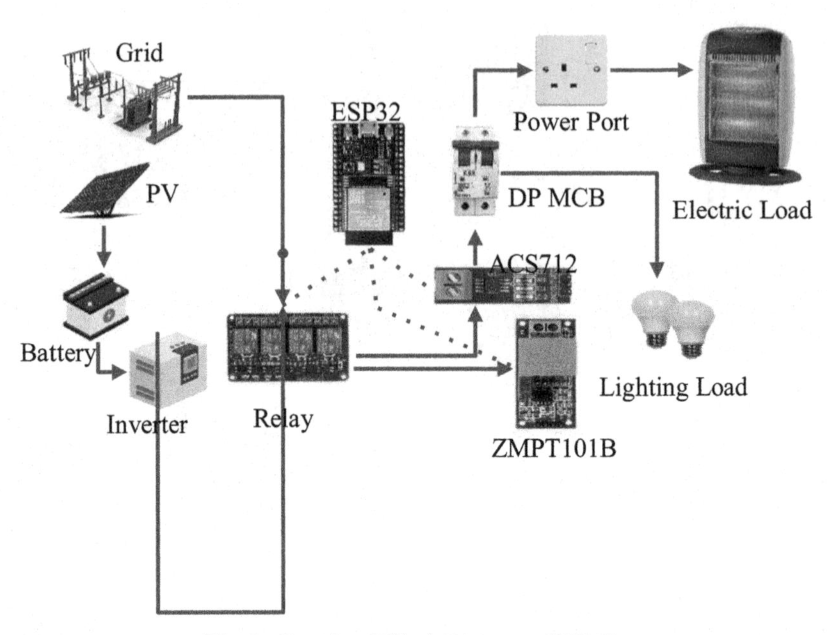

Fig. 2. Functional Block Diagram of PLMD.

of the ZMPT101B voltage sensor was somewhere around 57 mV/A obtained from its datasheet.

In addition to these, the DHT11 sensor was used to measure the ambient temperature and humidity as well which are some of the prime factors on which the demand curve is dependent. However, it's worth noting that the accuracy of the sensor depends on the quality of the ADC and the external circuitry used to sample the voltage and current signals. To get the most accurate measurements, it is recommended to use an external ADC or a sensor with high resolution and accuracy.

4.2 Dataset Generation

With this, the real-time current, voltage, power, and energy consumption are obtained. For AI processing, the data needs to be stored continuously on the server. Due to some limitations in the project, the data is stored on a local network, a spreadsheet in this case. The data is updated on the local spreadsheet every second. The sheet provides the comma-separated values (CSV) of current, voltage, temperature, humidity, and such. Now, the AI model is ready to predict future demand consumption based on the previous CSV obtained from the spreadsheet. The peak time at which automatic renewable switching can be done and the off-peak time at which load scheduling can be suggested are already known a day ahead.

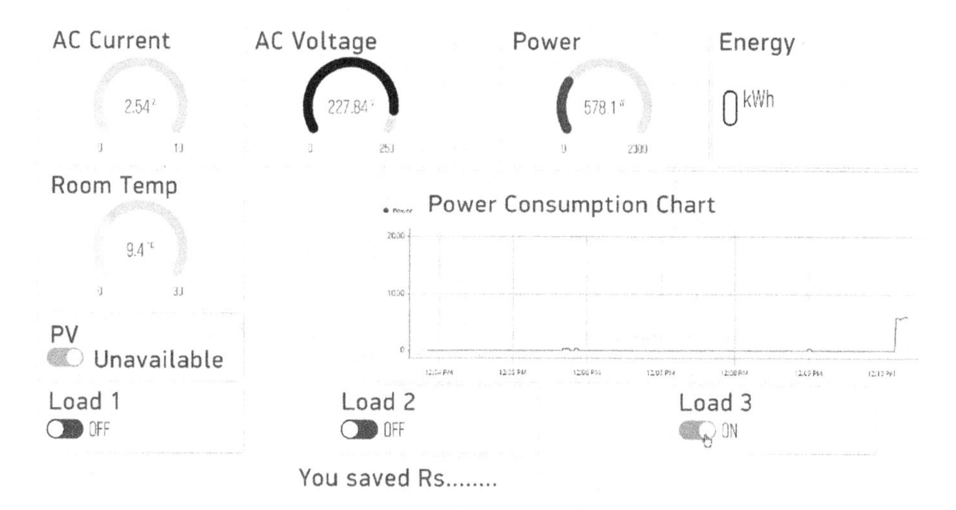

Fig. 3. Web Dashboard in Blynk for monitoring and controlling the electric loads.

4.3 Blynk IoT

Blynk is an IoT platform for controlling and monitoring the sensors remotely using smartphones or computers. The app for controlling the loads and monitoring the electricity consumption is registered in the Blynk IoT platform for the web dashboard (and mobile) is shown in Fig. 3. The real-time current, voltage, power, energy, and temperature keep on updating every second. Necessary switching can be automated as well as manually via the relay actions. The ESP32 communicates with the Blynk server over the internet to send and receive data from the app. The ESP32 uses the Blynk library to connect to the Blynk server and exchange data with the app. The library handles the low-level communication details and provides a high-level API for accessing the Blynk services.

Due to some limitations on the basic versions of Blynk IoT, the widgets available are only a few and not fully accessible. The solution to this problem will be to develop an IoT app communicating directly with ESP32 using Flutter.

4.4 Demand Prediction

The approach involves predicting the energy demand of individual households based on environmental factors such as temperature, humidity, and precipitation levels along with socio-economic aspects such as festivals, and holidays. For training and testing the model, the daily energy consumption data from Panama City was considered. To predict the 24-h demand, a multivariate LSTM network was used. As prediction was done based on the past 7 days' data, LSTM has a greater ability to capture patterns and trends compared to other RNN networks. The LSTM model architecture is shown in Fig. 4, where the memory cell state stores important patterns and trends in the demand curve.

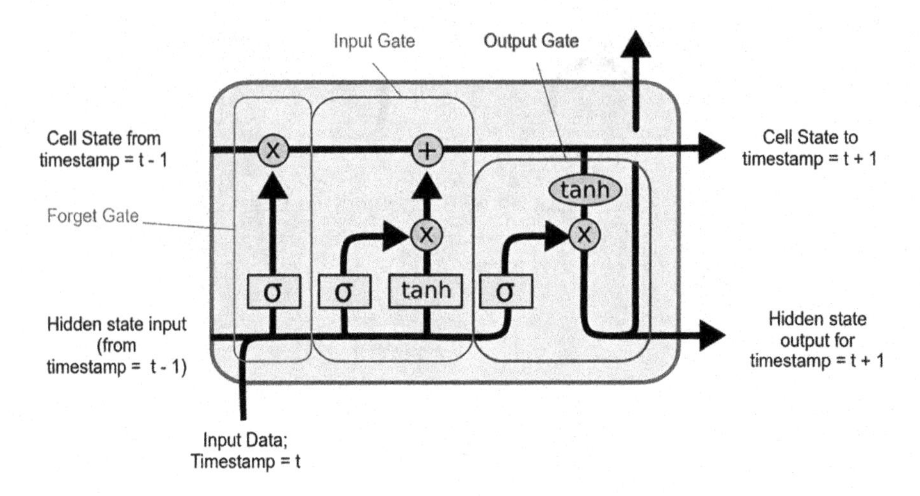

Fig. 4. LSTM Model Architecture [18].

The block diagram for the LSTM prediction system is shown in Fig. 5. The Panama-city dataset contains current energy demand along with temperature, humidity, precipitation, and holiday which are some of the variables affecting energy consumption. When the model was used to predict the demand curve, the necessary environmental conditions and the past 7 days' consumption pattern were fed.

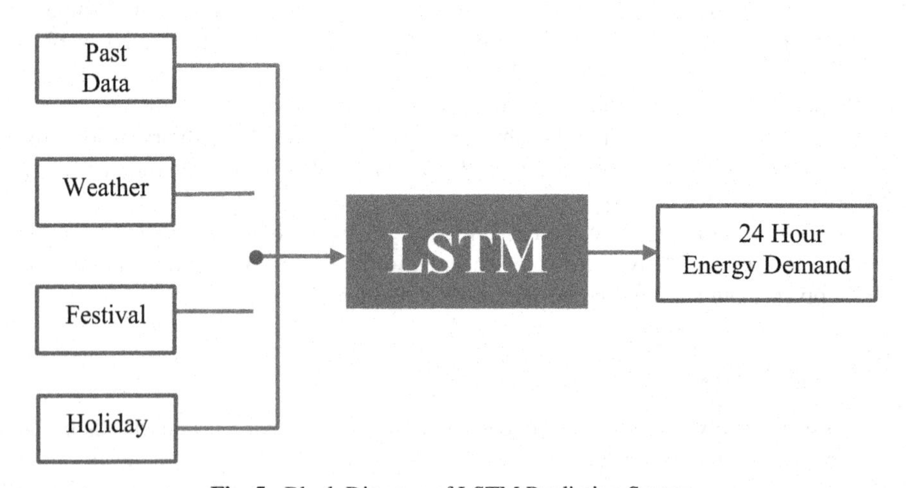

Fig. 5. Block Diagram of LSTM Prediction System.

The LSTM model was trained using the Panama City dataset for 100 epochs with a batch size of 100. Table 1 displays the model summary where the sequential model consists of two layers of LSTM with a single dense layer stacked after it.

The model was quite good at predicting the energy demand pattern and finding the peak demand time. The comparison between the actual power consumption and the

Table 1. Sequential Model consisting of two layers of LSTM.

Model: "sequential_1"		
Layer(type)	Output Shape	Param #
lstm_2 (LSTM)	(None, 168, 64)	17920
lstm_3 (LSTM)	(None, 32)	12416
dropout_1 (Dropout)	(None, 32)	0
dense_1 (Dense)	(None, 1)	33
Total params: 30,369 Trainable params: 30,369 Non-trainable params: 0		

predicted power consumption is presented in Fig. 6. The power consumption on the y-axis has been scaled down using a min-max scaler.

4.5 Load Scheduling and Dynamic Tariff Structure

Once the demand graph is obtained, a load optimization technique is applied to flatten out the demand curve at the individual level. This is known as decentralized demand flattening. To do this, the loads are categorized into deferrable and non-deferrable loads. Keeping the use of non-deferrable load constant as it is, since it has maximum priority for use at any time. The deferrable loads can be shifted to reduce the peak demand, PLMD switches the lighting load during peak hours to the PV source (if available) and uses the grid for only non-deferrable loads. Figure 7(a) shows the predicted energy demand, from the demand predicted, an average data line is drawn, this helps to find the peak region.

Now, to flatten out the curve, the best time to shift the deferrable loads is suggested when the user's demand curve hits the lowest minima. A variable tariff rate structure depending on the user's energy consumption is developed. This is shown in Fig. 7(b) and Fig. 8 respectively.

Until and unless consumers are not motivated, DSM cannot be achieved. One way to motivate consumers is by providing incentives. For that reason, a dynamic tariff structure should be implemented where consumers are charged by the utilities based on time of peak and non-peak hours. Three dynamic tariff rates can be compared based on annual average cost [13]. Figure 9 shows three dynamic tariff structures.

Figure 9(a) displays a 2-tier ToU tariff structure. In this model, each time period is assigned to a price level. Within one day, two price levels are possible for peak time and off-peak time. Figure 9(b) shows a 3-tier ToU tariff structure. In this model, each time period is assigned to a price level. Within one day, three price levels are possible; peak time, off-peak time, and normal time. Figure 9(c) shows a real-time tariff structure in which electricity price is allocated based on the real-time power consumption of the user. From these three tariff structures, the real-time tariff is found to be more scientific and efficient. If a 2-tier and 3-tier structure is applied, then there is a chance of shifting

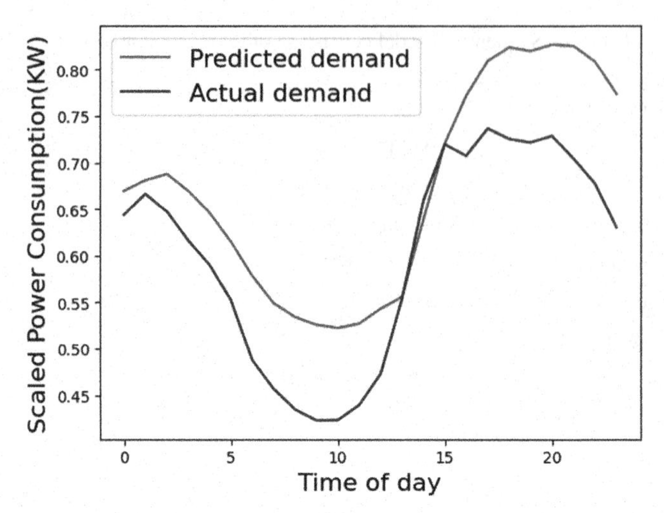

Fig. 6. Comparison of Actual vs Predicted Demand Curve obtained from the model.

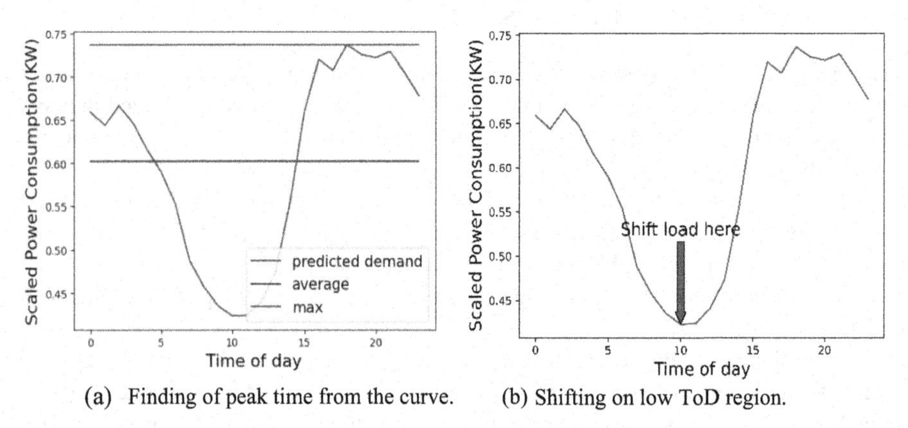

(a) Finding of peak time from the curve. (b) Shifting on low ToD region.

Fig. 7. Load Optimization Techniques for Load Flattening. (a) Finding of peak time from the curve. (b) Shifting on low ToD region.

peak demand to the time having lower electricity cost. This does not lead to effective DSM as the load curve is still not flattened. For this reason, a real-time tariff structure should be implemented.

To create an interface between regular consumers and load time schedule, an IoT-based app is developed which provides the best time to use specific loads, the charge associated with that time period, and total net energy consumption with a status of PV source. The final prototype model of PLMD is shown in Fig. 10.

Figure 10(a) is the compact model of PLMD and Fig. 10(b) shows the internal circuitry consisting of ESP32 microcontroller, current sensor, voltage sensor, relay modules, etc. as explained earlier. For demonstration purposes, on the left-hand side of PLMD, there are two lighting loads, and near that, there is a power socket for powering the large

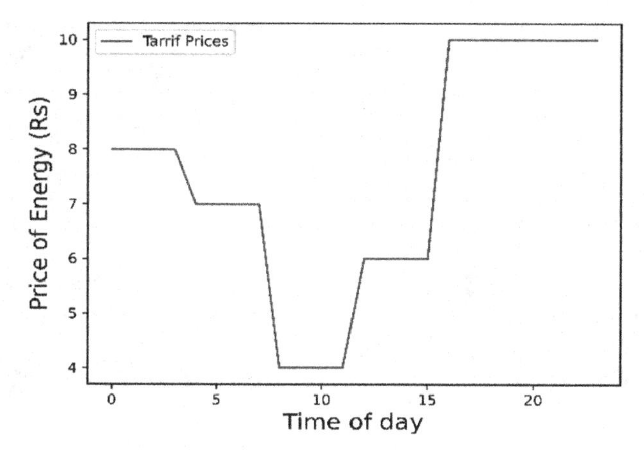

Fig. 8. Variable Tariff Rate.

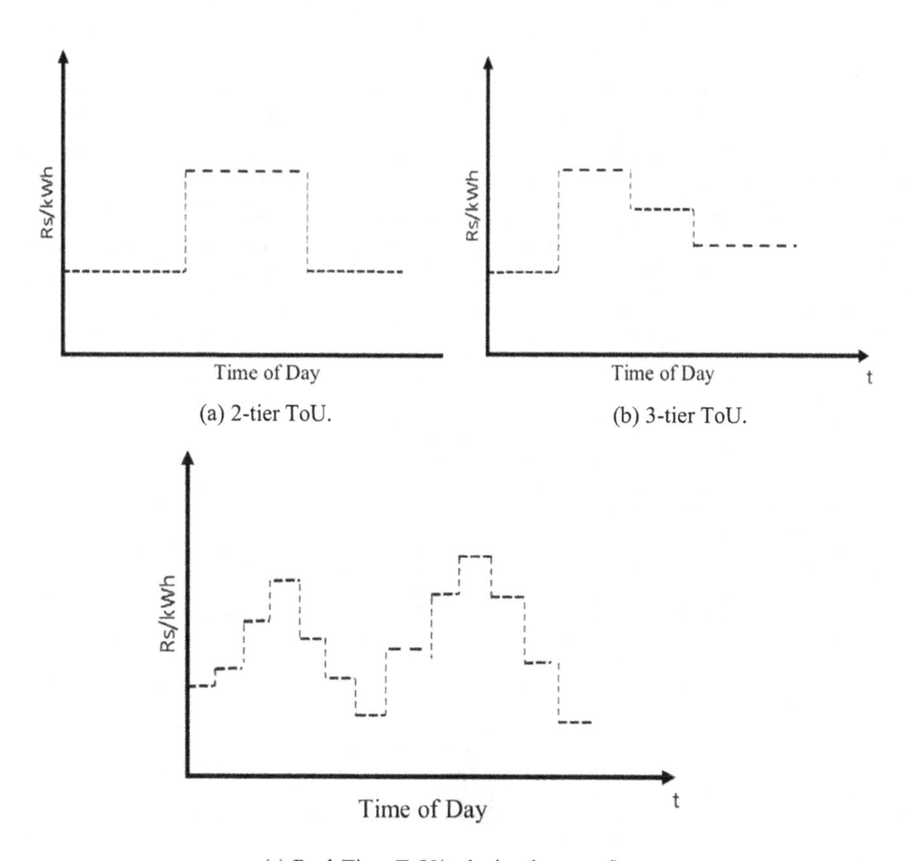

(a) 2-tier ToU.

(b) 3-tier ToU.

(c) Real-Time ToU(to be implemented).

Fig. 9. Different types of ToU tariff structures.

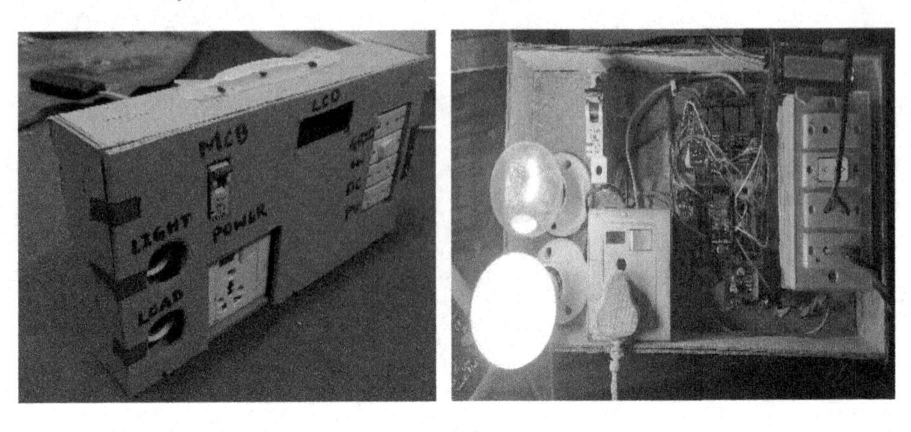

(a) Compact Model. (b) Internal Circuitry.

Fig. 10. Prototype Model PLMD. (a) Compact Model. (b) Internal Circuitry.

devices. And on the right-hand side of it, there are two sources one grid and another PV source or battery. ESP32 fetches the data into the local spreadsheet and Blynk IoT continuously. With the peak time already sensed by the AI model, the relay module can switch the incoming power from the grid to any (available and capable) renewable sources (either automatically or manually through the app). Similarly, when the off-peak time is identified, various suggestions and recommendations can be provided to users via application to shift the load to off-peak time with some attractive dynamic tariffs incentivizing.

Hence, the prototype PLMD tries to act as an interface between utility and consumer to realize the smart-grid concept and achieve DSM as shown in Fig. 11.

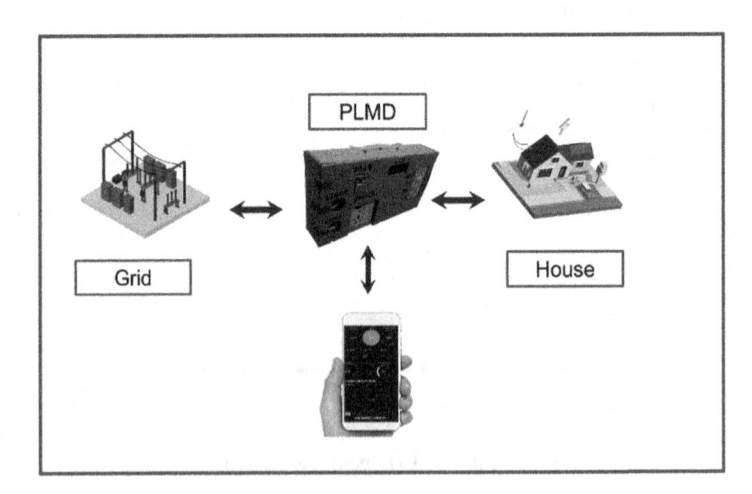

Fig. 11. PLMD as an interface between Grid and House.

5 Conclusion

In this work, the future electricity demand of households was predicted by using LSTM based network and incorporating the smart monitoring device PLMD, and different techniques of DSM were applied for decentralized demand flattening. Also, with the provision of a real-time dynamic tariff structure from utility companies, consumers can get incentives and be motivated to support DSM programs. With the exploitation of renewable energy in peak hours, the peak demand can further be flattened out. For shifting loads in off-peak times where the price of a unit is relatively low, the deferrable loads are shifted there as much as possible. However, there are still some better algorithms such as Heuristic Algorithms and other many computational algorithms for load optimization that can perform better. All in all, the idea was to create a data acquisition system (DAS) that predicts the energy demand and achieves decentralized demand flattening using AI and load optimization techniques.

References

1. Rodrigues, S., Faria, F., Ivaki, A., Cafôfo, N., Chen, X., Dias, M.: The Tesla powerwall: does it bring something new? A market analysis. In: Proceedings of the Engineering & Technology, Computer, Basic & Applied Sciences (ECBA-2015), Bangkok, Thailand, 9–10 December 2015 (2015)
2. Chatfield, C.: Time-Series Forecasting (2000). https://doi.org/10.1201/9781420036206
3. Gellings, C.W., Chamberlin, J.H.: Demand-Side Management: Concepts and Methods (1987)
4. Philippou, N., Hadjipanayi, M., Makrides, G., Efthymiou, V., Georghiou, G.E.: Effective dynamic tariffs for price-based Demand Side Management with grid-connected PV systems. In: 2015 IEEE Eindhoven PowerTech (2015). https://doi.org/10.1109/ptc.2015.7232387
5. He, W.: Load forecasting via deep neural networks. Procedia Comput. Sci. **122**, 308–314 (2017). https://doi.org/10.1016/j.procs.2017.11.374
6. Khan, Z.A., Jayaweera, D.: Smart meter data based load forecasting and demand side management in distribution networks with embedded PV systems. IEEE Access **8**, 2631–2644 (2020). https://doi.org/10.1109/access.2019.2962150
7. Fan, L., Li, J., Zhang, X.P.: Load prediction methods using machine learning for home energy management systems based on human behavior patterns recognition. CSEE J. Power Energy Syst. **6**(3), 563–571 (2020)
8. Saglam, M., Spataru, C., Karaman, O.A.: Electricity demand forecasting with use of artificial intelligence: the case of Gokceada island. Energies **15**(16), 5950 (2022). https://doi.org/10.3390/en15165950
9. Rahman, Md.A., Rahman, I., Mohammad, N.: Demand side residential load management system for minimizing energy consumption cost and reducing peak demand in smart grid. In: 2020 2nd International Conference on Advanced Information and Communication Technology (ICAICT) (2020). https://doi.org/10.1109/icaict51780.2020.9333451
10. Zhu, Q., Li, Y., Song, J.: DSM and optimization of multihop smart grid based on genetic algorithm. Comput. Intell. Neurosci. **2022**, 5354326 (2022). https://doi.org/10.1155/2022/5354326
11. Adejumobi, I.A., Adesina Adeoti, J.: Efficient utilization of industrial power: demand side management approach. In: 2019 IEEE PES/IAS PowerAfrica (2019). https://doi.org/10.1109/powerafrica.2019.8928817

12. Javor, D., Raicevic, N.: Two-steps procedure in demand side management for reducing energy costs. In: 2020 19th International Symposium INFOTEH-JAHORINA (INFOTEH) (2020). https://doi.org/10.1109/infoteh48170.2020.9066311
13. Stute, J., Kühnbach, M.: Dynamic pricing and the flexible consumer – investigating grid and financial implications: a case study for Germany. Energ. Strat. Rev. **45**, 100987 (2023). https://doi.org/10.1016/j.esr.2022.100987
14. Maharaja, K., Balaji, P.P., Sangeetha, S., Elakkiya, M.: Development of bidirectional net meter in grid connected solar PV system for domestic consumers. In: 2016 International Conference on Energy Efficient Technologies for Sustainability (ICEETS) (2016). https://doi.org/10.1109/iceets.2016.7582897
15. Anjana, S.P., Angel, T.S.: Intelligent demand side management for residential users in a smart micro-grid. In: 2017 International Conference on Technological Advancements in Power and Energy (TAP Energy) (2017). https://doi.org/10.1109/tapenergy.2017.8397265
16. Tabassum, Z., Shastry, B.S.C.: Peak power management of residential building using demand side management strategies. Int. J. Health Sci., 8978–8997 (2022). https://doi.org/10.53730/ijhs.v6ns2.7333
17. Allegromicro: ACS712: fully integrated, hall-effect-based linear current sensor IC. https://www.allegromicro.com/en/products/sense/current-sensor-ics/zero-to-fifty-amp-integrated-conductor-sensor-ics/acs712
18. Olah, C.: Understanding LSTM Networks. http://colah.github.io/posts/2015-08-Understanding-LSTMs/

AI and IoT for Other Engineering Applications

A BERT Classifier Approach for Evaluation of Fake News Dissemination

Tushar Rana, Darshan Saraswat(✉), Akul Gaind, Rhythem Singla, and Amit Chhabra

Chandigarh College of Engineering and Technology, Sector-26, Chandigarh 160019, India
falsesaraswatdarshan@gmail.com, amitchhabra@ccet.ac.in

Abstract. The intake of information via media has changed from newspapers to social networking sites as technology has evolved. Accessibility and availability are two important aspects that have led to this trend in media usage. Users share thousands of posts, articles, and videos as the internet penetration grows. These postings are done on a variety of social networking sites such as Facebook, Instagram, YouTube, Twitter, and others. It is now widely acknowledged that disinformation may often create tensions and has a substantial impact in the country. Stemming the tide of false news articles via social networking sites and the Internet is critical. This problem has indeed been handled in this study using various algorithmic approaches that could help us to control this type of hazardous thing. Including the findings, a contrast of how various classification functions are offered. This article gives a thorough assessment of numerous fake news detection strategies employed by various other authors, databases they had engaged with, and the multiple analytical people were using to evaluate the effectiveness of their respective algorithms. This research analyses the troubles and challenges of detecting this type of news. This article examines documents from 2017 through 2022, as well as various fake news detecting tools. This study provides a complete evaluation of present and past studies on false news recognition leveraging various ML and DL models. In this research work, BERT Classifier has been used which uses a deep learning technique and has the highest accuracy of all the methods described.

Keywords: BERT · Machine learning · Fake news · Deep learning · Artificial Intelligence · Regression · and classification models

1 Introduction

All of us have been getting news through the Online websites and apps rather than conventional Television channels because of simplicity of disseminating news articles through different social media networks [1]. Obtaining knowledge over the Internet, on the other hand, is a two-sided knife; the news quality we are getting on the Internet is poorer than among conventional TV channels. All facts and reports surrounding the spreading of the worldwide upsurge such as Corona, such as of infection, temperature-based spreading, and vaccination; the speech of convention of people when there is any

R. K. Challa et al. (Eds.): ICAIoT 2023, CCIS 1930, pp. 171–184, 2024.
https://doi.org/10.1007/978-3-031-48781-1_14

political speaker around or if there is any rally for political support and their unconfirmed statements on land intervention, producing and just doing essential utilities; inaccurate and defamatory photos of individuals to disparage or laud them; and modification of movies and voice recordings are among the incidents and instances of fake news [2]. This isn't really a modern problem, although its mass transmission is, and several news sources are actively promoting it. Social Media Networks like Meta, WhatsApp (Meta), vlogs, Instagram, Tweets, YouTube, and, tragically, news stations are also on the list of platforms that distribute fake news. This cannot be argued to be unknown to these mediums and internet platforms [3].

A person's viewpoint provides knowledge for someone else, and even those individuals construct their environment depending on biased and unconfirmed facts. Depending on just this perspective, great rise in knowledge caused a community to run with incorrect ideals. Individuals are seldom able to verify information since they are preoccupied with their personal and social lives. However, a community founded on erroneous and prejudiced ideas is a disaster waiting to happen anytime a new concept meddles and threatens the supremacy of existing notion that was not desirable for a person nor a community. Several actions were taken to manage this problem, one of which is to identify them and prevent their proliferation. There was previous research suggested that apply ML principles to extract such media articles [4]. The K-Nearest Neighbor was developed to segregate news as false or authentic, however due to the nature of the textual information accessible on the internet; this approach could not produce believable precision. These instances demonstrate how false news influences and shapes intellectual, social, religious as well as other views and relationships. Whereas the major benefactors of that kind of false information are the elected government for elective benefit, everyone else is gathering some traction. Here are some cases of fake news [4]:

- For unrelated information, use a fictitious image or title.
- Misunderstanding of data
- Information that is untrue
- Uninformed followers propagate rumors.

2 Fake News Detection

Fake News Detection involves different steps as represented in Fig. 1. Data captured from the news story out of a dataset or Web can be used as an input which can be processed further. This news item data is required to train the models' machine learning (ML) methods. Lemmatization, stop words, Regex, Stemming, Tokenization, and other methods can be used to convert text input into a format upon which modeling may be conducted. The following steps are outlined:

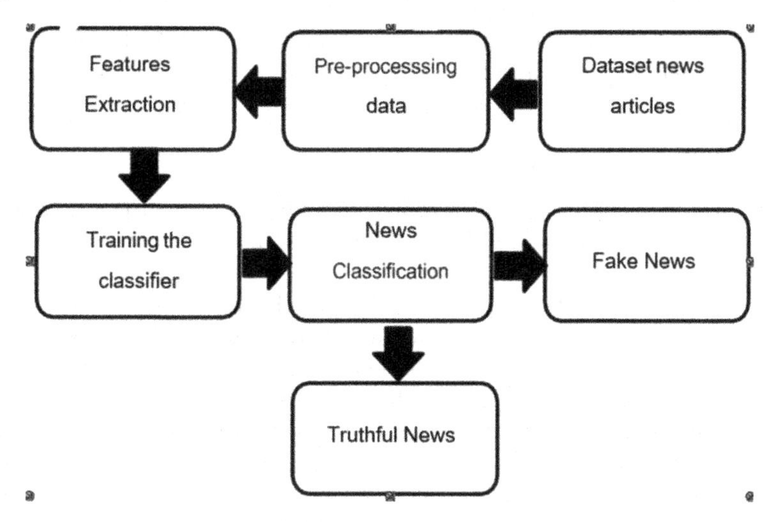

Fig. 1. Steps including the flow of fake news detection.

2.1 Data Preprocessing Techniques

Before implementing ML to textual information, it is necessary to eliminate specific special letters out from textual data. It is critical to do the analysis process, which may be accomplished through a variety of approaches [4].

Regex: Is a data analytical method that is utilized to eliminate commas, exclamation, and other punctuation symbols from a message.

Tokenization: This is a data analytical method that divides up into discrete parts called tokens. These are discrete components which can be characters, sub-words, words.

Stop Words: These are by far the most frequently utilized terms in any human speech. These words do not transmit any delicate relevant knowledge for classifier preprocessing. Instead, it causes perplexity for the classifier [4]. These words must be eliminated because these add little worth towards the understanding of the page.

Stemming: This is an essential Natural language processing method for converting a phrase to its fundamental form by eliminating some prefixes, suffixes, or substrings [5]. For example, if this method is used on phrases 'consuming,' 'consumes,' and 'consumed,' the core phrase created is 'consume.' While stemming, two types of mistakes happen: too much and the understemming. Too much stemming happens when a bigger portion of a phrase is deleted, leading to two or even more phrases getting truncated to the very same base phrase. Understemming happens when two or maybe more phrases are mistakenly simplified to even more than one base word rather than decreasing to the same base word.

Lemmatization: One type of above- mentioned technique is lemmatization, which involves eliminating some prefixes, suffixes, or substrings to get the base word [5]. This technique transforms each word to its logical original state, whereas stem may result in a nonsensical original state. Following data pretreatment, extracted features are required, which seems to be a crucial stage approach model development [4, 5]. Textual data could only be conducted once the actual text has been translated in nominal attributes.

2.2 Feature Extraction Techniques

Raw text must be translated into quantitative characteristics before feature extraction techniques on the textual data can be performed. The Bag of Words (BoW) and the term frequency-inverse document frequency (TF-IDF) are the basic extracting features approaches. Bag-of-words is a phrase - based approach which informs more about the presence of terms inside a page. It computes the frequency analysis of each sentence [4]. Its frequency analysis is used to generate the quantitative description of the information. The TF determines a phrase's local value based on its appearance in a paper. IDF, on alternate side of the coin, involves the use of statistical value that indicates the relevance of a certain phrase in a type of collection [4]. These input characteristics are critical for training different ML models. Typically, the information is partitioned into training datasets in 70% and 20% ratio [4]. Secondly, we must create an algorithm that can categorize the results into several essential entities. A variety of different classifiers can be used to perform this categorization. The following are a few categorization methods that could be utilized to identify bogus updates.

2.3 Classification Algorithms

There are different classification techniques, such as Gradient Boosting, Naive Bayes, KNN, DT-J48, RF, LR, MNB, SVM, SDG, and so on [6].

Naïve Bayes (NB) classification technique is an algorithm based on Naïve Bayes rule. It is a powerful classification method that aids in developing rapid ML models capable of generating good estimates. In Granik M. et al. [6], the classifying system is employed to recognize lies and misinformation, with an efficiency equaling roughly 74%.
Support Vector Machine (SVM) algorithm is an algorithm used to solve classification issues [7]. This is currently used to build an outcome limit that categorizes pieces of data consisting of a set of criteria.
Logistic Regression (LR) algorithm employs a function to predict a discrete outcome variable. LR employs a curve to convert the result into a significance level. This model's major goal in attempting to get an ideal chance is to reduce the computational complexity [8].
Decision Tree (DT) algorithm employs a splitting strategy to represent all and every potential consequence of a choice. Since every branch of the tree reflects a specific variable's trial, every route indicates the outcome of study [9].
Random Forest (RF) algorithm is an approach that outperforms simple ML algorithms. It incorporates many classification trees, which each operates independently to determine the future. Total number of seats gathered by a group affects the actual estimate. This technique employs a mixture of different techniques to provide the greatest classification or regression results. This method begins with such a "tree structure" with feed at the tip. The information is then divided down into manageable groups based on certain factors as it progresses down the tree. The benefit of this over other algorithms is their decreased failure rate [10].

3 Related Work

H. Ahmed et al. [11] presented bogus news identification utilizing the n-gram model to distinguish between similar and different news utilizing ML approaches. Different research works are conducted using experiments with linear and non - linear classifiers and made comparison over six different data mining techniques that are effective at detecting fake news. The researchers have exhibited their research observations utilizing assembled datasets from true and false internet sites; therefore, they must accomplish outcomes with high expectations. They employed five-doubling cross- authentication in their research, such that all data verification is being castoff for trained data, with the remaining 20% for testing datasets. The scientists used unigram approaches and LSTM (Long short-term memory) network classification to get the best accuracy of 92%. Table 1 represents previous research work related to News Article and ML Models.

Many ML methods have been applied by researchers to improve accuracy, including LSTM, CNN, KNN, SVM, and NB. Agarwal et al. [12] examined SVM with an efficiency of 73% while using NB, they obtained an accuracy of 91% which was far superior to SVM. The researchers ran their models using data from huge databases including 25680 rows of information. By employing LSTM, the precision improved to 97%. The purpose in this research was to study brief phrases and information and create the dependability count with media by sequential combining extraction of features and credibility ratings. Ultimately, researchers were able to get high efficiency of 94% using a mixture of three techniques: NB, CNN, and LSTM.

Jain et al. [13] presented an SVM classifier to demonstrate a technique that leverages ML. Experts discuss Chile disaster propaganda and the US National campaign. NLP has been suggested as method for predicting bogus news. The researchers utilized five distinct ML methods i.e., NB, Linear Regression (LR), SVM, RNN, and LSTM and found that SVM was the right choice for identifying bogus information.

Reis et al. [15] presented a technique as they identified social networking sites as a major issue today. Anyone could sign up as a media writer on online platforms and raise awareness. It quickly misled culture. As a result, the researchers claim that online communication is a medium for propagating disinformation. The researchers had provided several analyses to identify bogus information which are absent from media articles, sources of news, and surroundings. Linguistic elements are also employed in the detection of bogus media. Text is extracted from photos and movies using the image processing approach. The overall number of literary characteristics examined by the researcher is 141. A model is made up of semantic characteristics, semantic features, vocabulary items, and cognitive and meta cognitive features. The researchers utilized a classification to assess the strength of characteristics. KNN, NB, RF, and SVM are the classifiers.

B. D. Horne et al. [16] presented that headlines of misleading information contain extra phrases. Using these characteristics, the SVM classification model was developed. The researchers measured their approach using data sets. The very first collection is a real-world news database culled from BuzzFeed as well as other media websites, while the latter is Burfoot and parody database of Baldwin. When matching real journalism to parody items, they obtained 91% overall accuracy. Yet, it has been discovered when bogus news is forecasted in contradiction of actual news, the reliability increases to 71%.

Ahmed et al. [17] presented a technique to identify bogus information which is dependent on n-gram characteristics. In this methodology, N- gram attributes and ML classification algorithms are applied for content analysis. They examined several controlled classification methods, including K-Nearest Neighbor (KNN), SVM, SGD, LSVM, LR, DT. The results illustrate that employing Term Frequency and Inverse Document Frequency as extracting features approach with the classification method LSVM yields best results.

Baly et al. [18] presented research that predicts the selection bias and truthiness of various media sources. The project makes use of a huge variety of content portals as well as a diverse set of choices. The study results demonstrate the importance of each attribute category.

Yang et al. [19] presented a new approach. Rather than utilizing trained algorithms, they employed untrained analysis to identify bogus information. Researchers took up a difficult topic of analyzing the identities of official networking sites for misleading information, like Facebook. They set the basis for using user data to identify bogus news. UFD (Unique factorization domain) uses a probability statistical approach to pattern the veracity of information and the trustworthiness of individuals. Researchers utilized objective reality labeling like BuzzFeed and PolitiFact, a well-known realization service, for the database. Journalist specialists assign these designations. The researcher's suggested method performed the best on the LIAR database, outperforming the second highest conducting pathways by 18.4%. Except for recall, UFD is becoming the top scoring method on the BuzzFeed dataset.

Ahmad et al. [20] conducted tests to evaluate the results of their training approach with various algorithms, and found that the efficiency of probabilistic models was far worse than the suggested method.

X. Zhou, et al. [21] presented a hypothesis approach for detecting false propaganda that focuses on article content to detect bogus tales until they are circulated on social networking sites. Through the execution of an interdisciplinary approach, the model employs a collection of subjective attributes to describe news stories, preserving both text structure and content throughout language levels. They ran trials using two databases including news stories from BuzzFeed and PolitiFact, correspondingly.

Gupta et al. [25] presented a model using Twitter, which acquired more than 90% right results i n identifying bogus photos, which pounded the US. They ran a morality narrative on over 10,000 photos on Twitter at the same time to explore the influence patterns of misleading images. They focused on NB and DT algorithms at this time. With these two ML techniques, they get a decent outcome, with a DT reliability of 97%.

Arkaitz Zubiaga et al., [26] presented sequential classifiers to categorize rumor attitudes on social networking sites. The researchers used LSTM, LCRF, and TCRF, on eight databases, each of which were related to important news. They demonstrate that consecutive classifiers that make use of the reciting possessions outperform non- sequential predictors in social media activity. LSTM also outperforms other serial classifications.

Table 1. Previous research work related to News Article and ML Models.

Source	Classifier	Description\Dataset	Accuracy
Agarwal et al. [12]	NB, CNN, and LSTM	Collected news articles from the World Wide Web and Kaggle.com	94%
Jain et al. [13]	NB, SVM, and NLP	Various news websites, RSS Feeds	93.50%
Abdullah et al. [14]	SVM, NB, LR, LSTM, and RNN	20,360 data have been collected from the Chile earthquake 2010 dataset	89.34%
Julio et al. [15]	KNN, SVM and NB	News articles from Buzzfeed	89%
Horne BD et al. [16]	SVM	Buzzfeed; Political News Data	78%
		Burfoot and Baldwin dataset	71%
Ahmed H, et al. [17]	SVM, SGD, LR	Kaggle Dataset	92%
Ramy Baly, et al. [18]	SVM	Articles	64.35%
Yang, et al. [19]	UFD	LIAR	75.9%
		Buzzfeed	67.9%
X. Zhou, et al. [21]	SVM, NB, RF	PolitiFact	89.2%
	LR, XGBoost	BuzzFeed	87.9%
A. Gupta & H. Lamba [25]	NB	Twitter feeds (total 1,782,526 tweets)	91%
Arkaitz Zubiaga & Elena Kochkina [26]	SVM	Supporting tweets on twitter	65.7%
	Linear CRF		60.3%
	Tree CRF		55.2%
A. Jain & A. Kasbe [28]	NB (on title)	From GitHub (11,000 articles)	80.6%
	NB (on text)		91.2%
	NB (on title with n-grams)		80.7%
	NB (on text with n-grams)		93.1%
Oluwaseu Ajao, & Shahrzad, Zargari [29]	LSTM	5800 tweets on five rumored stories	82%

(continued)

Table 1. *(continued)*

Source	Classifier	Description\Dataset	Accuracy
	LSTM- DROP		73%
	LSTM- CNN		80%
Wenlin Han & Varshil Mehta [30]	NB	Data in form of multimedia, Text, audio, hyperlink	67%
	RF		56%
	LSTM		82%
	LSTM-DROP		73%
	LSTM-CNN		80%
J. C.S. Reece, A. Correia, F. Murai, & A. Veloso [15]	KNN	2282 Buzzfeed news article	80%
	NB		72%
	RF		85%
	SVM		79%
	XGBoost		86%
C. Yuan, Q. Ma, W. Zhou & J. Han [31]	SMAN	Twitter 15	92.9%
	SMAN	Twitter16	93.5%
	SMAN	Weibo	95.6%
P. Bharadwaj, and Z. Shao. [32]	RF (Bigram) RNN	Kaggle Dataset	95.66%
	(GloVe)		92.70%
	NB		90.77%
B. Kwadwo Osei. [34]	LR		91.71%
	CART		79.87%
	PAC		92.89%
	LR		91.9%
			81.28%
M. Granik, M ykhailo, and V. Mesyura. [1]	NB	BuzzFeed News	75.4%
A. Farzana Islam, et al. [9]	NB (TF-IDF)	Kaggle Dataset	87.4%
	NB (count vector)		87.2%

Koteti et al. [27] presented a model which concentrated on enhancing the observation of fake news using attributes of data. To make the performance better, the writers used a unique data preparation approach to top up the lost value in the raw database. They applied data modeling to quantitative and organizational features with missing values. They choose the most common value in columns as tiers, and which are numerical for the column's overall mean.

Jain, A. et al. [28] presented a model which worked on identifying bogus news and proposed a way for adopting this technology on social media network Facebook. He used

NB to predict. The writers used a Github collection of eleven thousand articles organized by category. Aside from political news, the information also contains scientific and economic news. They employed the title and content as their primary implementation sources, as well as certain n-gram sources. He then examined the outcomes and determined that NB (10 n-grams) had an efficiency of 93.1%, as well as numerous techniques to improve this system.

Ajao et al. [29] presented a prototype that detects fake tweets of news from the social media platform Twitter postings by merging (Convolution Neural Network) and (Recurrent Neural Networks) algorithms. For the database, they accumulated 5800 tweets concentrating on 5 rumor plots: Charlie Hebdo, Sydney S, Crash of German wing, Shooting of Ottawa, and Shooting of Ferguson. The work presented on a CNN-RNN mix recognizes essential qualities associated with bogus news plots intuitively without any background knowledge of news and achieves better than 80% precision.

Mehta et al. [30] presented a review on evaluating the efficacy of misleading news detecting strategies. Researchers split the lies and misinformation database into two groups. The first one is this news model, the second one is the socio-cultural model, and both split information between visible (photo, clip) and language (word, headline) divisions. Researchers measured the effectiveness of machine learning techniques (NB, RF) with the most recent algorithms (LSTM DROP, LSTM-CNN). The goal of any such study would be to give a foundation for individuals to select one amongst two strategies. Researchers discovered that combined CNN-RNN model performs and produces superior outcomes.

Reis et al. [15] presented a model which looks for aspects inside the articles of news, blogs, and tales that help anticipate bogus news more accurately. They illustrated the importance of these new characteristics in recognizing fake news. Some of these qualities include biasing, dependability/trust, commitment, and patterns which are temporal. The writers used a database including 2082 BuzzFeed posts (articles of news). The writers used KNN, NB, RF, SVM, and Gradient Boosting algorithms to analyze and explain the merits and drawbacks of this technique, and determined that Gradient Boosting beats the others with the precision of 86%.

Yuan et al. [31] presented the structure of model which is a SMAN-based technique for detecting bogus news. This technique is conceptualized on the credibility of the two publications. They put this model through its paces on three different datasets (Twitter 15, Twitter 16, and Weibo) and observed that it was quite accurate.

Bhardwaj et al. [32] used RNN, NB classifier, and RF classifiers using six feature extraction algorithms on the databases from the Kaggle website which had false news. By merging bigram characteristics with the random forest classifier, this team produced high-quality results [32].

Osei et al. [34] presented the model by merging seven machine learning classifiers and presented a new approach for assessing bogus news. Islam et al. [1], presented a model that recognized false news using the NB classifier method. It used two characteristic removal algorithms to two datasets which have fake news obtained by the Kaggle community. Different Classifiers in Machine Learning have been explained in [35, 36] which can be used to extract valid and invalid information related to fake News Detection.

4 Methodology

The Process of Fake News detection has been implemented over Python by importing the python libraries that are required. The stop words available in English are collected and printing of these words is done. The number of missing values in the dataset are evaluated using appropriate algorithms. Null values are replaced with empty string and the author name and news titles are merged. Separation of data and label is done using data pre-processing. Stemming of the data is the process of reducing a word to its root word. Data and labels are separated using train test split. Textual data is converted to numerical data using TF/IDF vectorizer. Further training of the BERT classifier is done and afterwards accuracy is evaluated using test data. Figure 2 represents steps in BERT model to check the fake news dissemination.

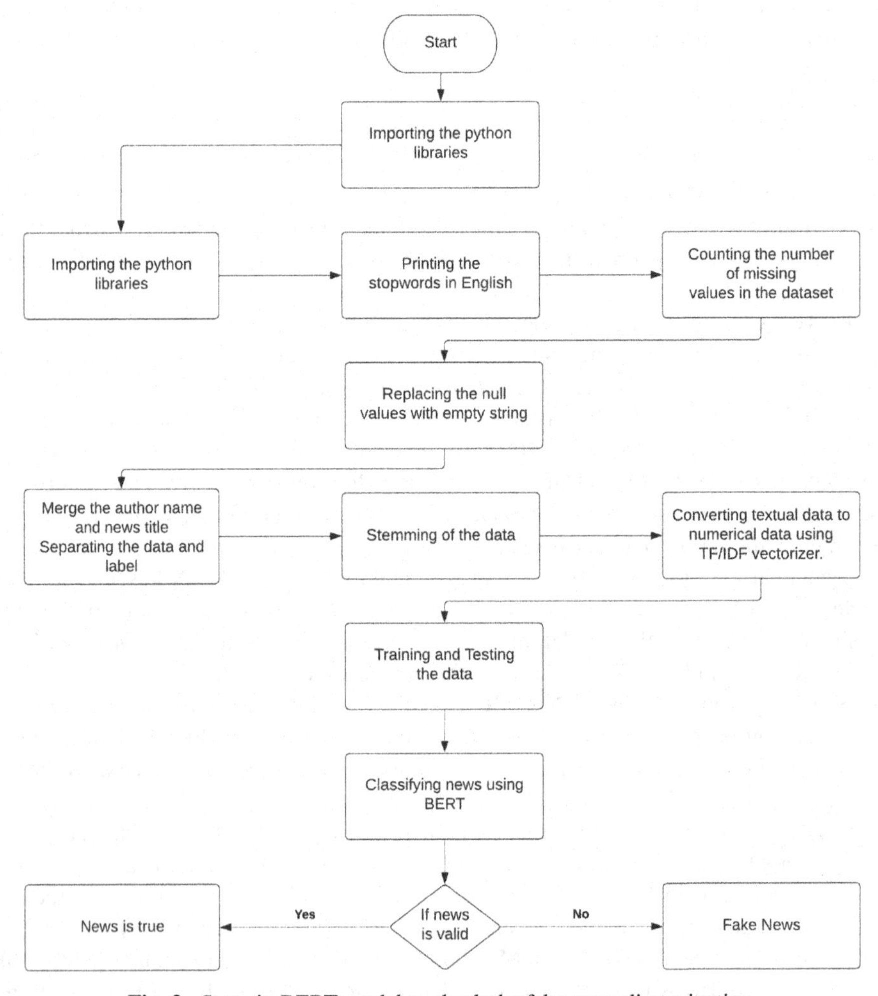

Fig. 2. Steps in BERT model to check the fake news dissemination.

5 Results and Discussion

In this study, the BERT deep learning technique has outperformed other models over this news article dataset which gives the best accuracy of 97.9%. Figure 3 represents comparison of accuracy for different models over a particular dataset. The model has been tested on a dataset of different types of news like political, geographical etc. with 21418 data articles. BERT is based on Deep Learning Technique based on AI Technology. BERT is a DL model where each result piece is connected to each intake piece of data and the relative importance with them are calculated periodically based on their connection. Earlier, models could only analyze text input consecutively, whether Left to right or Right to left, but cannot be together. BERT is the only device that can scan across both directions simultaneously. The BERT-based model has 1.2 billion characteristics and 12 levels (transformer units). A pre-trained BERT-based model performs wonderfully on small datasets. Their superior performance on datasets like Fake or Real News, which outperforms other models, illustrates this. BERT Frameworks beat other algorithms not just on large datasets, but also on tiny database samples. Even with a limited sample of just 500 data points, the BERT model achieves remarkable accuracy (about 90%). Therefore, when a wide variety of labelled data is not practicable, such techniques could be used to recognize bogus news in several languages.

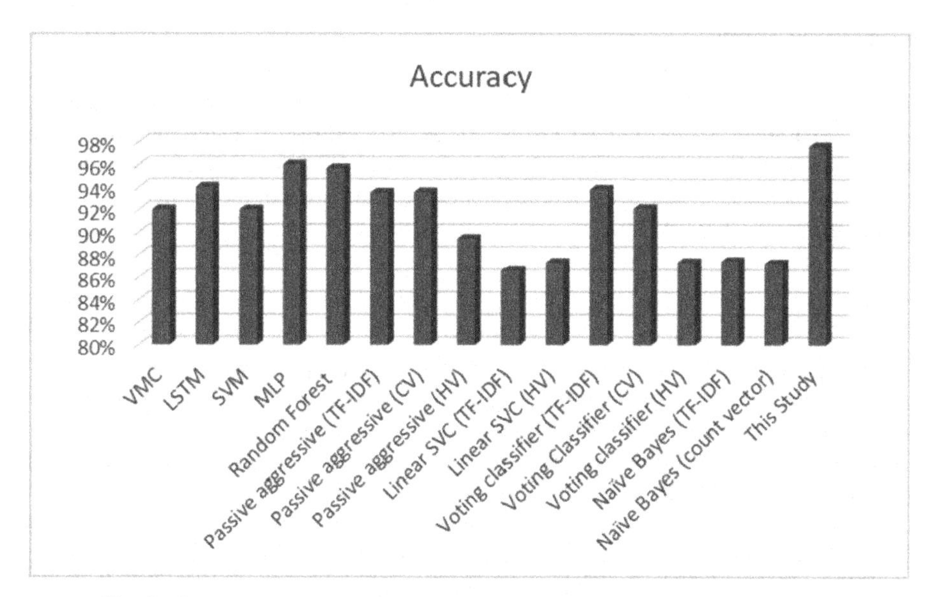

Fig. 3. Comparison of Accuracies for different models over a particular dataset

This research compares past work done in this area to our own research. In this research work, various classifying algorithms like VMC, SVM, LSTM, MLP, Random Forest Classifier, Naïve bayes, Voting Classifier, etc. are applied over Kaggle dataset to classify news and compared the results of all these classifiers against BERT Classifier. The results of this comparison are presented in the form of an index in this paper. BERT classifier provided maximum accuracy of 97.9%.

6 Conclusion

Previously, data was distributed through conventional media sources including newspaper, broadcast, and television. Every one of these media networks were subject to scrutiny from upper management, and this issue of false news, which is becoming more prevalent by the day, complicates matters and has the potential to affect people's opinions and attitudes regarding the use of electronics. This research analyzed few published articles by diverse researchers that used a variety of strategies for fake news identification in this systematic review. Several databases in conjunction with all these methodologies on different data sets have been examined and extensive examination of these published information was doubly checked for validity. When more people utilize these internet services, the problem of fake news grows, complicating matters and perhaps leading to the suppression of Social Conflict. This research shows that the BERT approach is the best for fake media classification, with an accuracy of 97.9% when evaluated on a dataset of diverse types of news stories. On contrary to other approaches, this accuracy is substantially greater. As a result, this strategy is suggested for detecting bogus news. Furthermore, for the benefit of the society, appropriate optimal model can be used which can help predict fake news at appropriate time.

7 Future Work

In future work, the accuracy of the BERT classifier will be improved by fine tuning it. Different parameters of BERT Classifier can be changed over transfer learning which in turn can improve the accuracy of the model. Also, BERT model can be executed on some other datasets which are available publicly for validation of the results obtained in the current research work.

References

1. Granik, M., Mesyura, V.: Fake news detection using naive Bayes classifier. In: 2017 IEEE First Ukraine Conference on Electrical and Computer Engineering (UKRCON) (2017)
2. Helmstetter, S., Paulheim, H.: Weakly supervised learning for fake news detection on Twitter. In: 2018 IEEE/ACM International Conference on Advances in Social Networks Analysis and Mining (ASONAM) (2018)
3. Agudelo, G.E.R., Parra, O.J.S., Velandia, J.B.: Raising a model for fake news detection using machine learning in Python. In: Al-Sharhan, S.A., Simintiras, A.C., et al. (eds.) I3E 2018. LNCS, vol. 11195, pp. 596–604. Springer, Cham (2018). https://doi.org/10.1007/978-3-030-02131-3_52
4. Ahmad, I., Yousaf, M., Yousaf, S., Ahmad, M.O.: Fake news detection using machine learning ensemble methods. Complexity **2020**, 1–11 (2020). https://doi.org/10.1155/2020/8885861
5. Wakefield, K.: A guide to the types of machine learning algorithms and their applications (2021). https://www.sas.com/en_gb/insights/articles/analytics/machine-learning-algorithms
6. Meesad, P.: Thai fake news detection based on information retrieval, natural language processing and machine learning. SN Comput. Sci. **2**(6), 425 (2021)
7. Cristianini, N., Shawe-Taylor, J.: An Introduction to Support Vector Machines and Other Kernel-Based Learning Methods. Cambridge University Press (2000)

8. Hofmann, T., Schölkopf, B., Smola, A.J.: Kernel methods in machine learning. Ann. Stat. **36**, 1171–1220 (2008)
9. Jadhav, S.S., Thepade, S.D.: Fake news identification and classification using DSSM and improved recurrent neural network classifier. Appl. Artif. Intell. **33**, 1058–1068 (2019)
10. Gregorutti, B., Michel, B., Saint-Pierre, P.: Correlation and variable importance in random forests. Stat. Comput. **27**, 659–678 (2016)
11. Ahmed, H., Traore, I., Saad, S.: Detection of online fake news using N-gram analysis and machine learning techniques. In: Traore, I., Woungang, I., Awad, A. (eds.) ISDDC 2017. Lecture Notes in Computer Science, vol. 10618, pp. 127–138. Springer, Cham (2017). https://doi.org/10.1007/978-3-319-69155-8_9
12. Agarwal, A., Dixit, A.: Fake news detection: an ensemble learning approach. In: 2020 4th International Conference on Intelligent Computing and Control Systems (ICICCS) (2020)
13. Jain, A., Shakya, A., Khatter, H., Gupta, A.K.: A smart system for fake news detection using machine learning. In: 2019 International Conference on Issues and Challenges in Intelligent Computing Techniques (ICICT) (2019)
14. Abdullah-All-Tanvir, Mahir, E.M., Akhter, S., Huq, M.R.: Detecting fake news using machine learning and deep learning algorithms. In: 2019 7th International Conference on Smart Computing & Communications (ICSCC) (2019)
15. Reis, J.C., Correia, A., Murai, F., Veloso, A., Benevenuto, F.: Supervised learning for fake news detection. IEEE Intell. Syst. **34**, 76–81 (2019)
16. Horne, B., Adali, S.: This just in: fake news packs a lot in title, uses simpler, repetitive content in text body, more similar to satire than real news. In: Proceedings of the International AAAI Conference on Web and Social Media, vol. 11, pp. 759–766 (2017)
17. Ahmed, H., Traore, I., Saad, S.: Detecting opinion spams and fake news using text classification. Secur. Priv. **1**, 1–15 (2017)
18. Baly, R., Karadzhov, G., Alexandrov, D., Glass, J., Nakov, P.: Predicting factuality of reporting and bias of news media sources. In: Proceedings of the 2018 Conference on Empirical Methods in Natural Language Processing (2018)
19. Yang, S., Shu, K., Wang, S., Gu, R., Wu, F., Liu, H.: Unsupervised fake news detection on social media: a generative approach. In: Proceedings of the AAAI Conference on Artificial Intelligence, vol. 33, pp. 5644–5651 (2019)
20. Baarir, N. F., Djeffal, A.: Fake news detection using machine learning. In: 2020 2nd International Workshop on Human-Centric Smart Environments for Health and Well-being (IHSH), vol 1, pp. 125–130. IEEE (2021)
21. Zhou, X., Jain, A., Phoha, V.V., Zafarani, R.: Fake news early detection. Digit. Threats Res. Pract. **1**, 1–25 (2020)
22. Sudhakar, M., Kaliyamurthie, K.P.: A machine learning framework for automatic fake news detection in Indian News 1(1), 1–14 (2022)
23. Castillo, C., Mendoza, M., Poblete, B.: Information credibility on Twitter. In: Proceedings of the 20th International Conference on World Wide Web, pp. 675–684 (2011)
24. Shu, K., Wang, S., Liu, H.: Beyond news contents. In: Proceedings of the Twelfth ACM International Conference on Web Search and Data Mining (2019)
25. Gupta, A., Lamba, H., Kumaraguru, P., Joshi, A.: Faking sandy. In: Proceedings of the 22nd International Conference on World Wide Web (2013)
26. Zubiaga, A., et al.: Discourse-aware rumour stance classification in social media using sequential classifiers. Inf. Process. Manage. **54**, 273–290 (2018)
27. Kotteti, C. M. M., Dong, X., Li, N., Qian, L.: Fake news detection enhancement with data imputation. In: 2018 IEEE 16th International Conference on Dependable, Autonomic and Secure Computing, pp. 187–192. IEEE (2018)
28. Jain, A., Kasbe, A.: Fake news detection. In: 2018 IEEE International Students' Conference on Electrical, Electronics and Computer Science (SCEECS) (2018)

29. Ajao, O., Bhowmik, D., Zargari, S.: Fake news identification on Twitter with hybrid CNN and RNN models. In: Proceedings of the 9th International Conference on Social Media and Society (2018)
30. Han, W., Mehta, V.: Fake news detection in social networks using machine learning and Deep Learning: performance evaluation. In: 2019 IEEE International Conference on Industrial Internet (ICII) (2019)
31. Yuan, C., Ma, Q., Zhou, W., Han, J., Hu, S.: Early detection of fake news by utilizing the credibility of news, publishers, and users based on weakly supervised learning. In: Proceedings of the 28th International Conference on Computational Linguistics (2020)
32. Bharadwaj, P., Shao, Z.: Fake news detection with semantic features and text mining. Int. J. Nat. Lang. Comput. **8**, 17–22 (2019)
33. Kaur, S., Kumar, P., Kumaraguru, P.: Automating fake news detection system using multi-level voting model. Soft. Comput. **24**, 9049–9069 (2019)
34. Bonsu, K.O.: Weighted accuracy algorithmic approach in counteracting fake news and disinformation. Econ. Reg. Stud. **14**, 99–107 (2021)
35. Gupta, S., Chhabra, A., Agrawal, S., Singh, S.K.: A comprehensive comparative study of machine learning classifiers for Spam Filtering. In: Nedjah, N., Pérez, G.M., Gupta, B.B. (eds.) International Conference on Cyber Security, Privacy and Networking (ICSPN 2022), pp. 257–268. Springer, Cham (2023). https://doi.org/10.1007/978-3-031-22018-0_24
36. Verma, R., Chhabra, A., Gupta, A.: A statistical analysis of tweets on covid-19 vaccine hesitancy utilizing opinion mining: an Indian perspective. Soc. Netw. Anal. Min.Netw. Anal. Min. **13**(1), 12 (2022)

Emoji Based Sentiment Classification Using Machine Learning Approach

Parul Verma[1] (✉) and Roopam Srivastava[2]

[1] Amity Institute of Information Technology, Amity University, Lucknow, UP, India
pverma1@lko.amity.edu

[2] Mahatma Gandhi Post Graduate College, Gorakhpur, UP, India

Abstract. Online media platforms like Facebook, Twitter, and Instagram continue to influence our world. People today are more closely connected than ever before, and they exhibit such sophisticated personas. Ongoing studies have revealed a link between excessive social media use and depression. This research focuses on the development of features using Term frequency – Inverse Document Frequency (TF-IDF) and Bag-of-Word (Bow). The work will also put light on the generation of models utilizing a machine-learning technique. The dataset for this research using Twitter API. Only the English context was kept from the Tweets after filtration. It focuses on categorizing users' mental health at the tweet level. About 20000 reviews make up this dataset. Using these reviews, emoji sentiment classification has been developed. Data is cleaned using a pre-processing approach before BoW and TF-IDF were used to extract features. Following to it, classifier deployment, training and assessment were carried out. Metrics for evaluation are used to gauge classifier accuracy. MultinomialNB (MNB) fared best in the field of Bag-of-words features among the three classifiers used to evaluate the accuracy, but Random Forest (RF) outperformed TF-IDF. In Bag-of-Words, we are able to classify data with an accuracy of 86% using TF-IDF Random Forest and 89% using multinomial NB and BoW.

Keywords: TF-IDF · Supervised · Bag of Word · Multinomial NB

1 Introduction

Due to the widespread use of social media in the twenty-first century, it is safe to say that practically everyone on earth is linked to at least one platform. In particular, Facebook and Instagram have developed into "Status Symbols." As a result, society frequently views those without Social Media profiles as outcasts. We have forgotten that conversations on social media will never be a substitute for interactions with people in person. Your hormones are triggered by in-person interaction, which reduces stress and leaves you feeling happier, healthier, and more optimistic every day. Constant sadness and feeling down are symptoms of depression. Social media is not entirely bad. We didn't say it. However, too much of anything is harmful. Here are some signs that social media is having a negative impact on our lives rather than a positive one. The effect on your relationship and work, unable to enjoy real-world happenings, mental addiction to digital

R. K. Challa et al. (Eds.): ICAIoT 2023, CCIS 1930, pp. 185–195, 2024.
https://doi.org/10.1007/978-3-031-48781-1_15

personas. There are several symptoms it could cause, such as losing interest in activities that used to make you happy, being easily annoyed or frustrated, and overeating or undereating. Oversleeping or under-sleeping. Having trouble focusing or remembering things. Suffering from bodily aches or discomforts, such as a headache, weariness, stomachache, or sexual dysfunction, and considering harming or dying oneself. These factors cause depression. In this investigation, it is determined whether person is depressed or not. The NLTK library is used for data preprocessing. Python language is used to implement Machine Learning (ML) models. Various ML models are implemented to increase the accuracy of the model prediction.

2 Related Work

Our daily lives and activities now include social media. We all spend a lot of time on Facebook, WhatsApp, Instagram, and other social media platforms. Although there are numerous benefits to social media, excessive use can also be harmful to mental health. Additionally, it can exacerbate feelings of loneliness, despair, and worry. Researchers observed that Twitter data may be a useful tool for assessing and forecasting severe depressive illness in people. First, a dataset of public self-professed cases of depression was assembled in collaboration with Johns Hopkins University. After that, they suggested using a Bag of Words method to quantify the data, and they produced an 846,496-dimensional feature space that served as an input vector. Finally, researchers used these distinctive characteristics to create, contrast, and evaluate several statistical classifiers that may be able to predict a person's chance of developing depression. They sought to develop a technique for detecting depression by examining extensive archives of user language history on social media and produced results that were both promising and accurate [1]. They predict when depression and post-traumatic stress disorder would start to manifest in Twitter users. Researchers used computer models. From 204 people, Twitter information and specifics on depression history were gathered, they extracted predictive characteristics assessing effect, linguistic style, and context, then used these features to build models utilizing supervised learning methods. The generated models, however, in a different demographic, compared favorably with general practitioners' average success rates in detecting depression and accurately differentiated between depressing and healthy material [2]. This study made use of actual campus Internet data that was continually, discretely, and privately gathered. The CES-D scale was used to assess 216 students for depressive symptoms. Then, using Cisco Net Flow logs, they gathered information on their on-campus Internet usage. Following an investigation, it was discovered that several aspects of Internet usage, including average packets per flow, peer-to-peer chat octets, mail (packets and duration), and depressive symptoms statistically significantly correlate with FTP duration and distant file octets [3]. By using a statistical model to determine which individuals may go from mental health discourse to suicidal thoughts, this work fills in gaps in the literature. They estimate the likelihood of these transitions using covert data sources from Reddit's semi-anonymous support forums. For this reason, they developed statistical methods based on propensity score matching as well as linguistic and interactional metrics. Their methodology enables them to identify specific indicators of changes in suicidal thoughts and to address the social and ethical consequences of this study [4]. This study's goals include determining the relationship between years, frequency of use, and hours per day of the Internet and psychopathological symptoms in an adult Portuguese population as well as how lonely they perceive themselves to be. The findings imply that unhealthy Internet usage habits that are common among young people are also present among adults. There was also evidence of a connection between the amount of time spent online and psychopathological symptoms, as well as a link between loneliness and online usage. Effect of Internet use people's well-being must be recognized since they should be seen as a public health concern in our increasingly digitally disconnected offline environment. The target audience of this study is what gives it its originality. Portuguese Internet users over 18 for whom there isn't a dedicated study, highlighting the problem's interdisciplinary character [5]. The fact that those people who had enough Facebook data limited to only six months before the first formal diagnosis of depression gave a greater forecast accuracy and highlights the need to broaden the use of the present screening processes during this study [6]. Six focus groups with 54 teenagers, aged 11 to 18, who were recruited from schools in Leicester and London over the course of three months were placed as a result of this article's contribution (UK). According to three themes that emerged from a thematic analysis, teenagers' mental health was seen as being at risk due to social media because it was thought to contribute to certain of those youngsters' anxiety and mood disorders. It was seen as a venue for cyberbullying, and social media use itself was frequently described as an addiction [7]. Adolescents who use social media may be more susceptible to mental health issues. A small number of longitudinal research has examined this link, but none have quantified the percentage of teenage mental health issues owing to social media use [8]. This review gathered data on the effects of social media use on teenage depression, anxiety, and psychological discomfort. 13 appropriate studies of which 12 were cross-sectional were found after a search of the PsycINFO, Medline, Embase, CINAHL, and SSCI databases. The results were divided into four categories related to social media: time spent, activity, investment, and addiction. Depression, anxiety, and psychological discomfort were associated across all dimensions [9]. The results of this study imply that potential social media causative variables should be taken into account when working with individuals who have been given an anxiety or depression diagnosis. Additionally, if other links with another construct were explored using the study's findings, this may strengthen the conclusions and help lower rates of anxiety, despair, and suicide [10]. They talked about how social media can isolate and test communication theories in a way that helps us understand how people connect with computers. Overall, this essay offers a common paradigm to support and enable communication scholarship going forward [11]. The purpose of the study was to evaluate the efficiency of official social media communication plans during the COVID-19 epidemic. Egypt was the primary case study country for the methodology. The results of the survey demonstrated that while the Egyptian government performed a good job of informing the populace, transparency was absent and more effort was needed to promote ethical behavior on the part of residents [12]. The study points out that the majority of research so far has been correlational, has concentrated on adults rather than teenagers, and has produced a variety of occasionally contradicting weak positive, weak negative, and null relationships. The most recent and thorough large-scale preregistered studies report weak correlations between the amount of daily digital technology use and adolescents' well-being that do not provide a means of separating cause from effect and, as estimated, are unlikely to be of clinical or practical significance [13]. The purpose of this systematic review was to locate and compile studies that looked at depression and anxiety about social networking site (SNSs). In addition, it looked for studies that assess well-being in addition to mental illness and examined moderators and mediators that contribute to the environment's complexity [14]. Despite the importance of mobile technology and social media (MTSM) for teenagers, it is still unknown how these platforms relate to depressive illnesses in this group. While some studies have linked the use of MTSM to an increased risk of depression, other studies have suggested that they may also have a preventative impact [15]. Over the past two decades, there have been rises in adolescent depression and suicidal behavior that correspond with the introduction of social media (SM) (platforms that allow contact via digital media), which are heavily used among teenagers. The relationships between teen depression and suicidality and the usage of SM, particularly social networking sites, were examined in this scoping review. Through thematic analysis of the papers under consideration, four key topics in SM usage emerged: amount and quality of SM use, social networking aspects, as well as the admission of mental health problems [16]. This paper stated that current research has ignored the reverse causation theory, which holds that social media use is driven by depression. The dynamics governing depressed symptoms and unhelpful social media use can be better understood by highlighting the need for longitudinal and experimental approaches to determine directionality [17]. By showcasing the research that has been done in this area over the past 20 years, they were able to throw light on the field's research patterns. This study outlines a methodical division of labor among researchers in the field as well as the problems and difficulties associated with emoji-based sentiment analysis [18]. Emoticons are expressive; therefore, we may give them more nuanced emotional characteristics like anger, happiness, or sadness as well as some superficial semantics like events, places, or things to look at [19]. This work looked at the value of sentiment analysis in predicting stock prices using stock market indices like the Sensex and Nifty [20].

3 Methodology

The methodology is explained in this section. The flow of work shown in Fig. 1.

We used the "Mental-Health-Twitter" dataset in this work. Twitter API was used to gather the information, which is in an unclean format. Only the English context was kept from the Tweets after filtering. It focuses on the classification of the user's mental health at the level of the individual tweet and has 20k reviews gathered from Mental-Health-Twitter, which was acquired from kaggle.com. It features a lot of columns. For reviews, the "post-text" column is used, and for features, the "label" column. In our analysis of the dataset, only positive and negative evaluations were included. Comma-separated value (.csv) file format. For implementing the system, the NLTK toolkit is used for text processing and the Sickit-Learn machine learning framework in Python. We perform text preprocessing, emoji sentiment analysis, feature generation with a supervised model, and classifying with the classification procedure.

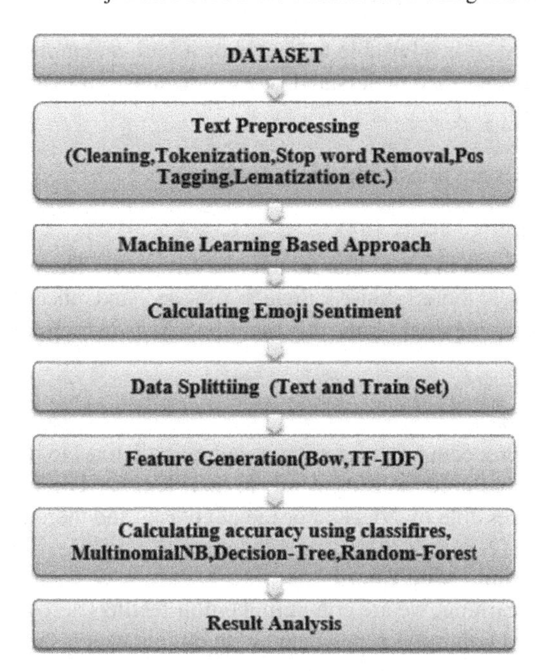

Fig. 1. The flow of the system work

3.1 Text Pre-processing

In the process of feature engineering and modeling, it is one of the first steps. The data is cleaned during the pre-process, and the text is standardized into a standard form, which includes phrases and words. This makes it possible to standardize document corpora, which helps with the development of important features and noise reduction brought on by undesirable items. The NLTK toolkit was used to preprocess the data. The step of text preparation, detailed as follows-

Text-Cleaning

HTML tags and other unnecessary information frequently show up in the content, preventing sentiment analysis from being effective. Consequently, before extracting features, it is essential to remove them.

Lowering Case

Whenever a phrase has the same case as letters, a computer easily understands the words since it views lower case and upper case differently. Hence as the text should be made the same case, with lowercase being the ideal choice, to prevent issues like this.

Special Characters and Digits Removal

The words 'hurray' and "beautiful!" as well as square10 can be handled by this text preparation technique. It is advisable to get rid of it with an empty string because this kind of word is challenging to understand. We use regular expressions for this.

Word Tokenization
During this, sentences are transformed into words.

Remove Stopword
The most frequent words in a text that do not add any value. There are several phrases in it, including "they," "there," "this," "there," "a," "an," and "the". While maintaining the most crucial and contextual terms, this function assists in removing stopwords from a corpus.

Text Lemmatization
It follows the same procedure as stemming in removing affixes to reveal a word's basic form. It is a technique for condensing words to their lemma. The WordNetLemmatizer program is made accessible by NLTK. For the next tasks and there is a clear review.

Sentiment Calculation Using Emoji
During supervised learning, we are computing emoji feelings across the dataset's post-text column. Our text column is represented with emojis which shows the positive and negative sentiment as seen in Table 1.

Table 1. Emoji Sentiment Expression

Text	Sentiment
Can't be bothered to cook, take away on the way 😁👍	positive
Had a headache so took a nap. For three hours! Woke up a bit confused as to the time and day...😴	negative

3.2 Using TF-IDF and a Bag-of-Word Feature Engineering for Supervised Machine Learning Models

The act of turning unstructured data into attributes aids prediction models understanding the environment better, improving the accuracy of previously unidentified data. Feature engineering is another name for this. Feature selection methods aim to minimize the dimensionality of the dataset by removing characteristics that aren't necessary for classification [21]. By counting how many times each word appears, the bag-of-words (BOW) model converts any text into fixed-length vectors. Vectorization is a common term used to describe this procedure. Although BOW is a very simple model, it is frequently employed for Natural Language Processing (NLP) applications including text classification. Its simplicity is its strength since it is cheap to compute and often simpler is preferable when placement or contextual information isn't important.

The Term Frequency-Inverse Document Frequency Measure
By looking at how frequently a specific phrase is used with the document, word frequency is calculated. Numerous frequency definitions or measurements exist-

- Documents containing a particular term frequently.
- The count of occurrences in a document is adjusted for its length and term frequency.
- logarithmically scaled frequency, such as log (1 + raw count).

A word's usage within the corpus is examined using inverse document frequency (IDF) is calculated as in Eq. (1), where N indicates the number of documents (d) in the corpus (D) and t is the term (word) whose frequency we need to estimate. The denominator is the number of texts that contain the letter "t".

$$\text{idf}(t, d) = \log (N / \text{count} (d \text{ € } D: \text{€}d) \tag{1}$$

A term's relevance is inversely correlated with its frequency across documents, according to the fundamental concept behind TF-IDF. IDF provides details regarding a term's relative scarcity among the documents in the collection, whereas TF provides information about how frequently a term appears in a document. Our final value may be obtained by multiplying these numbers as shown in Eq. (2) collectively.

$$\text{tfidf}\ (t, d, D) = \text{tf}\ (t, d) \times \text{idf}(t, D) \tag{2}$$

3.3 A Classifier for Classifying

After preprocessing and feature selection, the usage of classification techniques comes next. Numerous text classifiers have been proposed in the literature [22]. Machine learning techniques such as Multinomial NB, Decision Tree, and Random Forest are used.

Multinomial NB. It is a probabilistic learning technique mostly employed in natural language processing (NLP). The method anticipates the tag of a text and is based on the Bayes theorem. The tag with the highest probability of occurring is output by an algorithm after it has calculated the chance of each label in a collection. With situations involving numerous classes and text data processing, it is a potent method. Understanding the Bayes theorem principle, on which the Naive Bayes theorem is founded, is essential to comprehend how the latter works. Based on previously known event-related factors, Bayes theorem determines the probability that something will happen.

Random Forest. It is a widely used technique that makes use of supervised learning. Classification and regression-related ML issues can be resolved with it. Its foundation is the idea of ensemble learning, a method for merging several classifiers to solve complex issues and enhance model performance. A set of different tree predictors, each of which is dependent on the results of a distinct sample taken at random from a vector with the same distribution as all the other trees in the forest. The strength of each tree in the forest and the correlation between them determine the generalization error of a forest of tree classifiers.

Decision Tree. It is a supervised learning that may be applied to classification and regression problems, but is typically chosen to deal with classification difficulties. A dataset's nodes represent its properties, branches represent the decision-making process, and each leaf node represents the categorization outcome in these tree-structured classifiers. The Leaf Node and the Decision Node are the two nodes in this. Unlike Leaf nodes, which are the results of decisions and do not have any additional branches, these nodes are used to make decisions. A tree is constructed using the Classification and Regression Tree Algorithm, or CART algorithm.

4 Result and Discussion

Using the above-mentioned data set, out of ten columns, we worked over two columns from the dataset that is 'post_text' and 'label'. An imbalanced dataset is used with two classes ('positive' and 'negative') and here positive label shows depression and vice-versa. This is shown in Fig. 2. Seventeen thousand sample reviews are taken from the dataset for a model. Reviews are preprocessed using the NLTK tool. Tweets are used for the calculation of emoji sentiment. Training and testing sections of the dataset are separated, with test data making up 25% of the total for the supervised learning model. For classification, a variety of machine learning techniques are utilized. To create features, the 'label' column from the target dataset is used. For accurate computation, a model is created utilizing classifiers such as multinomial NB, decision tree, and, random forest. Feature extraction was carried out using

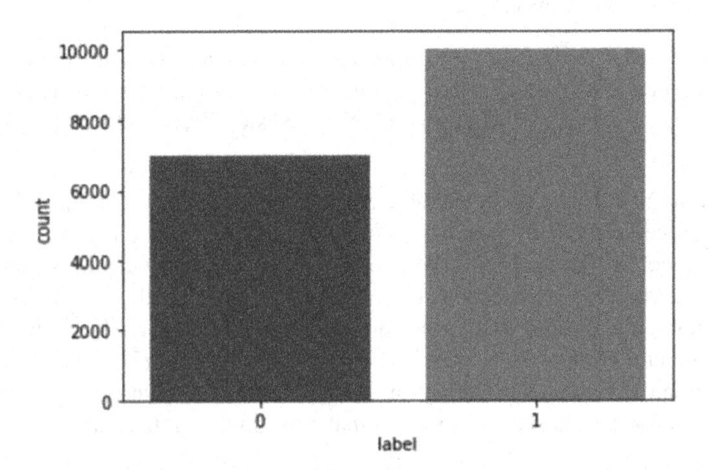

Fig. 2. Data Visualization

BoW and TF-IDF. A fit function is applied to the train set to fit the model, and a prediction function is applied to the test set. Objects were created for these functions.

For accuracy calculation, MNB, DTR, and RF are applied as classifiers via a bag of word features in the Table 2(a). TFIDF characteristics are employed by these classifiers for accuracy, as shown in Table 2(b). As can be shown by comparing the results from Tables 2(a) and 2(b), the Multinomial NB model outperforms the Random Forest classifier utilizing TF-IDF features for a BoW characteristic. Therefore, it is found that the model performs best in supervised learning when employing a bag of word characteristics.

Table 2(a). Comparative Result Table of Supervised learning approach using Bag of Word

Bag of Word (Bow)				
MultinomialNB				
Class	Precision	Recall	F1- measure	Accuracy
0	0.83	0.93	0.87	**0.89**
1	0.94	0.87	0.91	
Decision Tree				
0	0.82	0.79	0.81	0.84
1	0.86	0.88	0.87	
Random Forest				
0	0.87	0.75	0.81	0.85
1	0.84	0.92	0.88	

Table 2(b). Comparative Result Table of Supervised learning approach using TF-IDF

TFIDF				
MultinomialNB				
Class	Precision	Recall	F1- measure	Accuracy
0	0.99	0.59	0.74	0.83
1	0.78	0.99	0.87	
Decision Tree				
0	0.82	0.79	0.81	0.84
1	0.86	0.88	0.87	
Random Forest				
0	0.89	0.76	0.82	**0.86**
1	0.85	0.94	0.89	

5 Conclusion

In this research, a classification model is developed utilizing a supervised approach. A key objective of text mining is text categorization. Using tweets, emoji sentiment is produced. Operations related to general text analytics, such as tokenization, normalization, stop word removal, etc., were carried out. For feature creation in this study, Bag-of-Words and TF-IDF are employed. N-gram words are employed in this study, which uses the bag of words, one of the simplest methods for extracting characteristics from the text. Our approach achieves the best outcome, with about 89% accuracy. There are a great number of additional potential causes of anxiety and depression that need to be looked into. These findings are significant because they will facilitate future studies on social media and mental health.

References

1. Nadeem, M., Horn, M., Coppersmith, G.: Identifying depression on Twitter (2016). arXiv arXiv:1607.0738
2. Reece, A.G., Reagan, A.J., Lix, K.L.M., Dodds, P.S., Danforth, C.M., Langer, E.J.: Forecasting the onset and course of mental illness with Twitter data. Sci. Rep. **7**(1), 13006 (2017). https://doi.org/10.1038/s41598-017-12961-9
3. Katikalapudi, R., Chellappan, S., Montgomery, F., Wunsch, D., Lutzen, K.: Associating internet usage with depressive behavior among college students. IEEE Technol. Soc. Mag. **31**(4), 73–80 (2012). https://doi.org/10.1109/MTS.2012.2225462
4. Choudhury, D.M., Kiciman, E., Dredze, M., Coppersmith, G., Kumar. M.: Discovering shifts to suicidal ideation from mental health content in social media. In: Proceedings of the SIGCHI, pp. 2098–2110 (2016). https://doi.org/10.1145/2858036.2858207.PMID:29082385
5. Leite, A., Ramires, A., Amorim, S., E Sousa, H.F.P., Vidal, D.G., Dinis, M.A.P.: Psychopathological symptoms and loneliness in adult internet users: a contemporary public health concern. Int. J. Environ. Res. Pub. Health **17**, 856 (2020). https://doi.org/10.3390/ijerph17030856
6. Eichstaedt, J.C., et al.: Facebook language predicts depression in medical records. Proc. Nat. Acad. Sci. **115**, 11203–11208 (2018). https://doi.org/10.1073/pnas.1802331115
7. O'Reilly, M., Dogra, N., Whiteman, N., Hughes, J., Eruyar, S., Reilly, P.: Is social media bad for mental health and wellbeing? Exploring the perspectives of adolescents. Clin. Child Psychol. Psychiatry **23**(4), 601–613 (2018). https://doi.org/10.1177/1359104518775154
8. Riehm, K.E., et al.: Associations between time spent using social media and internalizing and externalizing problems among US youth. JAMA Psychiatry **76**, 1266–1273 (2019). https://doi.org/10.1001/jamapsychiatry.2019
9. Keles, B., McCrae, N., Grealish A.: A systematic review: the influence of social media on depression, anxiety and psychological distress in adolescents. Int. J. Adolesc. Youth **25**, 79–93 (2020). https://doi.org/10.1080/02673843.2019.1590851
10. Karim, F., Oyewande, A.A., Abdalla, L.F., Chaudhry, E.R., Khan, S.: Social media use and its connection to mental health: a systematic review. Cureus **12**, e8627 (2020). https://doi.org/10.7759/cureus.8627
11. Carr, C.T., Hayes, R.A.: Social media: defining, developing, and divining. Atlantic J. Commun. **23**, 46–65 (2015)
12. El Baradei, L., Kadry, M., Ahmed, G.: Governmental social media communication strategies during the COVID-19 pandemic: the case of Egypt. Int. J. Pub. Adm. **44**, 907–919 (2021). https://doi.org/10.1080/01900692.2021.1915729

13. Odgers, C.L., Jensen, M.R.: Annual research review: adolescent mental health in the digital age: facts, fears, and future directions. J. Child Psychol. Psychiatry **61**(3), 336–348 (2020). https://doi.org/10.1111/jcpp.13190
14. Seabrook, E.M., Kern, M.L., Rickard, N.S.: Social Networking sites, depression, and anxiety: a systematic review. JMIR Ment. Health **3**, e50 (2016). https://doi.org/10.2196/mental.5842
15. Arias-de la Torre, J., et al.: Relationship between depression and the use of mobile technologies and social media among adolescents: umbrella review. J. Med. Internet Res. **22**, e16388 (2020). https://doi.org/10.2196/16388
16. Vidal, C., Lhaksampa, T., Miller, L., Platt, R.: Social media use and depression in adolescents: a scoping review. Int. Rev. Psychiatry **32**, 235–253 (2020). https://doi.org/10.1080/09540261
17. Hartanto, A., Quek, F., Tng, G., Yong, J.C.: Does social media use increase depressive symptoms? A reverse causation perspective. Front. Psychiatry **12**, 641934 (2021). https://doi.org/10.3389/fps.2021.641934
18. Vashist, G., Jalia, M.: Emoticons & emojis based sentiment analysis: the last two decades! Int. J. Sci. Technol. Res. (IJSTR) **9**(03), 366–371 (2020). ISSN 2277-8616
19. Kralj Novak, P., Smailović, J., Sluban, B., Mozetič, I.: Sentiment of emojis. PLoS ONE (2015). https://doi.org/10.1371/journal.pone.0144296
20. Bhardwaj, A., Narayan, Y., Vanraj, P., Dutta, M.: Sentiment analysis for Indian stock market prediction using Sensex and NIFTY. Procedia Comput. Sci. **70**, 85–91 (2015)
21. Forman, G.: An experimental study of feature selection metrics for text categorization. J. Mach. Learn. Res. **3**, 1289–1305 (2003)
22. Breiman, L.: Classification and Regression Trees. Routledge (2017)

Comparative Analysis of Structural Characteristics of Social Networks and Their Relevance in Community Detection

Shano Solanki[1](✉), Mukesh Kumar[2], and Rakesh Kumar[3]

[1] Department of Computer Science and Engineering, NITTTR, Chandigarh, India
shano@nitttrchd.ac.in

[2] Department of Computer Science and Engineering, UIET, Panjab University, Chandigarh, India

[3] Department of Computer Science and Engineering, Central University of Haryana, Mahendergarh, India

Abstract. Social Network Analysis based on structural properties is relevant to many application domains including influence maximization, product recommendation, and viral marketing to name a few. The study of structural properties has a vital role in effective community detection and researchers have suggested several algorithms in this direction. The research paper not only gives details of various structural characteristics and their relevance for community detection but it also covers community visualization tools and their useful features. Along with the mapping of structural properties and community detection, efforts have also been made towards understanding the essential community evaluation metrics to know the goodness of communities resulting after applying various algorithms. Network characteristics vary for disjoint and overlapping community detection, so, social network datasets are analyzed to produce different network statistics for a better understanding of their suitability for disjoint and overlapping community detection. The research paper also covers applications of community detection to motivate the researchers for exploring the possibility of using characteristics of social networks for detecting communities for reducing computation time.

Keywords: Social Networks · Community Detection · Structural characteristic of social networks · Algorithms for Community Detection · Clustering · Community Visualization

1 Introduction

The structure of Social Networks (SN) revolves around nodes and edges and the way nodes are connected to other nodes in a network. Online Social Networks(OSNs) are not only providing a platform to connect and interact with friends, family members, relatives, and professionals but they are popular for a lot of other services as well, namely, recommending friends, products, professional links, and marriage partners; question and answers forums; job opportunity platforms; multimedia content sharing [1], etc.

R. K. Challa et al. (Eds.): ICAIoT 2023, CCIS 1930, pp. 196–217, 2024.
https://doi.org/10.1007/978-3-031-48781-1_16

A network can be partitioned into communities that are made up of strongly connected nodes but relatively few nodes in between those communities. There are numerous cutting-edge approaches like centrality-based methods, statistical inference methods such as spin models, random walks, and synchronization, as well as block modeling, bayesian inference, model selection, and information theory. Most of the Community Detection (CD) approaches are based on static networks, so, it is the need of the hour to identify communities in dynamically evolving networks. Moreover, it's still challenging to identify Overlapping Communities in Social Networks (SN). Another challenge for researchers is to understand the community definition in a better way as still researchers are trying to define what actually a community means. The research work in this paper is primarily focused on using structural characteristics of networks to identify communities whether weak or strong and further it is used for different application areas like influential node identification, generation of recommendations, etc. The applications of CD in different emerging areas are discussed in Table 1.

Table 1. Applications of Community Detection

S. No.	Application Area	Description
1	Recommendation Systems	Community-based recommendation systems can be developed for sending recommendations to a group of persons related to books, movies, music, food, and different interest-specific products. In addition to this, friendship recommendations can also be generated at the community level
2	Influential Nodes Identification	The community-based approach can also be used to find experts in a particular domain/ region when the community represents persons in a particular domain or region. It is also quite popular in election campaigns for finding influential persons to spread the party agenda among community members
3	Criminal Activities Detection	CD is very useful in studying the possibilities of criminal activities across social networks or different groups
4	Disease Spread	The study of disease spread across communities is also one of the important application areas of CD
5	Targeted Marketing through customer segmentation	CD plays a major role in identifying customer segments by looking into their profiles for doing targeted marketing rather than viral marketing to increase sales and business profits

This research paper is organized in different sections: Section 2, describes the related work in the direction of community detection that is based on structural characteristic of networks. Section 3, explains in detail about different structural characteristics and their relevance in CD. Section 4, provides features and comparison of various network visualization tools. Section 5, discusses about CD algorithms based on structural characteristics and python libraries available for CD. Section 6, describes various evaluation metrics for effective CD. Section 7, concludes the understanding gained from the comparative analysis done on social networks based on different structural characteristics.

2 Related Work

The research work related to the structural analysis of SN and their impact in several application areas has been studied by researchers [1–3]. Efforts have been made to emphasize how even a weak tie in a network can contribute to information diffusion and the spreading of innovative ideas [4, 5]. The structural analysis contributes to both kinds of CD i.e. disjoint and overlapping. Several CD algorithms are based on network statistics computation like network density, max-clique, centrality measures, etc. In [6], an empirical study has been made by the author to validate the results of CD with ground-truth communities.

CD is highly dependent on similarity measures between nodes in a graph or network. Nodes' similarity can be measured using "structural similarity and attribute similarity" [7]. Structural similarity-based community detection primarily refers to network topology-based extraction whereas attribute-similarity-based CD refers to similarity computation using the characteristics of the individual nodes such as age, gender, language spoken, region, organization affiliation, interests, and occupation. There can be several parameters that can be computed using the structural information of a social network to detect communities [7, 8].

In network science, Overlapping Community Detection (OCD) is one of the challenges w.r.t handling large-scale datasets including SN for analysis [9]. Moreover, due to belongingness of one node to more than one community makes it more difficult to clearly identify community structures. Overlapping communities can be traditionally identified using a node-clustering-based approach but it suffers from the drawback of not being able to depict pervasive overlaps, and on the other hand, the link-clustering-based approach is not effective in terms of computation time. Ding et al. proposed a new algorithm based on a network decomposition approach for OCD and it results in less computation time and more accuracy [10].

3 Structural Characteristics of Social Networks

Network analysis is highly dependent upon structural characteristics of network where nodes represents different entities such as persons, organization, city, product, etc. and edges (ties) simulates information regarding different kind of interactions among entities. Network may vary from small to large-scale networks, sparse to dense graphs, etc. Generally, graphs are of different types such as directed, undirected, directed and unweighted, undirected unweighted, directed weighted, and undirected weighted. Different network statistics can be generated based on the type of graph. The network structure can be analyzed at different levels, namely, "small-scale or micro-scale", "medium-scale or meso-scale", and "large-scale" network structure analysis [11].

The small-scale structural analysis helps us in understanding the role played by individual nodes and their position in the network which makes them influential. The specific properties to identify such influential nodes are referred as centrality measures. The metrics which can be computed are local clustering coefficient, closeness centrality, betweenness centrality, degree centrality, and eigenvector centrality. The betweenness centrality, closeness centrality, and eigenvector centrality depends upon their relation to other nodes in the network but local clustering depends upon relationship between its neighboring nodes i.e. whether node's neighbors tends to connect each other or not.

Large-scale network analysis helps us in differentiating between real-world networks. Different type of analysis that we can perform at the network level are the diameter and shortest path; Global clustering helps in the quantification of how neighbors of neighbors of a node are interconnected; density and minimum cut helps in quantification of error and attack tolerance i.e. resilience of a network etc. Further, analysis of global structure of a network helps in measuring equality or inequality of networks. The structural analysis of networks can be done on the basis of the following parameters:

3.1 Network Order

It refers to the *number of nodes* present in a network. For example, a network consisting of nodes A, B, C, and D would have a network order 4.

3.2 Length (Size) of a Graph

It means how many edges exist in a graph say "G".

3.3 Distance Between Two Vertices

The distance between two vertices say u and v in a graph "G", is measured as number of edges present in a shortest path and there is a possibility that graph may contain more than one shortest path between pair of two vertices.

3.4 Network Connectedness (Network Density)

Network density refers to calculation of how many actual number of edges present in a network out of the maximum number of possible edges. In a network consisted of 3 nodes, say [x, y, z], node x is connected to node y, and node y is connected to node z, the density of a network is 2/3. Further, it can also be find out that whether the graph is connected or not, count of connected components, nodes in the connected components, the list of nodes or edges which needs to be deleted for making a graph disconnected. Connectivity also refers to as cohesion i.e. it tells about how edges of the graph are distributed [12].

3.5 Eccentricity of a Graph

Eccentricity of the vertex is defined as the maximum distance between a vertex 'V' to rest of the vertices in the graph say 'G' and it is denoted by e(V). Eccentricity of a graph is defined as the maximum eccentricity of nodes in the graph [13, 14].

3.6 Diameter and Periphery

Among different network metrics based upon shortest path lengths, one of the most popular measure is diameter. It refers to the longest of all shortest paths between every possible pair of nodes in the network or it is the length of the path between the two nodes that are furthest apart. This metric tells about size of the overall network. It is also referred to as the maximum eccentricity value among all nodes in a network [14]. Further, the periphery is defined as the set consisting of nodes having their eccentricity value equal to the diameter of a graph.

3.7 Radius and Centre of a Graph

Radius is defined as the minimum eccentricity value of a node [14]. The centre of a graph is the set of nodes whose eccentricity is equal to the radius of the Graph.

3.8 Degree Distribution

Different networks including social networks such as the internet, biological networks, communication networks, and email networks, exhibit different degree distribution. The graph can be plotted between unique degrees of nodes in the graph and number of nodes having those degree value as shown in Fig. 1, and Fig. 2.

3.9 Strength of Ties

The strength of ties can be weak or strong depending upon how closely two nodes (persons), say A and B are connected with each other in a social network. Moreover, the nodes which are either connected to node A or node B are mostly going to connect with both A and B. On the other hand if connection between node A and node B is weak then such connections may not occur. Weak ties play an important role in applications areas like information diffusion and they may be the only connection between two group of nodes in a graph [4]. The strong connections between nodes also indicates more similarity between those nodes.

3.10 Degree Centrality

This centrality measure helps in computing popularity of a node by counting number of neighboring nodes to whom it is connected. In undirected graphs, it simply refers to a node's degree but in directed graphs, this centrality can be computed as "in-degree" and "out-degree" of a node [11]. The node with highest degree may be termed as most influential or central node. Such nodes are of high importance in many application domains like, viral marketing, spreading awareness regarding disease spread, running successful advertisement campaigns and many more.

3.11 Closeness Centrality

It is defined as a node that is close to the remaining nodes in the network and it is also referred to as the reciprocal of farness. The farness is the sum of distances between that node and all the other nodes [11].

3.12 Eigenvector Centrality (Hubs)

It usually refers to how well connected a node is to other highly connected persons and it is also referred as hubs [11].

3.13 Betweenness Centrality

The nodes which connects many parts of the networks plays a vital role in information dissemination and there is a high probability of information being changed through such nodes. Such nodes are not only crucial for SN but also in other type of complex networks like flow networks. In flow networks such as water pipes and telecommunication networks, such nodes may interrupt the amount of flow and it is needed to protect such nodes from attacks and failures. Overall, such nodes are referred as broker nodes and edges that spans distant parts of networks are called as bridges.

Betweenness centrality helps in identifying broker nodes and bridges [11] and its computation is based upon finding, how many number of shortest paths between a pair of nodes in a graph goes through a particular node (broker) or edge (bridge). The sum of paths that goes through a particular node is called as node betweenness and number of nontrivial paths that goes through an edge, add 1 to that for the edge itself will give us edge betweenness.

3.14 Local Clustering (LCC)

This refers to whether a foaf (friend of a friend) is your friend also or we can say consider a scenario where a node, say x, is connected to node y and z then finding out whether there is a possibility of an edge (representing friendship) between nodes y and z. The connectedness between 3 nodes forms triangles and this refers to clustering. The strong clustering between nodes in a network generally refers to robustness and redundancy and this also helps in finding a path between two nodes even if the edge which is directly connecting two nodes disappear. The LCC defined as the fraction of the all pairs of node's neighbors that have an edge between them. The low value of LCC refers to low connectedness between neighbors of a node.

Using LCC we can measure clustering for a single node by finding how many pairs of a node's friends are also friends with each other out of the all possible pairs of a node's friends. It can also be understood as measuring how close the node's neighbors in becoming a complete graph.

$$\mathrm{LCC} = \frac{No.\ of\ pairs\ of\ node's\ friends\ that\ are\ friends\ (POFTAF)}{No.\ of\ pairs\ of\ the\ node's\ friends\ (POF)} \tag{1}$$

$$\text{POF} = \frac{d(d-1)}{2} \quad (2)$$

In POF, "d" is a number of neighboring nodes representing friends of a node say "v" and a possible number of pairs of friends is calculated using (2). The numerator in (1), specifies an actual number of friends' pairs that exist. In order to compute the clustering of the whole graph, we can compute an average over all the LCC of individual nodes i.e. the sum of the LCC of all nodes divided by the number of nodes [13].

3.15 Global Clustering/Transitivity/Triadic Closure

Global Clustering is the measurement of the tendency of the edges in a graph to form triangles (also referred to as Triadic Closure). This metric also understood as the extent to which nodes in a graph tend to form clusters [15]. Alternatively, triadic closure can also be defined as nodes having common neighbor tends to get connected through an edge [13]. A *triangle* is consisted of set of three nodes which are connected by three edges whereas an *open triad* is defined as a set of three nodes such that they are connected by only two edges. So, we can say, each triangle contains three open triads. The structural characteristic named as *transitivity* refers to a measurement of the percentage of open triads that are actually triangles. In other words, it is three times the ratio of count of closed triads to count of open triads.

$$\text{Transitivity} = \frac{3 * No.\ of\ triangles\ in\ the\ graph}{No.\ of\ all\ the\ open\ triads\ in\ the\ graph} \quad (3)$$

3.16 Clique

A clique C of an undirected graph say "G = (V, E)" is a subset of the vertices i.e. "C ⊆ V". The induced subgraph of G i.e. C is a complete subgraph and all of its vertices are linked to rest of the other vertices in the clique. The Max-Clique problem is quite challenging as it requires huge computation. A clique in friendship network identifies a group of people knowing each other. Other metrics that can be computed are number of maximal cliques in a graph, clique of largest size, clique containing particular node of a graph, [16] etc.

3.17 Node Cover

In graph theory, node cover is defined as a subset of nodes of graph G such that all the edges in graph has at least one of their node in this subset. On the other hand, line cover or edge cover is defined as a set of edges which covers all the nodes of a graph G.

3.18 Structural Holes

The presence of structural holes (empty space) in a social network refers to the absence of a direct tie (edge) between two or more nodes representing persons in SNs or organizations, etc. The position of the nodes helps us in deciding whether the node is central to network (or community) or it is present on the boundary of the network. The structural holes in a network is consisted of subset of nodes present on the boundary of communities [6]. Further, they are different from the concept of "Strength of Weak Ties" as introduced by Granovetter (1973) as it represents lack of link rather than a presence of link [2]. The between-group brokers i.e. people who are present near the structural holes in a social network are more likely to express innovative and valuable ideas that are generally assumed as different from homogeneous ideas expressed by nodes within a group [17]. The applications of bridging structural holes can be in the area of advertising campaigns, solving business problems creatively.

3.19 Boundary Spanners

It refers to nodes on boundary of graph or subgraphs representing communities and they provide critical links between two or more graph or subgraphs [18]. The identification of bridge nodes (bridges) in a complex network is still a challenging problem as it requires high consumption of computational resources. Further, they exhibit the characteristic that, if they are removed then the network will split into two or more connected components (subgraphs). The concept is also similar in nature to weak ties which also play vital role for maintaining connectivity among different connected components of a network [5]. Hence, a bridge can also be defined as an edge whose removal results in more number of connected components. Such kind of edges can also be referred as cut edges or cut arcs and they should be a part of any chain or cycle.

3.20 Cut Ratio

It is calculated as the ratio of existing edges leaving the cluster and count of possible edges [19].

Betweenness Centrality, Clique/ k-clique/max-clique, Network Density, Clustering coefficient, Transitivity are some of the important measures that employed for selection of dataset having more probability of detecting community structures. In summary, structural characteristics of networks are significant for community detection and in Table 2, a comparative analysis of network characteristics is done for two social networks namely Zachary Karate Club (in-built dataset in NetworkX) and Facebook Dataset.

Facebook dataset consists of friendship information between persons on Facebook and it also includes features of nodes. Dataset is anonymized w.r.t political affiliation and person id for maintaining privacy. Zachary karate club is well-known social network dataset used by several researchers to establish their empirical results. In this dataset, each node is a club member, and every edge is a tie between two club members. Both of the networks used in experimentation are undirected and often used by researcher for studying their community detection algorithm results. Datasets are selected in such a manner that network statistics can be derived for small-scale and medium-scale datasets. The Degree Distribution of both Network Datasets is given in Fig. 1 and Fig. 2.

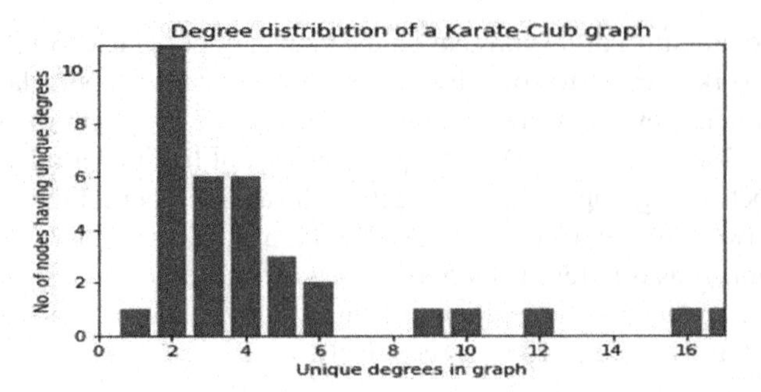

Fig. 1. Degree Distribution of Zachary's Karate Club Network Dataset.

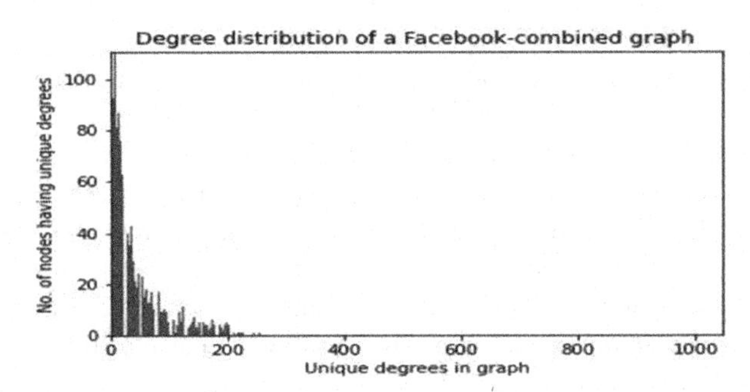

Fig. 2. Degree Distribution of Facebook_combined Network Dataset.

Table 2. Comparative Analysis of Network Characteristics using different Datasets.

Network Properties	Karate Club	Facebook_combined
Network Order and Length of a Graph	Nodes: 34 Edges: 78	Nodes: 4039 Edges: 88234
Network Density	0.139	0.011
Minimum and Maximum Degree of a node	Minimum Degree:1 Maximum Degree: 17	Minimum Degree:1 Maximum Degree: 1045
Average Degree of nodes in a graph	4.59	43.69
Radius and Diameter of a graph	Radius: 3 Diameter: 5	Radius: 4 Diameter: 8
Centre of a Graph	[0, 1, 2, 3, 8, 13, 19, 31]	[567]

(continued)

Table 2. (*continued*)

Network Properties	Karate Club	Facebook_combined
Node with Maximum Degree Centrality	Node 33 Degree Centrality 0.52	Node 107 Degree Centrality 0.26
Node with Maximum Betweenness Centrality	Node 0 Betweenness Centrality 0.44	Node 107 Betweenness Centrality 0.48
Node with Maximum Closeness Centrality	Node 0 Closeness Centrality 0.57	Node 107 Betweenness Centrality 0.46
Node with Maximum Eigenvector Centrality (EC)	Node 33 EC 0.37	Node 1912 EC 0.1
Avg. Clustering Coefficient	0.571	0.606
Global Clustering Coefficient/Transitivity/ Triadic Closure	0.26	0.52
Number of maximal cliques in the graph	36	Requires high computation power to generate results
Clique No. of a Graph (Largest clique size)	5	69
Graph connectedness (Graph having more than one component)	True	True
Node with minimum and maximum eccentricity	Minimum: Node 0 Maximum: Node 33	Minimum: Node 0 Maximum: Node 4038
Periphery of a Graph	[14, 15, 16, 18, 20, 22, 23, 26, 29]	[687, 688, 689, 690, 691, 692, 693, 694, 695, 696, 699, 700, 701, 702, 704, 705, 706, 707, 709, 710, 711, 712, etc.]

4 Network Visualization Tools

Network visualization is a part of data visualization and gives us an added advantage by providing interesting insights into network data represented using nodes and edges [20]. It helps in understanding complex relationships among different nodes in the network, highlighting the significant nodes by using different colors and size depending upon some network statistics like degree, detection and visualization of vulnerable nodes and edges, so that, they can be protected from any kind of attacks or for maintaining the connectedness of the network graph. Moreover, visualization of graphs with additional nodes and edge attributes gives us quick help in decision making and presentation of our ideas for solving graph based problems. As most of the problems can be represented in the form of graphs, so, it becomes very important to make use of network visualization tools to create, visualize, and analyze data of different datasets for better clarity and exploration of dynamics behind complex networks including SNs. The features of various tools for Network visualization are given below [21]:

Cytospace. It is capable of analysis and visualization of large-scale networks having nodes greater than 10,000 and it supports RESTful API for programmatic access. Originally it has been designed for visualizing molecular interactions in biological networks. It has native support for Delimited text and Excel Workbook (. Xls) and many other file formats such as, Graph Markup Language (GML or . Gml format), Simple interaction file (SIF or .sif format) [22], etc.

Gephi. It is capable of performing effective exploratory data analysis of networks with up to 100,000 nodes and 1,000,000 edges. It is the most versatile tool available for SNs analysis and visualization. It supports file formats including CSV, GDF, GEXF, GraphML, Pajek NET, PDF, SVG, etc [23]. Gephi Framework can be used for computing commonly used network metrics such as Centrality Measures, Clustering Coefficient, Shortest path, PageRank, Community detection (Modularity), Diameter, and Random graph generators [23]. It can also be used for analyzing dynamic graphs i.e. temporal networks that evolve over time. It provides an interactive interface to filter networks and produce new networks as an output in real-time. It makes use of interface similar to excel for data manipulation. It allows extensibility feature by allowing plugins to be used that are developed by community. No programming skills needed and provides high-performance due to built-in render engine.

NetworkX. It is very popular and well documented package for any kind of complex network analysis and visualization [24]. It allows network (graph) based algorithms to be implemented in easy manner. It required knowledge of Python language.

Pajek. It can be used for Large Network Analysis and Visualization including SNs, Communication Network Optimization, Chemical Technology, Bio-Technology, Genealogy, etc. It can be used from small to large-scale networks, multi relational networks, 2-mode networks, and temporal networks. It makes use of notepad as an editor and takes input and stores output as a .txt, .net, and .mat file. It makes use of six types of objects namely, Network, Partitions, Clusters, Permutation, Hierarchies, and Vectors [25, 26].

SocNetV (3.0.4). It is a Social Network Analysis and Visualization software and it supports file formats GraphML, GraphViz, EdgeList, GML, etc [26].

NodeXL. It makes easy exploration and visualization of network data including SN. It is an add-on to Microsoft Excel (compatible with MS Office 2007 or above)

R. It can be used for network analysis effectively in terms of manipulation of the graphs, statistical, and mathematical modeling using packages like igraph, network, tidygraph, and ggraph [27].

4.1 Comparative Analysis of Various Network Visualization Tools

The various existing network visualization tools vary in terms of open source or licensed, ease of use and learning, scalability i.e. capability to handle small-scale to large-scale networks analyses, etc. Complex Networks analysis and visualization helps in discovering structures and hidden patterns in a network for providing useful insights [28]. Some tools requires programming knowledge while some don't require. In addition to this, some tools run on desktop while others run using web-interface. Tools can also be differentiated based on file formats supported by them and cost-effectiveness. Tools may be compatible with single or multiple platforms. The network visualization tools are compared in Table 3.

Table 3. Comparative Analysis of Various Network Visualization Tools.

S. No	Name of Tool and latest version	Free / Licensed	Development Platform	O/S Support
1	Cytospace (3.9.1)	Free and Open Source	Java based desktop application	Windows, Linux, Mac
2	Gephi (0.10)	Free and Open Source	Cross-platform application developed in Java and Gephi current stable version works with Java 11	Windows, Linux, Mac
3	NetworkX	Free and Open Source	Python	Cross-platform
4	Pajek (5.16)	Free and Open Source	Delphi (Pascal)	Windows
5	SocNetV(3.0.4)	Free and Open Source	C + + and Qt (used for developing GUI's)	Windows, Linux, Mac
6	NodeXL	Free and open source	C#,.Net framework	Windows
7	R	Free and open source	C, Fortran	Windows, Linux

5 Community Detection (CD)

CD is defined as the identification of a subset (group) of nodes that are more connected with other nodes in the subset rather than with nodes outside the group. In other words, it can also be considered as a higher number of internal connections (intra community) as compared to external links (inter-community). The two main categories of CD are classified as disjoint (non-overlapping) and overlapping. The communities can also be detected based upon node similarity calculation using Jaccard Similarity and Cosine Similarity.

Overlapping communities consist of nodes that are members of more than one community and each node's belongingness or membership value varies in different communities. Overlapping communities detection in the SNs can be done using various modularity based approaches and each approach is having its own characteristics [29]. Overlapping region may consist of overlapping nodes or edges and they play significant role in applications like information diffusion [6]. Overlapping nodes are also considered important in structural analysis of networks as they participate in evolution of networks by forming new links.

In Python we have different libraries available for detecting communities and one of them is *"communities"* library and it is helpful in implementation of CD algorithms namely, "Louvain method", "Hierarchical Clustering", "Spectral Clustering", and "Bron-Kerbosh algorithm". The prerequisite for using these algorithms is to represent an input graph using 2D matrix where adjacency matrix is an representation of undirected graph and edges are either weighted or unweighted [30]. Python also provides other CD libraries like CDLIB [31]. Different CD algorithms based on structural characteristics of networks along with important highlights are given in Table 4.

Table 4. Algorithms for Community Detection based on structural characteristics.

Algorithm Name and its Complexity	Type of Input needed and Communities Detected	Important Points
Louvain's Method complexity is O(n.log^{2}n), where n is the no. of nodes in the graph	Undirected Graph, weighted graph (Take input as an adjacency matrix) and detects Disjoint Communities	It makes use of properties of edges in the graph and generates dendogram as an output. Its lowest level represents communities of smallest size It may yield badly (weakly) connected communities or disjoint communities and it is based on Modularity optimization (maximization of modularity) Improved versions of Louvain algorithm like NI-Louvain can also be used for overlapping community detection
Girvan-Newman Algorithm complexity is O(m^{2}n) time, where m is the number of edges and n is the number of nodes in the graph	Takes input as Undirected Graph and detects Disjoint Communities	It is based upon edge betweenness centrality and not applicable to directed graph CONGA (Cluster-Overlap Newman Girvan Algorithm) is a CD algorithm based on Girvan-Newman Algorithm

(*continued*)

Table 4. (*continued*)

Algorithm Name and its Complexity	Type of Input needed and Communities Detected	Important Points
Hierarchical Clustering Algorithm complexity is $O(n^3)$, where n is the no. of nodes in the graph	Takes input as Undirected Graph and detects Hierarchical Clusters	It is based upon the bottom-up approach for getting hierarchical clustering. Each node belongs to its own community and then, the most similar pairs of communities are merged until no further gains in modularity are obtained
Spectral Clustering Algorithm is NP-Hard algorithm	Takes input as Undirected Graph and detects Overlapping Communities	It makes use of the eigenvalues of the adjacency matrix which holds information about community structure. It requires information about the number of communities to be formed It works via embedding the vertices of a graph into a low-dimensional space using the bottom eigenvectors of the Laplacian matrix It transforms clustering into a graph-partition problem and it is time-consuming for dense datasets
Bron-Kerbosch Algorithm If input graph has less than $3^{n/3}$ maximal cliques then its time complexity is $O(3^{n/3})$	Takes input as Unweighted, Undirected Graph and detects Overlapping Communities	It is a max-clique-based community detection algorithm
NISE [6, 32] (Neighborhood Inflated Seed Expansion)	Takes input as Undirected Graph and detects Overlapping Communities	It is PageRank-based local expansion method, scalable and effectively finds highquality overlapping communities, similar to ground-truth communities

(*continued*)

Table 4. (*continued*)

Algorithm Name and its Complexity	Type of Input needed and Communities Detected	Important Points
NEO-K-Means (Nonexhaustive, Overlapping K-Means	Takes input as Undirected Graph and detects Overlapping Communities	This algorithm is an improvement to the traditional k-means clustering with an objective to capture the degrees of overlap and non-exhaustiveness
Distributed Neighborhood Threshold Model (DNTM) [33] O(|V|.b + |OCN|.|Nr|)	Takes input as Undirected, Directed, Weighted Graph and detects Overlapping Communities	This algorithm is based upon generating non-overlapping clusters first and then overlapping nodes are found to update the previous clusters result to achieve final overlapping clusters In complexity analysis, |V| is the no. of vertex in graph, b is the branching factor, |OCN| is the no. of overlapping candidate nodes, and |Nr| is no. of neighborhoods in other clusters

6 Evaluation Metrics for Effective Community Detection

The output of CD algorithms i.e. detected communities can be validated using groundtruth communities if it is available for the dataset we are using for community detection. The benchmark datasets available with ground-truth communities such as DBLP (Collaboration Network), friendster, Orkut, and LiveJournal (Online Social Networks) can be used for OCD [34]. The various measures to quantify the performance of CD are discussed below:

6.1 Modularity

Various metrics can be used to measure the strength of the community and the most popular of them is the modularity measure. Modularity defines the strength of the partition of the network as a community. It is the measure of the density of intra-community links as compared to intercommunity links. As per the initial idea given by Newman and Girvan, modularity "Q" can be defined as given in (4) [35]:

$$Q = \frac{1}{2m}\sum\nolimits_{uv}[A_{uv} - K_u K_v]\delta(C_u, C_v)$$

$$\text{where } \delta(C_u, C_v) = \begin{cases} 1, & (u, v) \in C \\ 0, & \text{otherwise} \end{cases} \tag{4}$$

A_{uv} represents the adjacency matrix entry at row "u" and column "v" (It is equal to 1 if there is an edge from node "u" to node "v" and otherwise it is 0.)

C_u and C_v are the communities to which node "u" and node "v" belongs.

K_u and K_v represents the degree of node "u" and "v" respectively.

m is the number of edges in the graph. [12]

Equation (4) of modularity is only applicable to an undirected and unweighted graph and at any instance produces only two communities. Belongingness factor refers to degree of association of a particular node in a community. This basic definition is further modified for directed graphs, overlapping community detection, and weighted graph etc. Further, many researchers have tried to optimize modularity and this is considered as NP-Hard problem. Modularity varies between -1 and 1. The higher value of Q i.e. 1 or close to it indicates stronger community structures and Q = 0 means no communities identified [29]. In other words, modularity tells about the density of edges within communities with respect to edges outside communities.

6.2 Density/Edge Density

This metric tells about how connected a network is. It is calculated as ratio of number of edges in the network and number of possible edges. The possible number of edges for a directed graph of "*n*" nodes can be calculated as "*n * (n-1)*" and for an undirected graph it can be "*(n * (n-1)) / 2*" [12]. So, the density for a network is computed as:

$$\text{For the directed graph,} \qquad \text{Density} = \frac{e}{n * (n-1)} \tag{5}$$

$$\text{Similarly, for an undirected graph} \qquad \text{Density} = \frac{e}{n * (n-1)/2} \tag{6}$$

In (5) and (6), e is the number of edges. The value of density varies between 0 and 1 where 0 is the minimum density, and 1 stands for the highest density representing a scenario where all nodes are connected to all the other nodes i.e. clique. It can be used for comparison of networks especially egocentric networks.

6.3 Total Number of Intra-edges and Inter-edges

The higher count of intra-edges refers to better quality of community whereas higher count of inter-edges refers to low quality of community.

$$\text{No. of Inter - edges} = |\text{Outgoing Edges from community C}| \tag{7}$$

Equation (7) refers to number of edges connecting nodes from one community to another community nodes or total sum of weights of such edges.

$$\text{No. of Intra-edges} = |\text{Edges inside community C}| \tag{8}$$

Equation (8) refers to number of edges connecting nodes from the same community or the total sum of weights of such edges.

6.4 Contraction

This metric gives the average number of edges per node inside the community C or the average weight per node of such edges. The higher value of contraction refers to the better quality of the community. It can be calculated as in (9) and (10):

$$\text{Contraction} = \frac{2 * |Edges\ inside\ community\ C|}{|Nodes\ in\ community\ C|} \text{ for undirected networks} \tag{9}$$

$$\text{Contraction} = \frac{|Edges\ inside\ community\ C|}{|Nodes\ in\ community\ C|} \text{ for directed networks} \tag{10}$$

6.5 Expansion

Expansion measures the average number of edges (per node) that point outside the community C or the average weight per node of such edges. The lower value of Expansion refers to the better quality of the community. It can be calculated as in (11):

$$\text{Expansion} = \frac{|Edges\ connecting\ nodes\ outside\ community\ C|}{|Nodes\ in\ community\ C|} \tag{11}$$

6.6 Conductance

The evaluation of community detection can also be done using conductance metrics and it is defined as the ratio between edges representing relationships that point outside community "C" and the total number of relationships of "C" as in (12) and (13). The lower value of the conductance refers to strongly connected community [36].

For Undirected Graphs (Networks)

$$\text{Conductance} = \frac{|\text{Outgoing Edges from community C}|}{2|\text{Edges inside community C}| + |\text{Outgoing Edges from community C}|} \tag{12}$$

For Directed Graphs (Networks)

$$\text{Conductance} = \frac{|\text{Outgoing Edges from community C}|}{|\text{Edges inside community C}| + |\text{Outgoing Edges from community C}|} \tag{13}$$

6.7 Purity

Purity is one of the generally used metrics for cluster analysis and it can be applied to network partitions (communities). It varies between 0 and 1, where 1 stands for maximum purity i.e. all the nodes with one community label say c1 are partitioned in one community and there is no other community node (say a node labelled with community c2) present in it. The main disadvantage of this metric is that all nodes need to be labelled with community label i.e. we need to know beforehand to which community a node belongs to. Purity increase with increase in number of clusters (communities).

Table 5. Confusion matrix to evaluate performance of classification.

		Predicted Class	
		1	**0**
Actual Class	**1**	True Positive (TP)	False Negative (FN)
	0	False Positive (FP)	True Negative (TN)

6.8 Metrics for Community Evaluation Based on Confusion Matrix

The detected communities can also be validated using ground-truth communities and the metrics used for comparison are given in Table 5.

In order to compare result of CD algorithms in the form of clusters (communities) with ground-truth communities, we try to compare for every possible pair of nodes, whether they belong to same community or different community as shown in Table 6.

Table 6. Confusion matrix to evaluate performance of community detection [37]

		Predicted Class (Clustering Results)	
		1 ($Cv_i = Cv_j$)	**0 ($Cv_i != Cv_j$)**
Actual Class (Ground-truth)	**1 ($Cv_i = Cv_j$)**	True Positive (TP)	False Negative (FN)
	0 ($Cv_i != Cv_j$)	False Positive (FP)	True Negative (TN)

In the Table 6, for a pair of nodes (vi,vj), **Cvi** means the community to which node vj belongs. Similarly, **Cvj** means the community to which node vj belongs. The community detection method results in communities and each node in the graph is labeled with the community number to which it belongs as per prediction.

Cvi = Cvj means both nodes vi and vj belong to the same community and
Cvi != Cvj means both nodes vi and vj do not belong to the same community

TP: It means pair of nodes belongs to the same community in ground-truth communities and they also belong to the same community in communities generated by CD algorithms.

FP: It means pair of nodes does not belong to the same community in ground-truth communities but prediction results of CD algorithms show them in the same community.

FN: It means pair of nodes belonging to the same community in ground-truth communities but the prediction results of CD algorithms show them in a different community.

TN: It means pair of nodes belongs to a different community in ground-truth communities and prediction results of CD algorithms also show them in a different community.

In classification models, we can also set a threshold value and if it is set to a low value then it will either result in an increased number of False Positives or decreased number of False Negatives and vice-versa [40]. Setting up of right threshold value is very important for classification model performance. ROC Curve can be plotted between TPR (along the y-axis) and FPR (along the x-axis).

On the basis of the confusion (contingency) matrix in Table 6, some metrics for CD algorithms performance evaluation is computed as shown in Table 7.

Table 7. Community Detection Performance Evaluation Metrics [38]

Evaluation Metrics	Mathematical Formula and Description
Precision	$\text{Precision} = \frac{TP}{TP+FP} = \frac{No.\,of\ nodes\ belongs\ to\ community\ detected}{Total\ no.of\ nodes\ detected\ in\ community}$ It refers to the fraction of instances positively labeled are truly positives. Denominator refers to nodes truly classified in community detected and nodes which are wrongly classified in this community but belongs to another community [39]
Recall / Sensitivity/ hit-rate / True Positive Rate (TPR)	$\text{Recall} = \frac{TP}{TP+FN} = \frac{No.\,of\ nodes\ belongs\ to\ community\ detected}{Total\ nodes\ of\ the\ community}$ It is the proportion of actual positives that are identified correctly. Denominator refers to nodes truly classified in community detected and nodes which are wrongly classified in another community
Specificity/ True Negative Rate (TNR)	$\text{Specificity} = \frac{TN}{TN+FP}$ It measures the proportion of actual negatives that are correctly identified and it is opposite of recall
Accuracy	$\text{Accuracy} = \frac{TP+FN}{TP+FP+FN+TN} = \frac{No.\,of\ correct\ predictions}{Total\ no.of\ predictions}$ It is defined as proportion of predictions that are correctly classified by the model i.e. true class predicted as true and false class predicted as false
F1 Score	$\text{F1 Score} = \frac{2TP}{2TP+FP+FN} = \frac{2 * Precision * Recall}{Precision + Recall}$ It is defined as harmonic mean of Precision and Recall
False Positive Rate	$\text{FPR} = 1 - \text{Specificity} = 1 - \frac{TN}{TN+FP} = \frac{FP}{TN+FP}$ It indicates the proportion of wrongly classified (predicted) instances in the true class and total negative class instances

6.9 Normalized Mutual Information (NMI)

It is a normalization of the Mutual Information (MI) score to scale the results between 0 (no mutual information) and 1 (perfect correlation). It is used to estimate the clustering quality.

7 Conclusion

This research was primarily focused on the structural characteristics of networks and how they are playing a significant role in CD. A network with dense connections, having more connected components, and a high value for centrality measures like betweenness and closeness centrality is more suitable for finding communities. Further, communities can also be identified in a network by taking seed nodes included in the centre of a graph. Such nodes are generally connected with more nodes and due to their high influence in the network may lead to better CD. Max-clique finding is an NP-Complete graph problem and there is no polynomial time deterministic algorithm for finding max-clique. This problem requires high computational power and hence it is difficult to execute such tasks on large-scale graphs. Further, this study also leads to the identification of peripheral nodes that are important in applications like information diffusion across communities or subgraphs in a graph. Peripheral nodes are also part of overlapping regions between more than one network. This study is also helpful in deciding influential nodes for viral marketing. The comprehensive details of community evaluation metrics are beneficial for researchers to choose algorithms with better performance and comparative analysis of network visualization tools will be helpful in selecting the right tool as per requirement for exploring more features of networks. In the future, more indepth structural analysis of large-scale networks can be extended to generate better understanding of network characteristics based on empirical results.

References

1. Bhattacharya, S., Sinha, S., Roy, S.: Impact of structural properties on network structure for online social networks. Procedia Comput. Sci. **167**, 1200–1209 (2020). https://doi.org/10.1016/j.procs.2020.03.433
2. Bentley, C.: Introduction to Structural Hole Theory (2018). https://medium.com/@agreenmoment/introduction-to-structural-holes-theory-124c51c3ae31. Accessed 17 Jan 2023
3. Baagyere, E.Y., Qin, Z., Xiong, H., Zhiguang, Q.: The structural properties of online social networks and their application areas. IAENG Int. J. Comput. Sci. **43**(2), 156–166 (2016)
4. Granovetter, Mark S.: The strength of weak ties. Am. J. Sociol. **78**(6), 1360–1380 (1973). https://doi.org/10.1086/225469
5. Musiał, K., Juszczyszyn, K.: Properties of bridge nodes in social networks. In: Nguyen, N.T., Kowalczyk, R., Chen, S.-M. (eds.) Computational Collective Intelligence. Semantic Web, Social Networks and Multiagent Systems, pp. 357–364. Springer, Heidelberg (2009). https://doi.org/10.1007/978-3-642-04441-0_31
6. Whang, J.J.: An empirical study of community overlap : ground-truth , algorithmic solutions , and implications, pp. 2363–2366 (2017)
7. Zarandi, F.D., Rafsanjani, M.K.: Community detection in complex networks using structural similarity. Physica A: Stat. Mech. Appl. **503**, 882–891 (2018). https://doi.org/10.1016/j.physa.2018.02.212
8. Ladd, J., Otis, J., Warren, C.N., Weingart, S.: Exploring and analyzing network data with python (2021). https://programminghistorian.org/en/lessons/exploring-and-analyzing-network-data-with-python. Accessed 17 Jan 2023
9. Vieira, V.F., Xavier, C.R., Evsukoff, A.G.: A comparative study of overlapping community detection methods from the perspective of the structural properties. Appl. Netw. Sci. **5**(1), 1–42 (2020). https://doi.org/10.1007/s41109-020-00289-9

10. Ding, Z., Zhang, X., Sun, D., Luo, B.: Overlapping community detection based on network decomposition. Sci. Rep. **6**(April), 1–11 (2016). https://doi.org/10.1038/srep24115
11. Platt, E.L.: Network Science with Python and NetworkX Quick Start Guide. First Edit (2019)
12. Golbeck, J.: Network structure and measures. Anal. Soc. Web **5**, 25–44 (2013). https://doi.org/10.1016/b978-0-12-405531-5.00003-1
13. Clustering, connectivity and other Graph properties using Networkx. https://www.geeksforgeeks.org/python-clustering-connectivity-and-other-graph-properties-using-networkx/. Accessed 22 Jan 2023
14. Graph measurements: length, distance, diameter, eccentricity, radius, center (2021). https://www.geeksforgeeks.org/graph-measurements-length-distance-diameter-eccentricity-radius-center/. Accessed 23 Jan 2023
15. Hades. Triadic Closure (Clustering) (2017). https://necromuralist.github.io/data_science/posts/triadic-closure/. Accessed 22 Jan 2023
16. Clique. https://networkx.org/documentation/stable/reference/algorithms/clique.html. Accessed 05 Feb 2023
17. Burt, R.S.: Structural holes and good ideas. Am. J. Sociol. **110**(2), 349–399 (2004). https://doi.org/10.1086/421787
18. Cruz, J.D., Bothorel, C., Poulet, F.: Community detection and visualization in social networks: integrating structural and semantic information. ACM Trans. Intell. Syst. Technol. **5**(1), 1–26 (2013). https://doi.org/10.1145/2542182.2542193
19. J. Yang and J. Leskovec, "Defining and evaluating network communities based on groundtruth," *Proc. - IEEE Int. Conf. Data Mining, ICDM*, pp. 745–754, 2012, doi: https://doi.org/10.1109/ICDM.2012.138
20. Network Visualizations: The links that bind us (2012). https://rockcontent.com/blog/network-visualizations/. Accessed 23 Jan 2023
21. During, M.: From hermeneutics to data to networks: data extraction and network visualization of historical sources. https://programminghistorian.org/en/lessons/creating-network-diagrams-from-historical-sources. Accessed 17 Jan 2023
22. Cytospace: Network Data Integration, Analysis, and Visualization. https://cytoscape.org/. Accessed 23 Jan 2023
23. Gephi : makes graphs handy. https://gephi.org/users/exported-file-formats/. Accessed 23 Jan 2023
24. NetwrokX: Network Analysis in Python (2023). https://networkx.org/. Accessed 23 Jan 2023
25. Batagelj, V., Mrvar, A.: Pajek—analysis and visualization of large networks. In: Mutzel, P., Jünger, M., Leipert, S. (eds.) Graph Drawing, pp. 477–478. Springer, Heidelberg (2002). https://doi.org/10.1007/3-540-45848-4_54
26. Mukherjee, S., Goswami, A.: Visualizing Graphs with Pajek (2010)
27. Sadler, J.: Introduction to network analysis with R (2017). https://www.jessesadler.com/post/network-analysis-with-r/. Accessed 23 Jan 2023
28. Faysal, M.A.M., Arifuzzaman, S.: A comparative analysis of large-scale network visualization tools. In: Proceedings - 2018 IEEE International Conference Big Data, Big Data, pp. 4837–4843 (2019). https://doi.org/10.1109/BigData.2018.8622001
29. Devi, J.C., Poovammal, E.: An analysis of overlapping community detection algorithms in social networks. Procedia Comput. Sci. **89**, 349–358 (2016). https://doi.org/10.1016/j.procs.2016.06.082
30. Communities (2021). https://pypi.org/project/communities/. Accessed 20 Jan 2023
31. Rossetti, G., Milli, L., Cazabet, R., Rossetti, G., Milli, L., Cazabet, R.: CDLIB : a python library to extract, compare and evaluate communities from complex networks To cite this version: Open Access CD LIB : a python library to extract, compare and evaluate communities from complex networks (2019)

32. Wang, J., Dhillon, I.: Overlapping Community Detection in Massive Social Networks. https://bigdata.oden.utexas.edu/project/graph-clustering/. Accessed 29 Jan 2023
33. Jaiswal, R., Bello, S.R.: Using Distributed Neighbourhood. Springer, Heidelberg (2020). https://doi.org/10.1007/978-3-030-52705-1
34. Networks with ground-truth communities. https://snap.stanford.edu/data/#communities. Accessed 04 Feb 2023
35. Newman, M.E.J., Girvan, M.: Finding and evaluating community structure in networks. Phys. Rev. E **69**(2),(2004). https://doi.org/10.1103/PhysRevE.69.026113
36. Chen, M., Nguyen, T., Szymanski, B.K.: On measuring the quality of a network community structure. Proc. Soc. **2013**, 122–127 (2013). https://doi.org/10.1109/SocialCom.2013.25
37. Tang, L., Liu, H.: Community Detection and Mining in Social Media. Morgan and Claypool, San Rafael (2010)
38. Shin, T.: Understanding the confusion matrix and how to implement it in python (2020). https://www.datasource.ai/en/data-science-articles/understanding-the-confusion-matrix-and-how-to-implement-it-in-python. Accessed 04 Feb 2023
39. Pattanayak, H.S., Verma, H.K., Sangal, A.L.: Community detection metrics and algorithms in social networks. In: ICSCCC 2018 - 1st International Conference on Security Cyber Computer Communivations, pp. 483–489 (2018). https://doi.org/10.1109/ICSCCC.2018.8703215
40. Pandey, P.: Simplifying the ROC and AUC curve (2019). https://towardsdatascience.com/understanding-the-roc-and-auc-curves-a05b68550b69. Accessed 05 Feb 2023

A Meta-data Based Feature Selection Mechanism to Identify Suspicious Reviews

Rajdavinder Singh Boparai[1(✉)] and Rekha Bhatia[2]

[1] Department of Computer Science and Engineering, Punjabi University, Patiala 147002, India
rajiboparai@gmail.com

[2] Punjabi University Centre for Emerging and Innovative Technology (Mohali), Punjabi University, Patiala 147002, India
rbhatia71@gmail.com

Abstract. Deceptive reviews on internet platforms are a harsh reality in today's world. Businesses and products are praised or vilified through reviews. In an online or web society, users get help from reviews before making a decision and in a similar way, web reviews are also very helpful for organizations to keep them updated as per customer needs. Many efforts through various algorithms have already been made to detect deceptive reviews. It has been observed that instead of focusing only on algorithms to find optimal predictions, it is also necessary to pay attention towards the selection of effective features. Datasets used for machine learning are loaded with a large number of additional features which are usually not required. This paper is presenting the significance of the meta-data-based features in order to predict the fake and genuine reviews. In the domain of finding suspicious reviews, various methodologies used for the selection of the optimal features are presented in the paper with the help of the amazon dataset. The target of this approach is to ensure the use of noise-free, relevant, non-redundant, optimal features of the dataset which will result in better predictions.

Keywords: Feature selection techniques · attributes · features · big data · data mining · meta-data · noise · redundancy

1 Introduction

There is a significant growth in amount of high-dimensional data that is available and accessible to the general public online. Because of this, machine learning techniques struggle to handle the abundance of input features, which presents an intriguing challenge for researchers. Pre-processing the data is crucial for the successful usage of machine learning techniques. One of the most used and significant data preprocessing techniques, feature selection, is now a mandatory step in the ML process [1]. As a result, data mining methods are expedited. Specifically, when modeling is to be done for the detection of deceptive reviews accounts [2, 3] which is one of the leading problems of machine learning, then it becomes very important to identify best features for better predictions. Initially, it was believed that machine learning/classification/clustering

R. K. Challa et al. (Eds.): ICAIoT 2023, CCIS 1930, pp. 218–227, 2024.
https://doi.org/10.1007/978-3-031-48781-1_17

is all about algorithms when somebody wants to find the optimal prediction. It was another belief that a better machine will give better results or predictions but in actuality, this is not the case; the selection of correct data and correct attributes make the output optimal. End of the day, it comes down to creating or identifying variables that utilize previously-stored business experiences and then pursuing the best decisions regarding which variable to choose for your predictive models! Unfortunately, or fortunately, both these abilities require a lot of training. There is likewise some workmanship engaged with making new elements - certain individuals have a skill of tracking down patterns where others battle. The following approaches can be used for the selection of features [3]:

- The subset with best commitment in terms of size and assessment measure.
- Reduce the size of the subgroup that complies with the limitations of the specific assessment measure.
- Number of features in the subset with the specified size that maximizes an evaluation metric.

There are two main methods for choosing features. Individual evaluation and subset evaluation are the first and second, respectively. Individual Evaluation is the term used to rank the features [4, 5]. An individual feature's weight in an individual evaluation is assigned based on how relevant it is. Candidate feature subsets are built using a search method in subset evaluation. A commonly used procedure for selecting features is given in Fig. 1 which is divided into four phases: generation of subset/group, evaluation of subset/group, criteria used for stopping and validation of results.

2 Earlier Findings

A suitable classification model was created by Fong et al. [6] for extracting a set of (precise) characteristics from HD data. Swarm Search (SS), a revolutionary method, is created to identify an ideal feature set by utilizing meta-heuristics. This search is seen to be beneficial since it has the freedom to use any classifier as its fitness function. To create a model with good high accuracy, a unique feature selection method is proposed by Fong et al. [7]. The novel and effective CCV technique is created to achieve the best possible balance between generalization and overfitting in order to increase classification accuracy. Finally, a quick discriminating approach called Hyper-pipe is used to examine the group that produces the best classification accuracy. An improved firefly heuristics was created by Selvi and Valarmathi [8] for efficient feature selection. For local searches, the Firefly Algorithm (FA) is exceptional. However, it might be forced towards local optima, making it unable to perform global searches with great precision.

This uses classifiers from naïve bayes (NB), k-nearest neighbor algorithm (KNN) and multiple-layer perceptron neural network (MLPNN) to categorize the selected characteristics. This particular type of firefly-based feature selection improves twitter data classification while simultaneously reducing time complexity. The results of the studies demonstrate that the novel FA-based feature selection approach enhances classifier performance. The kernel-based techniques were applied in an online situation by Kivinen et al. [9]. The three main problems encountered in the online setting are addressed

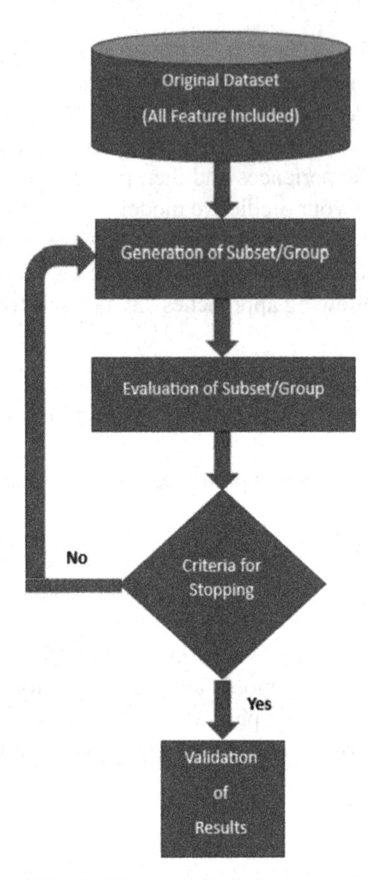

Fig. 1. Feature selection process.

by the examination since data is presented in this setting sequentially. The overfitting issue has been noticed among the typical online linear technique settings, kernel-based estimators, and training duration of batch. Grolinger et al. [10] proposed a method for classifying huge data using a MapReduce-based model. This paper focuses on the obstacles involved in implementing MapReduce for big data and proposes a classification model based on MapReduce to address them. Such a method determines the classes for samples and improves the parallelism observed in computing. With extreme learning machine (ELM), Huang et al. [11] looked into several learning strategies. Investigation demonstrates that ELM performs better in terms of big data categorization than proximal support vector machine (PSVM) and least square support vector machine (LS-SVM) and operates with more efficiency.

A variety of techniques presented in Fig. 2 were found in the existing literature. The set of representative features must be identified using the data class label, which is a prerequisite for all of the aforementioned procedures [12]. The majority of real-world applications, however, use unlabeled data, and identifying data can slow down an application. To the best of our knowledge, there is only one approach that is practical for applications that call for the selection of streaming features while not requiring data class labels for the feature selection procedure [13]. The disadvantage of the suggested approach is that it necessitates the initial establishment of link information. Furthermore, it is presumed that the link information is stable, which is undoubtedly false because it might alter dynamically [14].

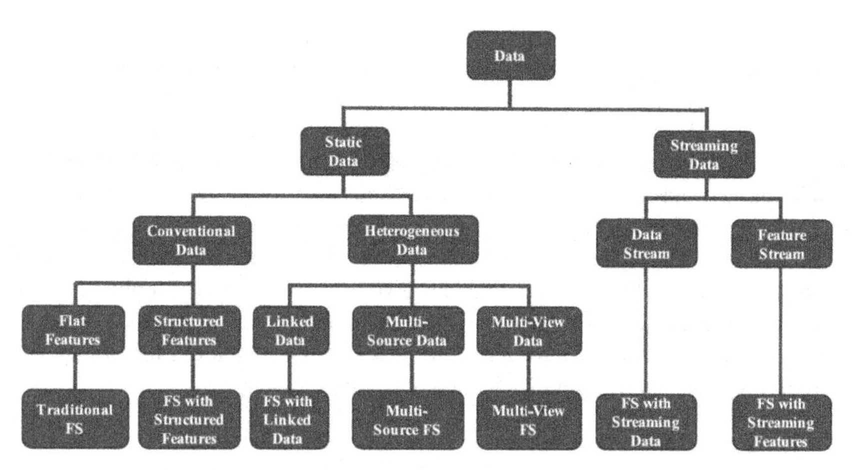

Fig. 2. Data extraction, collection or fetching strategies.

3 Features Selection Strategies

When there are a lot of features available, then it is not necessary to use every feature available while building an algorithm [15–17]. Supplying only the most crucial features may help the algorithm for better predictions, training of the model becomes faster, the complexity of algorithms is reduced and the problem of overfitting is minimized.

3.1 Filter Methods

The main criterion for variable selection by ordering in filter methods uses variable ranking approaches. Ranking techniques are utilized because they are straightforward and have a proven track record in actual applications. The variables are scored using an appropriate ranking criterion, and variables below the threshold are eliminated. Since ranking techniques are used prior to classification to weed out irrelevant factors, they are filter techniques. The ability to provide meaningful knowledge about the various data classes is the fundamental characteristic of a unique feature. This characteristic is best described as feature relevance [18], which gives an indication of how well a feature can distinguish between various classes. This method is used for preprocessing in most cases. The process lifecycle of the filter method is graphically demonstrated in Fig. 3. It used relationships and correlation, for example, chi-square, ANOVA, LDA and Pearson correlation etc.

Fig. 3. Process lifecycle of filter methods.

3.2 Wrapper Methods

An attempt is made to train a model using a subset of features and choose to add/ subtract features in reference to the earlier model. The wrapper method is highly expensive. The process lifecycle of the filter method is graphically demonstrated in Fig. 4. Due to the difficulty of evaluating 2^N subsets, poor subsets are discovered by using search algorithms that heuristically locate a subset. Tree structure was employed in the BB (Branch-bound) approach [19] to assess several sub-sets for the specified features. But if there were more features [18], the search would expand tremendously. For larger datasets, exhaustive search techniques can become computationally expensive. Because they are computationally practical and can give effective results, simplified methods like sequential search, swarm optimization [20] and genetic algorithm [21] are used. Commonly used examples are recursive and backward elimination, forward selection and algorithms such as sequential selection and heuristic search.

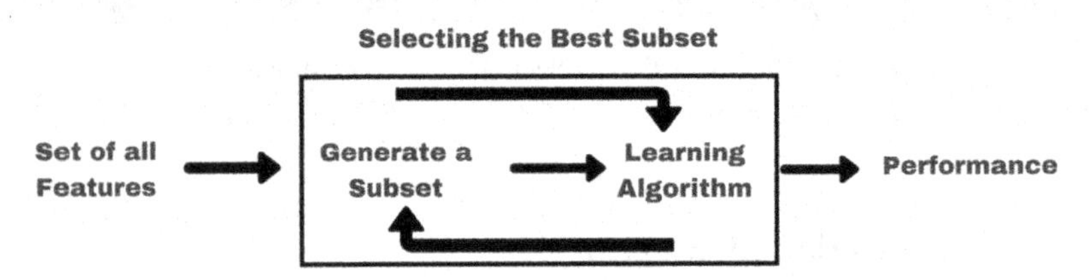

Fig. 4. Process lifecycle of wrapper methods.

3.3 Embedded Methods

To develop embedded methods, best features of filter and wrapper methods are combined. Built-in feature selection strategies are used in this method. Process lifecycle of the filter method is graphically demonstrated in Fig. 5. In contrast to wrapper approaches, embedded techniques [22–24] aim to shorten time required for re-classification. The main strategy is to include feature selection throughout the training phase. Although mutual information (MI) is a crucial concept, the ranking utilising MI produced subpar results and class output was taken into account. A greedy search strategy is employed in [25] to assess the subgroups. The objective function is created in a way that when a feature is selected, the mutual information between feature/class output is maximized, while feature and the subset of the features that have already been selected is minimized. LASSO and RIDGE reduce overfitting [26, 27]. Regularization L1 and L2 are carried out during lasso regression and ridge regression respectively.

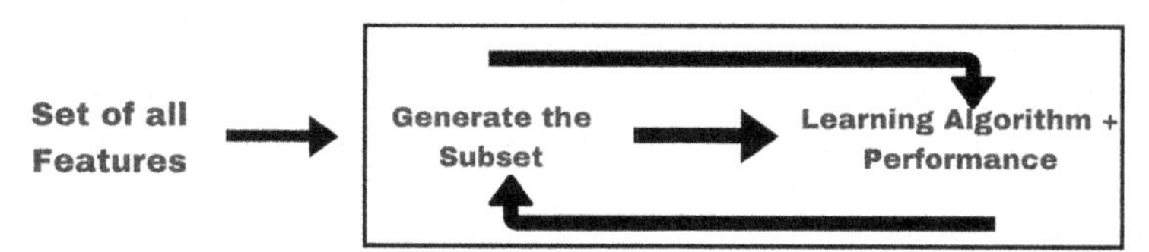

Fig. 5. Process lifecycle of embedded methods.

3.4 A Concatenation Approach Using Meta-data

In order to identify most suitable features for experiment purpose, various state of the art techniques falling under the filter, wrapper and embedded methods were explored experimentally. Techniques such as information gain, chai-square test, correlation coefficient were applied to select relevant attributes/features with respect to target attribute using correlation approach. Recursive feature estimation and generic algorithms were used to combine the attributes. Further, approach of combining attributes was also experimented using decision trees as well and it has been noticed that it is more cheaper or faster. Finally, meta-data of deceptive reviews was concatenated and performance of new approach was tested and same is presented in next section [28].

4 Experimental Setup and Result

This section will present an experimental setup of a meta-data-based approach used for feature selection, datasets used for experimental purposes, evaluation metrics and results. Only structured data was used for experimental purpose. Amazon datasets were used for experimental purposes having twelve features namely reviewer id, reviewer name, review time, product id, review text, summary, vote, image, overall, style, verified and unix review time. Further, performance of selected features is tested with the help of metrics such as accuracy, precision, recall, error rate and f1-score. Results show that it doesn't matter how many features are used as the independent variable; the property or characteristic of the features is more important than the count. Table 1 is presenting performance on various metrics with respect to count of features and type of features.

Table 1. Comparative performance of features selected using filter, wrapper and embedded methods.

Name of Features	Accuracy	Precision	Recall	F1-Score	Error Rate
reviewerID, reviewerName, reviewTime, asin, reviewText, summary, vote, image, overall, style, unixReviewTime	0.63450704	0.62505856	0.67450704	0.64884	0.36549296

(continued)

Table 1. (*continued*)

Name of Features	Accuracy	Precision	Recall	F1-Score	Error Rate
summary, vote, image, overall, style, unixReviewTime	0.73450704	0.80505856	0.70422535	0.75127	0.26549296
reviewerID, reviewerName, reviewTime, asin, unixReviewTime	0.85802817	0.87896184	0.84802817	0.86322	0.14197183
vote, image, style	0.59450704	0.42505856	0.53450704	0.47354	0.40549296
reviewerID, reviewerName	0.73802817	0.72896184	0.83802817	0.77970	0.26197183
reviewText, summary	0.83450704	0.82246701	0.83450704	0.82844	0.16549296
reviewTime, unixReviewTime	0.63450704	0.64775059	0.69545070	0.67075	0.36549296
reviewText	0.82394366	0.81276856	0.82394366	0.81832	0.17605634
Summary	0.72802817	0.77896184	0.76546817	0.77216	0.27197183

Performance of various approaches are presented graphically such as Fig. 6 is presenting a demonstration of comparative performance when different number of features are applied for each test, Fig. 7 presents the performance of embedded vs filter methods, Fig. 8 presents the performance of the direct feature and Fig. 9 is presenting performance of meta-data vs direct features.

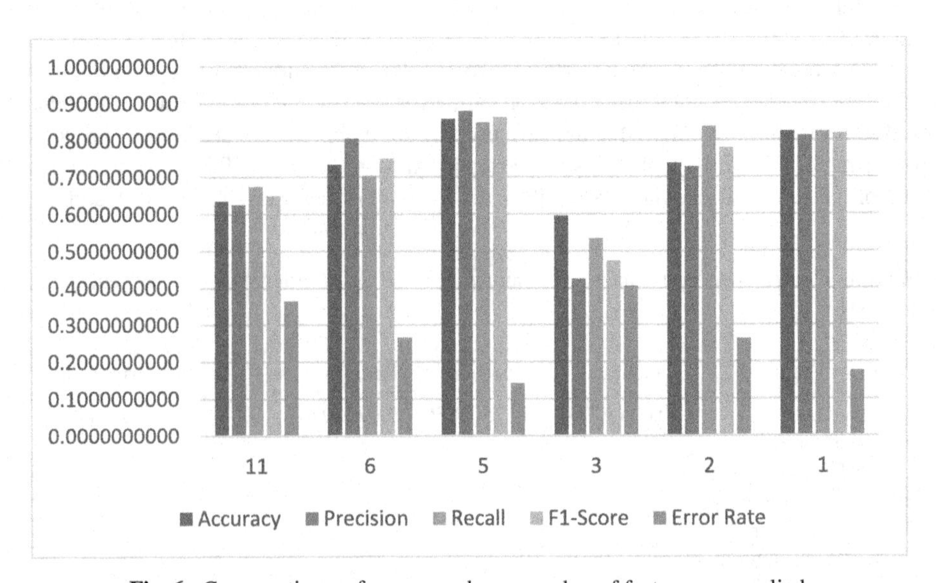

Fig. 6. Comparative performance when a number of features are applied.

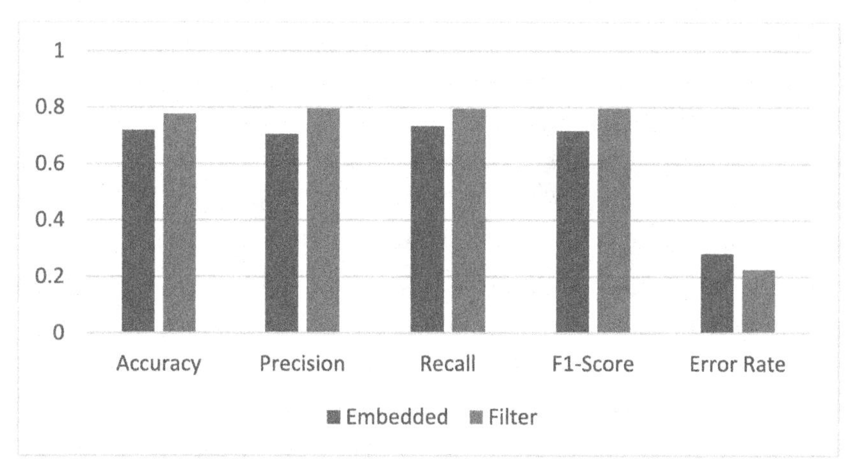

Fig. 7. Performance of embedded vs filter methods.

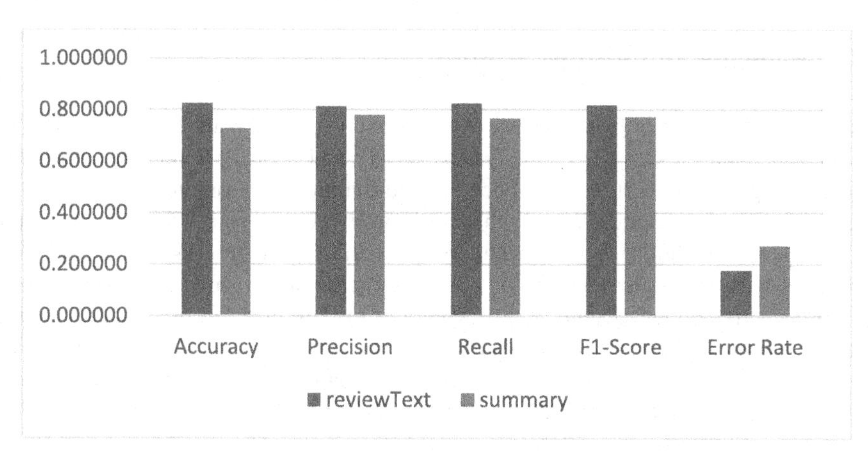

Fig. 8. Performance of direct features.

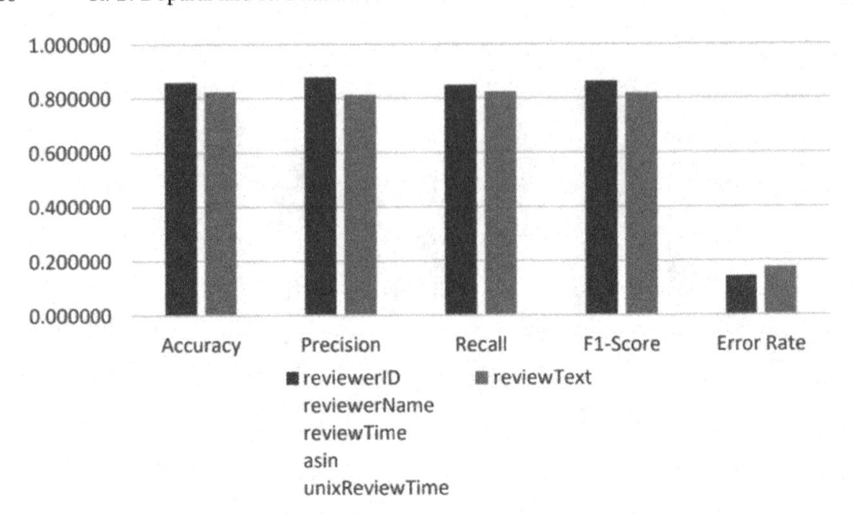

Fig. 9. Performance of meta-data vs direct features.

5 Conclusion and Future Scope

Fake reviews on internet platforms are sometimes considered spam and that's why day by day it becomes a very big challenge for society and people are relying upon these before taking decisions. It has also been noticed that products, organizations and even persons are praised/dispraised using these reviews which are sometimes being written by fake users or authors. This paper has tried to optimize the process of detecting fake reviews by focusing on the selection of attributes instead of focusing more on techniques to find deceptive reviews. State of art techniques are discussed in this paper which is further implemented on the amazon data set and their performance is tested and verified using various metrics. The leading outcome of this paper is that time and energy spent on the selection of features is really fruitful for better predictions.

References

1. Kalousis, A., Prados, J., Hilario, M.: Stability of feature selection algorithms: a study on high dimensional spaces. Knowl. Inf. Syst. **12**(1), 95–116 (2007)
2. Boparai, R.S., Bhatia, R.: Detection of fake profiles in online social networks – a survey. In: Proceedings of the Advancement in Electronics & Communication Engineering, pp. 52–57 (2022)
3. Molina, L.C., Belanche, B., Nebot, A.: Feature selection algorithms: a survey and experimental evaluation. In: Proceedings of ICDM, pp. 306–313 (2002)
4. Dash, M., Liu, H.: Feature selection for classification. Intell. Data Anal. **1**(1), 131–156 (1997)
5. Guyon, I., Elisseeff, A.: An introduction to variable and feature selection. J. Mach. Learn. Res. **3**(1), 1157–1182 (2003)
6. Fong, S., Yang, X.S., Deb, S.: Swarm search for feature selection in classification. In: Proceedings of 2nd International Conference Big Data Science Engineering, pp. 902–909 (2013)

7. Fong, S., Liang, J., Wong, R., Ghanavati, M.: A novel feature selection by clustering coefficients of variations. In: Proceedings of 9th International Conference Digital Information Management, pp. 205–213 (2014)
8. Selvi, S., Valarmathi, M. L.: An improved firefly heuristics for efficient feature selection and its application in big data. Biomedical Research (Special Issue), S236-S241 (2017).
9. Kivinen, J., Smola, A.J., Williamson, R.C.: Online learning with kernels. IEEE Trans. Signal Process. **100**(10), 1–12 (2010)
10. Grolinger, K., Hayes, M., Higashino, W.A., L'Heureux, A., Allison, D.S. Capretz, M.A.: Challenges for mapreduce in big data. In: IEEE World Congress on Services (SERVICES), pp. 182–189 (2014)
11. Huang, G.B., Zhou, H., Ding, X., Zhang, R.: Extreme learning machine for regression and multiclass classification. IEEE Trans. Syst. Man Cybern. Part B (Cybern.) **42**, 513–529 (2012)
12. Liu, X., Tang, J.: Mass classification in mammograms using selected geometry and texture features, and a new SVM-based feature selection method. IEEE Syst. J. **8**(3), 910–920 (2014)
13. Almusallam, N., Tari, Z., Chan, J., Fahad, A., Alabdulatif, A., Al-Naeem, M.: Towards an unsupervised feature selection method for effective dynamic features. IEEE Access **9**, 77149–77163 (2021)
14. Taşkın, G., Kaya, H., Bruzzone, L.: Feature selection based on high dimensional model representation for hyperspectral images. IEEE Trans. Image Process. **26**(6), 2918–2928 (2017)
15. Haq, A.U., Zhang, D., Peng, H., Rahman, S.U.: Combining multiple feature-ranking techniques and clustering of variables for feature selection. IEEE Access **7**(1), 151482–151492 (2019)
16. Gong, L., Xie, S., Zhang, Y., Wang, M., Wang, X.: Hybrid feature selection method based on feature subset and factor analysis. IEEE Access **10**(1), 120792–120803 (2022)
17. Feng, D., Chen, F., Xu, W.: Efficient leave-one-out strategy for supervised feature selection. Tsinghua Science and Technology **18**(6), 629–635 (2013)
18. Kohavi, R., John, G.H.: Wrappers for feature subset selection. Artif. Intell. **97**, 273–324 (1997)
19. Narendra, P., Fukunaga, K.: A branch and bound algorithm for feature subset selection. IEEE Trans. Comput. **6**, 917–922 (1977)
20. Kennedy, J., Eberhart. R.C.: Particle swarm optimization. In: Proceedings of IEEE International Conference on Neural Networks, pp. 1942–1948 (1995)
21. Goldberg, D.E.: Genetic Algorithms in Search, Optimization and Machine Learning. Addison-Wesley, Boston (1989)
22. Guyon, I., Elisseeff, A.: An introduction to variable and feature selection. J. Mach. Learn. Res. **3**, 1157–1182 (2023)
23. Langley, P.: Selection of relevant features in machine learning. In: AAAI Fall Symposium Relevance (1994)
24. Blum, A.L., Langley, P.: Selection of relevant features and examples in machine learning. Artif. Intell. **97**, 245–271 (1997)
25. Battiti, R.: Using mutual information for selecting features in supervised neural net learning. IEEE Trans. Neural Netw. **5**(4), 537–550 (1994)
26. Ashour, A.S., Nour, M.K.A., Polat, K., Guo, Y., Alsaggaf, W., El-Attar, A.: A novel framework of two successive feature selection levels using weight-based procedure for voice-loss detection in Parkinson's disease. IEEE Access **8**(1), 76193–76203 (2020)
27. Wang, Z., Xiao, X., Rajasekaran, S.: Novel and efficient randomized algorithms for feature selection. Big Data Min. Anal. **3**(3), 208–224 (2020)
28. Boparai, R.S., Bhatia, R.: Deceptive web-review detection strategies: a survey. Int. J. Intell. Eng. Inf. **10**(5), 411–433 (2022)

Fake News Detection Using Machine Learning

Abdul Samad[1], Namrata Dhanda[1](✉), and Rajat Verma[2]

[1] Department of Computer Science and Engineering, Amity University Uttar Pradesh, Lucknow Campus, Lucknow, India
ndhanda@lko.amity.edu

[2] Department of Computer Science and Engineering, Pranveer Singh Institute of Technology, Kanpur, India

Abstract. The news which is specially created to misguide as well as mislead the readers is termed fake news. Fake news can cause potential harm to both the individual and society. The problem of fake news has increased far more quickly in recent years. Social networks have significantly changed the scope and effect of their overall influence. Daily, a huge amount of information is released in print and online media, but it can be difficult to determine if the information is accurate or not. A significant amount of study has been done in this area in recent years with positive outcomes. In this modern era, machine learning is the answer to every issue. It can provide answers that people cannot readily think of and is applied in practically all disciplines, domains, and research. Many different algorithms in machine learning and artificial intelligence can help us in identifying as well as eliminating fake or false news. In this article, supervised learning is used. To determine the false news, the authors have used four machine learning algorithms in this study: Decision Tree (DT) classification, Logistic Regression (LR), Random forest (RF) and Passive Aggressive classifier (PAC). After comparing the output of each method, the one with the best accuracy is selected. The PAC outperforms the other classifiers and produces the best result.

Keywords: Machine learning · Confusion Matrix · Random Forest Classifier · Passive Aggressive Classifier · Decision Tree Classification · Logistic Regression

1 Introduction

The main aim of false news is deliberately mislead an individual or a group of individuals. It is a form of propaganda that is distributed under the disguise of actual news. More and more individuals are getting their news through social networks rather than from traditional media as a result of social networks' rising popularity. It is one of the biggest threats to democracy as well as the freedom of speech today. The general public's trust in the government has been damaged. The spread of false information will become more obvious over time. False information can spread widely and do great harm to both individuals and society.

Fake news, a form of yellow journalism, is news that might be a hoax and it is typically distributed through social platforms or any other kind of online source. This in turn is

R. K. Challa et al. (Eds.): ICAIoT 2023, CCIS 1930, pp. 228–243, 2024.
https://doi.org/10.1007/978-3-031-48781-1_18

frequently carried out with political agendas to advance or impose particular ideas. The rise of misleading websites has two main reasons: one is financial, as viral news earns significant advertising money; the other is ideological since false news outlets frequently try to influence public opinion on these topics [1]. It was noted that during the election of the US president, many of the circulated articles on social platforms were false. In recent, a fake video that was concerning Kerala fighting the floods went viral on social platforms like Facebook and claimed that Kerala's Chief Minister was enforcing the Indian army to cease conducting rescue operations in the parts of Kerala which were flooded [2]. The news that a particular user sees online may not always be accurate. When the news has an emotional connection to them, the probability that they will distribute it increases significantly [3]. It is becoming more and more difficult in making the difference between false and true news, which causes problems and misunderstandings. It is challenging to spot fake news manually; only when the individual doing it has a thorough understanding of the news's subject it is possible. Modern technology innovations have now made it quite easier to create as well as spread a false piece of information, but it is much more challenging to tell if the information is accurate or false. Based on their textual content, certain articles could be flagged as fraudulent using a variety of computer algorithms. Most of these methods are dependent on some of the fact checker websites like "PolitiFact" and "Snopes". There are several archives maintained by researchers that list the websites that have been considered suspicious and fake. The issue with this kind of resource is that the articles and websites that are fake must be identified by human expertise. More importantly, fact-checking websites only include articles from specific fields, like politics, and do not use a broad definition to distinguish between fake news from multiple fields, like entertainment, sports, and technology. The internet contains data in many different formats, such as documents, audios, and videos. Unstructured content that is released online, such as news, articles, videos, and audio, presents more difficulties for categorization and identification because it is purely a human-expertise task.

The Logistic regression, Passive aggressive classifier (PAC), Random Forest and Decision tree classification were employed as a supervised form of learning classifiers in this study for comparison. PAC is highly suited for news stories that are constantly being published. Online learning is more suited to applications where the training set typically arrives sequentially because predictive models are updated sequentially, compared to conventional batch learning that assumes training data is already available even before the learning phase starts. Even more, websites exist that nearly solely manufacture bogus news. They purposefully spread propaganda, disinformation, hoaxes, and half-truths disguised as news, and by frequent use of social media to boost web traffic and their impact. Given how challenging it has become to identify false news on social media, it is clear that an effective strategy must consider a variety of variables. Despite these drawbacks, machine learning has been crucial for classifying information. A significant amount of study in this field has been done recently with positive outcomes. The development and success of artificial intelligence and machine learning have relieved humans of needless labour. Machine learning is the study of how computers learn without being explicitly programmed. In this field computers or machines are allowed to learn and act like a human mind. Machine learning is one of the newest and most promising technology. Computers are most different from humans in that they can learn, as the name suggests. Machine learning is undoubtedly utilized widely in different kinds of applications today. In a variety of fields where it's difficult or not practical to develop conventional techniques which can complete the required tasks, such as computer vision, speech recognition, email filtering, medicine, and agriculture, machine learning algorithms are used. These technologies can protect society from unneeded upheaval and social unrest by detecting fake news. Machine learning uses many different kinds of algorithms and mathematical approaches to train and learn a system.

First, the authors receive user input in the form of a dataset, which is then handled by generating models based on a TF-IDF array. The next step is to extract the best features for TF-IDF vectorizer. For this, the most frequently used words are employed, with stop words such as "the," "when," and "there" removed, as well as the words which appear at least a certain number of times in the provided textual data set. The real goal is to develop a model that performs best with the TF-IDF vectorizer and choose which classifier gives us the best result among the four which were used. For comparison, supervised learning classifiers such as the Random Forest (RF), Decision Tree (DT) classification, Passive-Aggressive classifier (PAC) and Logistic Regression (LR) are employed.

Finally, the models' performance is determined by means of confusion matrix. It is a graph that is used to represent how effectively a classifier or classification model performs on a given collection of data. Depending on how well the classifiers work, the accuracy is received. Based on true and false positives and negatives, accuracy is determined. The formula used in the calculation of accuracy is written in (1).

$$Accuracy = \frac{TP + TN}{TP + FP + TN + FN} \tag{1}$$

Here TP shows "True Positive", TN is "True Negative", FP is "False Positive", and FN is "False Negative".

2 Literature Review

Since fake news has such a large negative influence on civic and social behaviour, research into fake news detection is a rapidly developing concern that is becoming more and more important every day.

One of the classification approaches used in [1] is the Passive-Aggressive classifier (PAC) for identification of the fake news. To determine the classifier with higher precision, comparative research is also conducted in which the efficiency of other classifiers is measured. The additional classifiers include the Random forest (RF), K-nearest neighbors (KNN), Decision Tree (DT) and Support Vector Machine (SVM). In addition to these classifiers, several additional deep learning, as well as a machine learning process, have been used to tackle the same issue. Here supervised form of learning is used. It is the process of developing a model using a data set that comprises input and output parameters. Classification is also an algorithm in machine learning. The process of classifying data elements involves predicting their class. Tags or categories are used to describe the classes. A classifier organizes data into predefined groups or categories.

One of these classifiers utilized in solving the problem of detection of false or fake news is the passive-aggressive classifier. With the confusion matrix, the first data set used for classification had a 98.87% accuracy rate, with true positives equaling 417, false positives equaling 6, false negatives equaling 1, and true negatives equaling 373. The accuracy of the second data set used in classification was 96.25% using a confusion matrix, with True positives equaling 1002, False Positives equaling 30, False Negatives equaling 45, and True Negatives equaling 924.

The three feature extraction strategies employed by the authors in the paper [2], include Hashing, Count as well as TF-IDF vectorizer. In the following phase, the produced features are passed to the chosen classification techniques. Different machine learning (ML) models, including PAC, Logistic regression (LR), Multinomial Naïve Bayes (MNB), Stochastic Gradient Descent (SGD) as well as a Multi-layer perceptron, Nu support vector, AdaBoost, SVM, DT, Gradient boost, Linear support vector and voting classifiers are utilized in learning and to recognize the patterns as well as their outcomes. These models are assessed using some of the performance indicators to select a better classifier that is effective. These models are then combined based on the analysis done to suggest a multilevel voting model achieves higher efficiency and then the previously performed work is compared to it.

In [3], it is dedicated to the handling of false news in both NLP and machine learning. In this paper, many data sets from various sources are used. On the data sets, the authors have decided to apply multinomial naïve bayes and PAC and also deep neural networks (DNN) are used. First, missing values and inaccurate data were eliminated from the data sets. Based on relevancy, some datasets had unnecessary columns that were eliminated. Following that, the data sets are split into different parts like training, development, and test set. The optimization of models was also done during training and development and then tested on the "test" set. Except for one data set, it was discovered that the DNN performed better than both the passive-aggressive as well as Naive Bayes classifiers. The success of DNN is a result of the better representation of complex and nonlinear structures by neural networks, which in this situation fits well.

In the study [4] the dataset is first preprocessed using text preprocessing techniques, then fed to machine learning classifiers, and finally assessed using performance metrics. The text preparation techniques make use of TF IDF and Count Vector. 20 percent of the total data of the data set was used for testing the model whereas the rest 80 percent was used to train the model. For comparison, they employed the scalable and efficient PAC, which is an online supervised kind of classifier. It is an online learning classifier that is utilized in situations like news, social media, and other situations where it is necessary to monitor the data constantly. Using a few of the performance measures, including recall, F1 score and also precision and accuracy. PAC has been compared to five distinct machine learning algorithms, which include the NB, SVM, the RF, LR, and Stochastic Gradient Descent(SGD). It was discovered that employing TF-IDF vectorizer and Naïve Bayes results in the poorest precision while PAC and SVM provide the highest precision. SVM and PAC provide the highest accuracy, whereas NB provides the poorest accuracy. The PAC as well as SVM are the best algorithms for this dataset when utilizing TF-IDF. The conclusion derived is based on the accuracy of each model. When TF-IDF was employed, both of these algorithms produced more accurate results, and when the count vector was used, the results were less accurate.

Paper [5] describes the procedure in three sections. The first section is static and focuses on a machine-learning classifier. They studied the model and trained it with four distinct classifiers before selecting the most effective one for the test run. The second part, which is dynamic, uses the user's keyword or text to do an online search to determine the probability that the news is accurate. The final stage validates the authenticity of the user-provided URL. In this experiment, the training of the model is done using 67% of the whole data, and it is then tested using the remaining 33% of the data. A confusion matrix is created to demonstrate the results of applying different extraction methods, like Bag of words, TF-IDF vectorizer, and N-grams, to these three classifiers, RF, LR and the NB. With a 65% accuracy rate, Logistic Regression was shown to be the best-performing model. The performance of LR was subsequently improved by the authors using gridded search parameter optimization, yielding a 75% accuracy.

In [6], the authors have chosen some machine learning techniques like NB, DT and passive-aggressive as well as NLP techniques. For feature extraction, the count, as well as TF-IDF vectorizer, are employed, and for classification, the DT along with the Naive Bayes and Passive aggressive (PAC). By analyzing the confusion matrices of other classifiers, they found that the PAC utilizing the TF-IDF offered the highest accuracy as compared to the other classifiers.

The authors in paper [7] employed the TF-IDF as well as the count vectorizer for text processing, and for the classification, they used NB and the PAC. This paper concludes that the PAC and TF-IDF vectorizer are efficient because authors attained the 90% accuracy with this model.

In [8], the authors employed frequency-based functions to train the algorithms Stochastic gradient classifier, multinomial NB along with SVM, LR and PAC. The results were deemed to be quite promising and provide scope for more study in the field. With a TF-IDF vectorizer, linear SVM and stochastic gradient classifier algorithms both obtained accuracy and the ROC area under the curve values of 90% and 95% respectively. The area under the ROC curve (AUC) is a binary classification performance statistic. If the value of the AUC is 1.0, then all of the predictions were accurate, whereas an AUC of 0.5 is considered to be good for random guesses. In two phases, this project has advanced. In the First Phase, it used the headlines and body of the articles to perform the highest posterior probability technique. The authors found that this strategy was not particularly effective because there was a risk that fake news may appear in a well-written piece, even though they saw greater accuracy results. In the second phase, they suggested a method for more precisely classifying false news by looking at how readers responded to such news pieces. In this paper, the author trained the model on each of two distinct datasets (LIAR and another dataset downloaded from GitHub) that were both of varying lengths. The author used Web scraping technologies in the name of Python, such as Selenium and Beautiful Soup, to extract the comments from each URL of the collected posts. With Selenium, the authors can extract the content from the server version of the website, but the Beautiful Soup library can only extract data from the client version of the page. Therefore, to scrape the necessary data, Beautiful Soup was combined with Selenium. The linear support vector machine with the TF-IDF vector performed better than other models in this experiment, with the greatest classification accuracy of 93.2%, the highest sensitivity of 92%, and the highest ROC AUC score of 97% documentedg.

The aim of paper [9] is to examine and compare several techniques to solve this problem, using a few of the traditional ML processes, such as Naive Bayes along with deep learning methods like hybrid CNN and RNN. For text mining and picture identification, CNN is frequently used. The hybrid strategy would improve the model's performance and produce better outcomes for content-based false news identification. These deep learning methods are used in solving issues like speech recognition, NLP as well as language translation and picture captions. In comparison to previous approaches and implementations shown in this paper, the TF-IDF along with the hybrid CNN as well as the RNN model has produced good outcomes.

NB, SVM, as well as the PAC, were utilized by authors in [10]. In this, the SVM outperforms both the PAC as well as the NB algorithms in terms of accuracy, but it takes longer to complete the task. On the test set, the PAC achieves an accuracy of 92.9%, NB classifier has an accuracy of 84.056%, and the SVM provides 95.05% accuracy. SVM is supervised learning algorithms. As a result, the model is created after training. SVM is mostly used to categorize fresh data. SVMs perform better with smaller data sets. SVM has a big disadvantage that while working on big datasets that are in training it takes a long time. Online learning techniques known as passive-aggressive algorithms are employed for both regression and classification. Compared to SVM, it is quicker and easier to use, but it does not offer the same level of accuracy. Large-scale data classification is its primary application. As a result, the SVM classifier is where the suggested model reaches its most remarkable accuracy. The accuracy score of 95.05% is the highest.

In the paper [11], the authors identify news items from diverse areas as genuine or false using ensemble approaches and different sets of linguistic features. A 70/30 split is used to divide each data set into training as well as testing sets. For maximizing accuracy for a particular data set with an ideal balance between the variance and the bias, the learning algorithms were then trained on several hyperparameters. To optimize the model for getting the best results, every model is trained several times on various sets of parameters, using grid search. To evaluate performance across numerous data sets, several ensemble approaches, including boosting bagging and the voting classifier, are examined. Authors employ two distinct voting classifiers made up of three learning methods: the very first voting classifier consists of an ensemble of the random forest along with K-nearest neighbor (KNN) and LR, while the next voting classifier is made up of the linear SVM, LR and classification as well as regression trees. Individual models are trained using the optimal parameters to train the voting classifiers, and each model is then evaluated based on the output label choice made using the top three models' votes. Using two booster set algorithms—AdaBoost and XGBoost, they trained a bagging set made up of 100 decision trees. For all of the students in the set, a K-fold (k10) cross-validation model is utilized. To lower the total error rate and enhance model performance, more than one model is trained using a specific approach, which results in ensemble learners, which typically have greater precisions. The learning methods that they utilized were the SVM, RF and the Multilayer Perceptron (MLP) along with the LR and KNN. Linear SVM,

CNN, and bidirectional long and short memory networks are the benchmark methods utilized in the paper. Here, recall, accuracy, F1 score, and precision are employed as performance indicators. In comparison to the individual learners, the ensemble learners have consistently outperformed them in all performance parameters.

The authors of [12] have utilized six algorithms to identify false news: XGBoost, NB, RF, DT as well as KNN and SVM. In creating this document, the LIAR PLUS master dataset was utilized. Two sections of data are presented: 75% of the data is in the first portion, which is trained data. Here, the algorithm distinguishes between genuine and false news, and the data is then labelled with a 0 or 1, with 0 denoting false news and 1, true news. After then, 25% of the remaining data will undergo testing to check if the news is authentic or not. Final results reveal that XGBoost has the greatest accuracy, over 75%, followed by SVM and RF, around 73%.

The authors of the paper [13] have proposed a hybrid system for detecting false news that makes use of the incident classification model of machine learning. The model of incident classification is made up of five NLP features and three knowledge check features. They put the system into practice by utilizing the News API for knowledge verification and Google's NLP API for NLP analysis. Their machine-learning technique for detecting fake news was the LR model in event categorization. The Kaggle Dataset of fake news was used for training and validation. Then created a separate test dataset of real news headlines from several verified news sources that adhered to strict journalistic standards, and they processed each example in the set using the mechanism they had already implemented. It displayed the system's estimated probabilities that several news headlines are fake.

The NB classifier, which is among the most utilized machine learning techniques, has been utilized by the authors in [14] to provide a straightforward fake news detection approach. The project aims to investigate the performance of NB for this specific task, provided with the manually labelled dataset, and to encourage the use of AI for false news identification. The authors also learned about the idea of web scraping, which offered a plan for regularly updating the dataset so the authors can verify the accuracy of newly updated Facebook posts. Web scraping is a method for extracting massive amounts of data from various websites. A dataset created by GitHub that contains 11000 news articles classified as true or fake was used to assess the model's efficiency. There are 4 columns and 6335 rows in it. Index along with the title, text as well as label are the four columns in the data set. The results of the simple NB implementation show that the AUC scores have risen as the amount of data already collected under a given label has increased, as in the case of Title and Text. Because the title was a condensed form of the news stories, and the text was a more detailed version. The authors observed that the second model's area under the ROC curve scores increased when the idea of n-grams was applied because the second model's larger number of vectors gave it a stronger capacity for judgement.

The authors of the paper [15] use a variety of traditional classifiers such as RF, XGBoost (XGB) along with the KNN, SVM with kernel RBF, and NB. To categorize fresh news stories as "fake" or "not fake," each classifier learns a model using a set of labelled data. The XGB and RF classifiers produced the best results, having 0.85 and 0.86 for AUC, respectively.

The writers of the paper [16] extracted significant features in the article using the TF-IDF feature extraction technique. The article headlines, article body, and publication name are the characteristics that were retrieved from the datasets. The classification of tweets uses the NB as well as the passive-aggressive algorithms. The PAC has produced the best result, by giving out an accuracy of 78%.

The authors of the paper [17] employed seven machine learning processes on the three standard data sets. They then used precisions and F1 scores to statistically confirm their results. The study indicates that through experimentation the XGB classifier, in particular, has helped in the effective identification of false or incorrect news with an average accuracy of 88% and an F1 score of 0.91, beating other ML classifiers.

In [18], the authors present an ensemble of multiple models and reclassify the problem as a binary classification instead of a multiple classification problem. Ensemble machine learning refers to a technique that integrates the output of multiple learners and applies it to a data set to make a prediction. When multiple base models are used to extract predictions that are combined into a single prediction, that prediction is likely to provide higher accuracy than individual base learners. They have used N-Grams, Bag of words, and TF-IDF vectorizer as textual feature extraction techniques. Here, the Bagging classifier and AdaBoost give 70% accuracy in precision, F1 score and recall.

The authors of [19] utilized five alternative categorization methods, including LR, long-term memory, SVM along with the NB and recurrent neural network (RNN). NB along with the SVM and the LR all have the same degree of precision. SVM outperforms other algorithms when a recall is considered, and SVM and Naive Bayes both have the greatest F1 scores of 0.94.

To classify the fake news dataset, the authors in [20] used ten different kinds of machine learning along with deep learning classifiers along with four conventional methods for extracting features from texts, including count vector, TF-IDF, N-gram level vector and along with the character level vector. The outcomes show that false news content can be identified better with the use of the convolutional neural network. Using several classifiers, this study got a precision range of 81 to 100%.

After conducting this extensive literature survey some of the machine learning techniques were analyzed which had outperformed other algorithms which were used along with them. Some of the best-performing algorithms which were used by many authors in their study are Logistic Regression (LR), Passive Aggressive Classifier (PAC), SVM, Random Forest (RF) and XGBoost.

3 Proposed Approach

Online news may come from many different kinds of places, including search engines, news agencies' homepages, social networking sites, and fact-checking websites. There are various publicly available datasets for identifying false news on the internet. These data sets have been widely utilized in numerous research projects to assess the trustworthiness of the news. The data set employed for detection of fake news was downloaded from Kaggle's official website. This dataset is also small in size and requires a memory space of only 29.2 MB. It has a size of 7796 x 4. Accordingly, there are 7796 rows and 4 columns. The news is defined in the very first column, the headlines, as well as its content, are present in the second as well as third columns respectively, and the news' TRUE or FAKE status is indicated by labels in the fourth column.

3.1 Data Preprocessing

Social media data is unstructured sources of information with typos, slang, bad language, etc. It is also relatively unstructured. Data needs to be cleaned before it can be utilized for predictive modelling to get better insights. This was accomplished by doing a fundamental preprocessing on the News training data. To transform the raw data into the necessary format, data preparation, data cleaning, data reduction, integration etc. is carried out. Datasets for this project are gathered from several sources with a variety of properties and formats. As a result, the data could be redundant and contain unhelpful qualities. So, using the necessary properties, the authors converted the data in desired format.

A text classification model performs best when the words or text of a corpus and features derived from those words are taken into account. Common terms, or "stop words," like "a," "the," and "they," are eliminated, and the words which are appearing only a few predetermined numbers of times in specific text datasets are kept. Before the data is ready for analysis, text preparation is a crucial step. With a smaller sample space for features in a noise-free corpus, accuracy is increased.

3.2 Feature Extraction

Feature extraction is the process which involves extracting clean and more precise information from the data. The data is initially collected and stored in CSV format. When solving real-world machine learning issues, the authors primarily used this approach to extract or get the necessary information from the whole set of raw data in a dataset. Here, in this model TF-IDF is the method which is employed.

TF-IDF Vectorizer. It is a technique which is used in converting the text into a meaningful representation of numbers. Based on their frequency, it is used to extract text characteristics from strings. For instance, if several words are repeated many times, then it means that the given text has higher importance. The author normalized the appearance of the word with the size of the document and that is why it is called frequency. The formula to calculate term frequency is written in (2).

$$TF = \frac{Number\ of\ times\ term\ appears\ in\ the\ document}{Total\ number\ of\ terms\ in\ the\ document} \tag{2}$$

When calculating the frequency of terms, each term is assigned the same weight. Some terms could appear often in the papers, which would make them less helpful in determining the meaning of the text. Such words, such as "a," "the," and others, may lessen the impact of words with more significance. The Inverse Document Frequency factor is applied to lessen this impact. The formula used for calculating it is shown below in (3).

$$IDF = loglog\left(\frac{Total\ number\ of\ documents}{Number\ of\ documents\ containing\ the\ term}\right) \tag{3}$$

The product of TF and IDF is then used to calculate TF-IDF as illustrated in (4). A higher TF-IDF score would be assigned to the most significant terms.

$$TF - IDF = TF \times IDF \tag{4}$$

3.3 Algorithms Used For Classification

The following algorithms have been used for the classification of fake news in the model.

Passive Aggressive Classifier (PAC). This is one of the few online learning methods which is used for large-scale learning. It is quick and also simple to put into practice. Such an algorithm is passive in the case of a successful classification and turns aggressive in the event of a calculation, update, or adjustment error. As opposed to batch learning when using the whole training data set at once, the online ML techniques make use of the incoming data in a sequence and also update the model of machine learning incrementally. When there is a lot of data and it would take too much computing power to train the complete data set, this is extremely helpful. To put it simply, an online learning algorithm acquires an example for training and updates the classifier and later on, it discards the sample. Finding fake news over social media platforms such as Twitter where a large number of new information is getting published each second, would be a very good illustration of this. It would be excellent to use an online-learning algorithm to dynamically read Twitter data continuously because the amount of data would be enormous.

Logistic Regression (LR). It is one of the most well-known ML methods used. This algorithm is used in the model because of some of its important features, it utilizes the predetermined set of independent factors and then it predicts a categorical dependent value. The output predicted by the logistic regression is a categorical dependent variable. The output of it must thus be a categorized or a discrete value. It either can be a true or a false, either 1 or 0, a yes or a no, etc. Besides generating a fixed value as an output between 0 and 1, it gives out the probability values which fall under the range of 0 to 1. It is used for estimation of the likelihood that an event will occur (0/1, True/False, Yes/No). The categorization issues are resolved using logistic regression because there are only two scenarios such as either news would be fake or true.

This is particularly a helpful machine learning algorithm since it can help in the classification of new data using both discrete as well as continuous data sets. It uses the sigmoid function, a mathematical tool, to convert the predicted values into probabilities.

Decision Tree Classification (DT). This algorithm is used in this paper for detecting fake news because it is a supervised form of learning algorithm which can be utilized in solving both classification problems as well as prediction problems but it is mostly utilized in handling classification problems. It is like a tree kind of structured classifier. In this, there are the decision as well as the leaf nodes present in it. Where leaf node represents the outcomes or results of the choices and they do not have any further divisions or branches while the decision nodes are those that are used in making decisions and may have a large number of branches. To run the test or make the decisions, the given dataset's features are used. It uses a graphical representation to find every solution that is possible under the given circumstances to a particular choice or a problem. It is termed a decision tree because it starts from a root node which develops further in multiple branches and forms a tree-like structure.

Random Forest (RF). This machine learning algorithm is also a type of supervised learning. Here this technique can be used for solving the problems of both regressions as well as classification. As the name suggests, this method employs multiple decision trees on the different subsets of a particular data set provided and then it takes their overall average which results in obtaining a better accuracy from the model. It takes the outputs of each decision tree and based on the majority of the predictions it predicts a particular output, as opposed to relying just on one decision tree. The more trees would be there, the better the results would be achieved in determining the true and fake news.

3.4 Metrics Used to Access the Performance of Model

Some of the most significant metrics used to evaluate a machine learning model's effectiveness are examined in this section. These metrics assess the accuracy with which the model can categorize or assess predictions. In this project, the metrics listed below are applied.

Confusion Matrix. It is the tabular form of representation that represents the performance of a model. It shows how many correct and incorrect predictions are made by an algorithm. The performance of the particular model may be determined by using a confusion matrix and getting different measures like accuracy, F1 score, recall and precision.

For a binary classification issue, a 2×2 matrix is preferred. The target variable can be given either positive or negative values. Its column displays the real values of the target variable whereas rows display the estimated or can say the predicted values for the target variable. The basic terms that are required to find the different performance metrics are as follows:

- True Positive (TP): It is the positively predicted numbers that were true.
- True Negative (TN): It is the accurately foreseen negative values.
- False Positive (FP): It is also called type 1 error where the prediction is positive, but the original value is negative.
- False Negative (FN): This is also referred to as type 2 errors, where a positive value is predicted as negative.

Classification Accuracy. In most cases, when something is claimed to be accurate, it has to be correct. It evaluates the proportion of the accurate estimates made relative to all the other predictions. The formula used to calculate accuracy is shown below in (5).

$$Accuracy = \frac{TP + TN}{TP + FP + TN + FN} \tag{5}$$

Precision. Precision is calculated by dividing the actual correct prediction by the total number of predictions the model produced. The formula which is used to calculate it is shown in (6).

$$Precision = \frac{TP}{TP + FP} \tag{6}$$

Recall. It is calculated as the number of TP divided by the total of TP and FN. The formula used for calculating the recall is (7).

$$Recall = \frac{TP}{TP + FN} \tag{7}$$

F1-Score. A weighted average of recall and accuracy is the F1 score. There are false positive and false negative in accuracy and recall, therefore it takes both into account. The formula to calculate the F1-score is written in (8).

$$F1 - Score = 2 \times \left(\frac{Recall \times Precision}{Recall + Precision} \right) \tag{8}$$

4 Results and Discussion

The library Pandas were used to load the data set from the CSV file. Scikit-Learn was used to import the necessary packages used for text preprocessing, algorithms, and performance metrics. The graphics are created using the NumPy and Matplotlib libraries. Python has many libraries and extensions that make it simple to use in machine learning. The Scikit-learn library is one of the good libraries to find machine learning algorithms because it makes every type of ML algorithm available to Python, allowing for quick and simple evaluation of ML algorithms. TF-IDF vectorization is the method used in this study for text preprocessing, and four algorithms—PAC along with the LR, RF and DT classification are used for the classification. The outputs of every algorithm are compared by the authors. Some metrics are used to calculate the outcome. Accuracy is one of the metrics to determine the performance of a particular model that has been used in analyzing the result. The confusion matrix is generated while accuracy is taken into account. The confusion matrix, which contrasts the actual classification with the anticipated classification, is a 2×2 matrix. The outcomes of each algorithm used in this paper are then contrasted. The outcomes of all the employed algorithms are displayed below.

Figure 1 shows the confusion matrix of the PAC which obtained an accuracy of 93% and LR gained accuracy of 92%.

Figure 2 shows the confusion matrix of the DT with an accuracy value of 80% and RF with an accuracy of 90%.

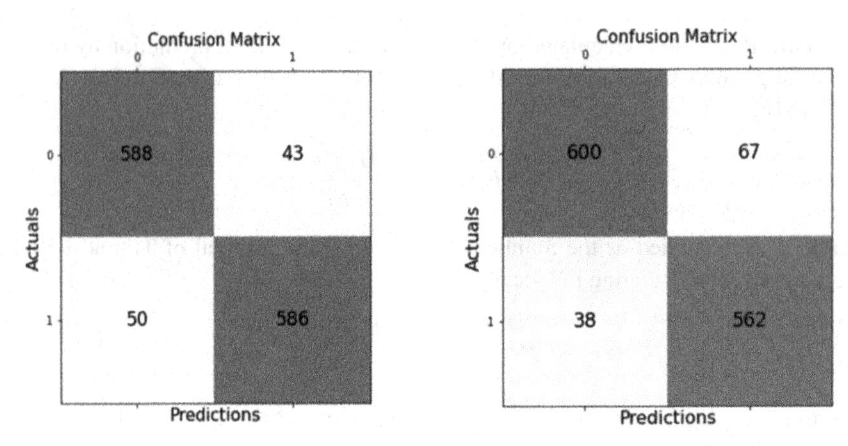

Fig. 1. Confusion matrix for the PAC and LR respectively.

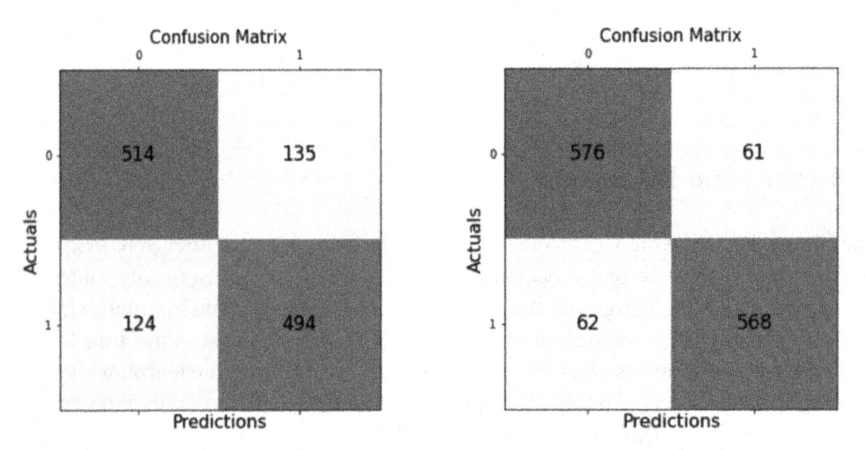

Fig. 2. Confusion matrix for the DT and RF respectively.

The PAC yields the best accuracy for this problem. It can handle big datasets because they only change the model's parameters in response to incorrectly categorized data. Results analysis shows that the Passive-Aggressive Classifier can significantly outperform traditional classification techniques to detect fake news for specific data within proven and effective machine learning algorithms. The Fig. 3 shows the chart of accuracies for all the four algorithms used. The detailed performance of each algorithm used is listed in the Table. 1.

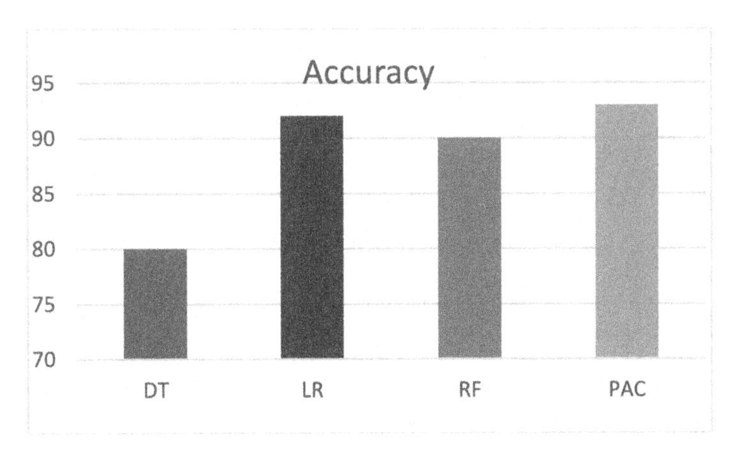

Fig. 3. Accuracies of used techniques.

Table 1. Performance metrics values for all algorithms used.

Algorithms Used	Accuracy (%)	Precision (%)	Recall (%)	F1 Score (%)
Passive Aggressive Classifier (PAC)	93	93	93	93
Logistic Regression (LR)	92	92	92	92
Decision Tree (DT)	80	80	80	80
Random Forest (RF)	90	90	90	90

5 Conclusion and Future Scope

The majority of tasks are being carried out online in the 21st century. Newspapers that were previously more popular in hard copies are now being replaced by online news articles and apps like Facebook and Twitter. Additionally, WhatsApp forward is a valuable resource. The growing issue of fake news only serves to complicate matters further and attempts to sway public opinion and attitude toward the use of digital technology.

Spreading information that is not accurate to its source is known as fake news. They sometimes have grave repercussions and can be deceptive. It is crucial to identify fake news because of this. In this study, the authors looked for the best algorithms for spotting fake news. Additionally, social media platforms have frequently been used to spread misleading or fraudulent information, which has a detrimental impact on both a specific user and also the society at large. This study aims to review, summarize, compare, and assess current research on fake news in its entirety. The major goals of this project were to propose a ML model which can accurately determine whether given news is real or not and to achieve high accuracy in doing so. The TF-IDF vectorizer was used to extract features from a political dataset, and then used four classification algorithms to compare their performance and choose the best one for the model. In the end, a maximum accuracy of the magnitude 93% was obtained. The model's performance is also compared with some existing models in the Table. 2. The proposed model has produced good results.

Table 2. Comparison of proposed model with the existing models.

Authors	Algorithms Used	Accuracy (%)
Thampi, Sabu M., et al. [4]	PAC	94.2
	NB	82.5
	RF	84.7
	LR	91.5
	SVM	94.2
	SDG	93.8
Khanam, Z., et al. [12]	XGBOOST	75
	SVM	73
	RF	73
Nikam, Shivani S., et al. [16]	NB	73
	PAC	78
Bali, Arvinder Pal Singh, et al. [17]	XGBOOST	88
Proposed Model	PAC	93
	LR	92
	DT	80
	RF	90

The project's implementation and the experiment that accompanied it revealed that the PAC and TF-IDF work best together and can accurately identify more than nine out of ten fake news articles, making them ideal for text classification tasks. The use of the PAC and TF-IDF vectorizer is efficient, as this document concludes, as the authors were able to obtain a good level of model accuracy. The classification in this study was done on a select few news stories. As a result, adding more data on the news in the dataset in the future will test its consistency of performance and use additional techniques to boost users' confidence in the system. Additionally, gathering real news that almost seems fake will also enhance model training.

References

1. Ranganathan, G., Chen, J., Rocha, A.: Inventive Communication and Computational Technologies (2021)
2. Kaur, S., Kumar, P., Kumaraguru, P.: Automating fake news detection system using multi-level voting model. Soft Comput. **24**(12), 9049 (2020)
3. Mandical, R.R., Mamatha, N., Shivakumar, N., Monica, R., Krishna, A. N.: Identification of fake news using machine learning. In: 2020 IEEE International Conference on Electronics, Computing and Communication Technologies (CONECCT), pp. 1–6. IEEE (2020)
4. Thampi, S.M., Gelenbe, E., Atiquzzaman, M., Chaudhary, V., Li, K.C.: Advances in computing and network communications. In: Proceedings of CoCoNet, vol. 1 (2020)
5. Sharma, U., Saran, S., Patil, S.M.: Fake news detection using machine learning algorithms. Int. J. Creat. Res. Thoughts (IJCRT) **8**(6), 509–518 (2020)
6. Indarapu, S.R.K., Komalla, J., Inugala, D.R., Kota, G.R., Sanam, A.: Comparative analysis of machine learning algorithms to detect fake news. In 2021 3rd International Conference on Signal Processing and Communication (ICPSC), pp. 591–594. IEEE (2021)

7. Kudari, J.M., Varsha, V., Monica, B.G., Archana, R.J.: Fake news detection using passive aggressive and TF-IDF vectorizer (2020)
8. Ranjan, A.: Fake news detection using machine learning (2018)
9. Han, W., Mehta, V.: Fake news detection in social networks using machine learning and deep learning: Performance evaluation. In: 2019 IEEE International Conference on Industrial Internet (ICII). IEEE (2019)
10. Shaikh, J., Patil, R.: Fake news detection using machine learning. In: 2020 IEEE International Symposium on Sustainable Energy, Signal Processing and Cyber Security (iSSSC). IEEE (2020)
11. Ahmad, I., Yousaf, M., Yousaf, S., Ahmad, M.O.: Fake news detection using machine learning ensemble methods. In: Complexity 2020 (2020)
12. Khanam, Z., Alwasel, B.N., Sirafi, H., Rashid, m.: Fake news detection using machine learning approaches. In: IOP Conference Series: Materials Science and Engineering, vol. 1099, no. 1, p. 012040. IOP Publishing (2021)
13. Ibrishimova, M.D., Li, K.F.: A machine learning approach to fake news detection using knowledge verification and natural language processing. In: Barolli, L., Nishino, H., Miwa, H. (eds.) INCoS 2019. AISC, vol. 1035, pp. 223–234. Springer, Cham (2020). https://doi.org/10.1007/978-3-030-29035-1_22
14. Jain, A., Kasbe, A.: Fake news detection. In: 2018 IEEE International Students' Conference on Electrical, Electronics and Computer Science (SCEECS). IEEE (2018)
15. Reis, Julio CS, André Correia, Fabrício Murai, Adriano Veloso, and Fabrício Benevenuto, Supervised learning for fake news detection. IEEE Intelligent Systems 34, no. 2 (2019)
16. Nikam, S.S., Dalvi, R.: Machine learning algorithm based model for classification of fake news on twitter. In: 2020 Fourth International Conference on I-SMAC (IoT in Social, Mobile, Analytics and Cloud)(I-SMAC). IEEE (2020)
17. Bali, A.P.S., Fernandes, M., Choubey, S., Goel, Mahima: Comparative performance of machine learning algorithms for fake news detection. In: Mayank Singh, P.K., Gupta, V.T., Flusser, J., Ören, T., Kashyap, R. (eds.) ICACDS 2019. CCIS, vol. 1046, pp. 420–430. Springer, Singapore (2019). https://doi.org/10.1007/978-981-13-9942-8_40
18. Waikhom, L., Goswami, R.S.: Fake news detection using machine learning. In: Proceedings of International Conference on Advancements in Computing & Management (ICACM) (2019)
19. Mahir, E.M., Akhter, S., Huq, M.R.: Detecting fake news using machine learning and deep learning algorithms. In: 2019 7th International Conference on Smart Computing & Communications (ICSCC). IEEE (2019)
20. Abdulrahman, A., Baykara, M.: Fake news detection using machine learning and deep learning algorithms. In: 2020 International Conference on Advanced Science and Engineering (ICOASE). IEEE (2020)

Tuning of Hyperparameters and CNN Architecture to Detect Phone Usage During Driving

Nishant Bhardwaj, Ayushi Yadav, and Sunita Daniel(✉)

Lal Bahadur Shastri Institute of Management,
Plot 11/7Sector-11, Dwarka 110075, New Delhi, India
sunitadaniel@lbsim.ac.in

Abstract. The primary goal of this paper is to classify images using Convolutional Neural Networks (CNN) to determine whether a driver is using their phone while driving or is paying attention to the road. There are a lot of road accidents happening on a day-to-day basis for which one of the main causes was found to be the driver behaviour. Talking on or texting on a smartphone while driving, along with other distractions, can be highly dangerous. Recognizing these poor driving habits can prevent many unfortunate events as well as save many lives. In this study, we propose a CNN model to categorize photos of drivers on a binary basis to ascertain whether the driver is using a phone while driving or is driving safely. Dataset was created using images from Google to conduct the analysis. For model training, equal proportions of the two image categories—driving safely and talking on the phone—were taken. Customized CNN models were created using different layers, and their accuracies were tested. Manual hyper parameter tuning was done on various parameters as well as the CNN architecture. The best model among all the models was the one which had 2 layers of convolution before 1 layer of pooling, along with dropout layers and was trained with 20 epochs. This model can be used in automobiles or in the traffic system to maximize the safety on the road.

Keywords: Deep learning · Convolutional neural network (CNN) · Image classification · Driver behaviour · Artificial Intelligence · Artificial Neural Networks (ANN)

1 Introduction

According to a report from the Ministry of Road Transport and Highways (MoRTH), there were a total of 4.12 lakh road accidents, which resulted in 1.53 lakh fatalities [1]. Driver behaviour can be a significant contributor to accidents, and so it is imperative that drivers remain focused at all times. Safety cameras can be used to record the behaviour of the driver, or the cameras which are built into the dashboard of the cars that can alert the driver if he is getting distracted. Connected and autonomous vehicles are the future of the world of transportation, but it will take a long time to reach complete autonomous

R. K. Challa et al. (Eds.): ICAIoT 2023, CCIS 1930, pp. 244–256, 2024.
https://doi.org/10.1007/978-3-031-48781-1_19

driving of vehicles. Till then it will be necessary for a person to stay behind the driving wheel for safety reasons. The driver will not only have to drive the vehicle safely, but he also needs to monitor if the vehicle is having any issues. The driver in this situation should always have his/her attention on the road and not get distracted by his phone or any other distraction. Therefore, the connected vehicles should be capable of detecting the driver behaviour and alert him in case he is distracted. This is where a machine learning model can be deployed to detect the driver's behaviour by using image classification.

Numerous researchers over the past ten years have sought to develop algorithms that can detect driver preoccupation. These investigations differ in several ways. Driving simulations were used by some while test tracks and/or realistic driving were used by others to collect data. These investigations' main objective, which is to pinpoint specific form or types of distraction, also varies (i.e., visual, combined and cognitive). Additionally, there is a massive concern with driver inattention since one of the primary factors in deadly traffic accidents is driver distraction. In order to create a sophisticated yet secure transportation system, it is crucial to be able to identify driver inattention. Many deep learning and machine learning techniques have been used to determine whether or not a driver is paying attention while driving.

In [2], the paper outlines the investigation into whether monitoring and evaluating driving behaviour alone can identify unfavourable health conditions of a vehicle driver in real time. The method bases its models of "regular" and "bad" driving behaviour on observation and uses machine learning to generate them. In the use case described, attention-deficit/hyperactivity disorder (ADHD) was the target condition. This condition affects many teenagers who are old enough to drive, and if left untreated with medication, it can be hazardous for safe driving. The exploratory character of the work reported in this study aims to demonstrate scientific viability. As a result of thorough testing, traces could be accurately characterized in up to over 82% of the test scenarios that were presented. In [3], the authors aim to provide an in-depth review of the methods used to identify driving distractions. Every study published between 2014 and 2021 was examined, and each was categorised based on the sensors they employed. A more straightforward structure for illustrating the detection flow was introduced, beginning with the sensors that were used, followed by the data that was gathered, measured, processed, and finally inferred behaviour and distraction type. Based on the studies, a condensed framework for displaying the detection flow is suggested and discussed using CNN in [4]. As regularized CNN are crucial for MLP-Multilayer Perceptions, CNN use a variety of regularization techniques and hierarchical patterns to achieve so. The success of CNN can be linked to the neurons' resemblance to those of living things. In this study, CNN for object identification and picture categorization were compared.

In [5], the authors suggest using the sensor fusion technology to enhance outcomes for identifying or detecting driving behaviour, and then discuss the performance matrix after going over several in and out of car sensors. In-car sensors in the automobile, measure things like the vehicle's and engine's speeds, the position of the throttles, and the load on engine while driving. Out car sensors are the GPS, magnetometer, and gyroscope. Risky driving behaviour was improved by the suggested integration of in-car and outdoor sensors using Machine Learning techniques. In [6], TensorFlow, a machine learning system, operates effectively and efficiently on a large scale in many different scenarios. It uses dataflow graphs to represent computation, shared state, and state-modifying operations, and distributes the graph's nodes across a cluster of machines. TensorFlow is particularly well-suited for neural network training and inference and can be utilized in a broad range of applications. As an open-source project, TensorFlow is widely used in machine learning research and is also employed in various Google services.

[7] aimed to develop algorithms for identifying distraction tasks that require concurrent visual, cognitive, and physical workload. Vehicle dynamics data was used to achieve this goal. The effectiveness of several data mining techniques for detecting distraction was examined using two linear (logistic regression and linear discriminant analysis) and two non-linear models (random forests techniques and support vector machines) to provide the most precise eating and texting distraction detection while driving. Random Forest (RF) algorithms were used to obtain the highest accurate texting and eating distraction detection, with detection accuracies of 81.26% and 85.38%, respectively. In [8], computer vision algorithms are used in most existing methods for tracking driver behaviour. Deep learning method are applied for analysing driver behaviour. This article identified five distinct driving behaviours, including normal, aggressive, distracted, drowsy, and drunk driving, using a range of driving signals, including throttle, acceleration, speed, gravity, and revolutions per minute (RPM). They trained a 2D convolutional neural network using deep learning techniques using driving signal images created by the recurrence plot method (CNN). According to [9], a driving behaviour detection algorithm was developed using a Long Short-term Memory (LSTM) Network and driver models from IPG's TruckMaker. Based on lateral and longitudinal acceleration restrictions, six driver models have been developed. The suggested method is practiced using driving cues from these drivers operating an actual truck model pulling five distinct trailer loads along a synthetic training route. Results demonstrate that even in short time intervals, the LSTM structure has a significant capacity to discern dynamic relationships between driving signals.

In [10] the authors propose a new method for real-time drowsiness detection. To identify the real-time drowsiness detection of the driver, the analysis method relies on deep learning techniques. A fundamental network structure is created by recognizing facial landmark key points and compressing a heavy baseline model into a lighter version. The suggested model based on this structure can attain an accuracy of over 80% for facial recognition. In [11], using facial images, faces are recognised and the eye region is extracted using Viola-Jones method. A novel eye movement-based method for identifying driving while intoxicated is presented in this article. To categorize the driver as asleep or awake, a SoftMax layer of the CNN classifier is used. The Viola-Jones Object Detection Framework is a highly efficient and accurate system for detecting objects in images, with a particular strength in recognizing human faces. The quick and precise Viola-Jones Object Recognition Framework develops an object detection system by integrating the ideas of Integral Images, AdaBoost Algorithm, Haar-like Features, and Cascade Classifier. This technology activates an alarm to let the driver know when they start to feel drowsy. The proposed work is evaluated using a gathered dataset, and it outperforms conventional CNN with an accuracy rate of 96.42%.

In [12], a system was developed to classify unsafe driving behavior among truck drivers by analyzing the data collected from 2000 individuals. The system identified nine specific types of unsafe driving behaviors as outputs. In this article four machine learning models - Classification and Regression Tree (CART), Random Tree (RT), Adaptive Boosting (AdaBoost), Gradient Boosting Decision Trees (GBDT) were used. Results revealed that various driving styles have various formation mechanisms, which have an impact on how accurately the model predicts outcomes (accuracy ranges from 64% to 95%). In [13], it was shown that attitudes, driving practices, and aggression levels of the driver have a big impact on vehicle control and energy economy. In the Advanced Driving Assistance System (ADAS), energy efficient control, or active safety system can successfully gather and evaluate the driving patterns. Using Artificial Neural Networks (ANN), to classify drivers into aggressive, normal, and calm states, a system employs three driving inputs: vehicle acceleration, speed, and throttle pedal angle. The generated models are 90% accurate in categorising vehicles based on different driving situations.

From the above, it is observed that the usage of phone during driving has not been studied so far. The use of different CNN architecture to determine if talking on a smartphone is safe or harmful is not yet covered in any of these studies. Hence, in this paper, we propose a Deep Learning (DL) model to classify the images into driving safely or talking while driving. To illustrate the results and demonstrate how the model operates on the test and training dataset, we utilize CNN techniques. Here is an overview of how the paper is organized, Sect. 2 gives a brief introduction about DL and some of the techniques available in DL for image classification. Section 3 describes the process to be followed and the dataset used for the analysis. Section 4 gives an overview of the different CNN architectures used and identification of the best one. Section 5 is the conclusion.

2 Theory and Formula

In this section, we recall the ANN and DL techniques needed for classification. A deep learning neural network is created for processing structured dataset arrays, such as images. For example, when a CNN learns from images, it tends to learn patterns, textures, edges, and brightness in the first few layers. These features from images are then used to do classification, clustering, and prediction.

Artificial Neural Networks (ANN) use mathematical models. It is based on the composition and operation of biological brain network. However, a flow of information has an impact on the ANN structure. As a result, adjustments to neural networks were depending on input and output. ANN are made up of several nodes those look like human brain's biological neurons. However, connections join these neurons together. They also communicate with one another by joining the neurons together. Nodes are used to collect input data and moreover to run basic operations on the data. These actions are consequently transferred to other neurons. The output of each node in a neural network can be referred to as its activation or node value, which is determined by the weights assigned to each link. This allows the network to learn and improve over time. By changing weight values, that is accomplished.

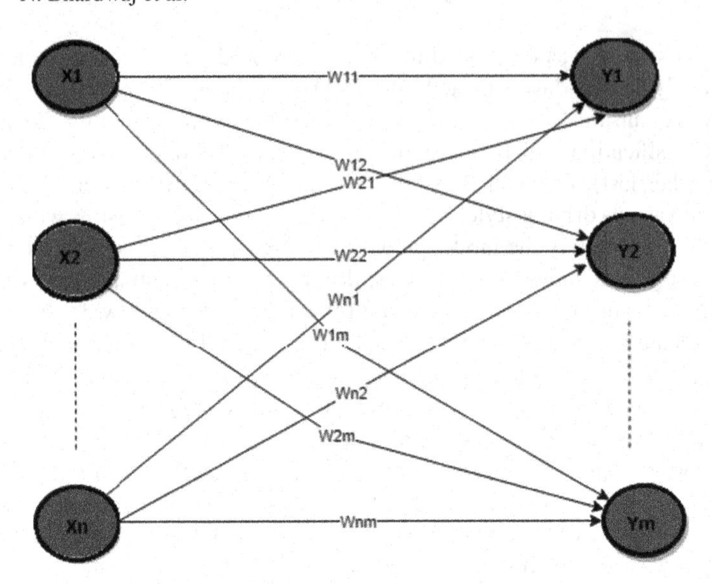

Fig. 1. Single layer forward propagating Network [14].

DL is a sort of ML where the model makes use of many artificial neural networks. Varying deep learning architectures can be created by connecting different numbers of neurons in various ways. These deep learning architectures handle both the classification and feature extraction.

Convolutional Neural Network (CNN) excel because of their layering. A three-dimensional neural network is employed to process the Red, Green, and Blue (RGB) components of an image simultaneously in a CNN. This approach reduces the number of artificial neurons required for image processing compared to traditional feed-forward neural networks. CNN is good at recognizing what's in images. It looks at the image and learns how to tell different objects apart. The way it does this is by using a special method called "convolution" instead of normal mathematics. This makes it better at understanding what's in an image. Convolutional networks typically have four different sorts of layers in their architecture:

(i) **Convolution** layer is used for feature extraction and creating feature maps.
(ii) **Pooling** layer decreases the size (height and width) of the image.
(iii) **Flatten** layer basically gives a single output as the CNN is not fully connected and
(iv) **Fully connected** layer sums up all the detections in the previous layers that were not fully connected which means each neuron is now connected to every activation of previous layer.

The CNN architecture is explained in Fig. 2.

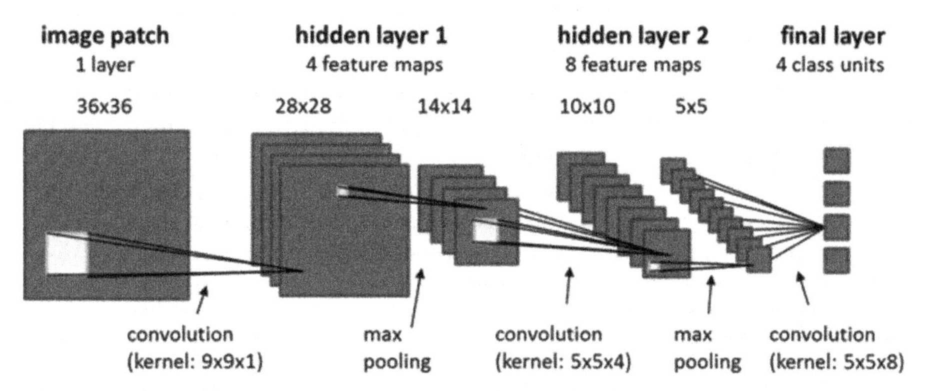

Fig. 2. Convolution Neural Network layers [15].

3 Proposed Framework

Figure 3 gives a complete framework of the experiment to be carried out. Firstly, the data is collected and loaded in the system. Then the pre-processing of the data is done – scaling of data, dividing it to training and test etc. The different CNN architectures are applied on the dataset and the accuracies are calculated and saved and these accuracies are compared to find the best model.

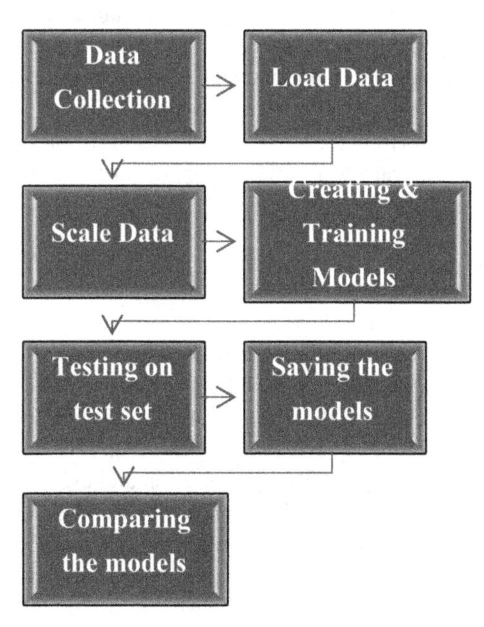

Fig. 3. Flowchart depicting the framework to be followed in analysis.

3.1 Data Collection

The dataset for analysis was gathered from the web by using terms like "Driving safely", "Man Driving Safely", "Woman Driving Safely", "Driving while talking", "Driving while using phone". A total of one hundred and ten images were selected based on the above search results and saved in two classes "Phone usage" and "Driving safely".

The images in the dataset were all in colour, but they varied in both shape and size. To address this challenge, pre-processing was done for the data.

3.2 Data Pre-processing

Since the amount and diversity of data was very less, the data was augmented. Random rotation, shifts, shear, and flips were performed on 55 images of each class and a total of 1100 images were created from them (550 of each class) which were saved in their respective folders.

The pixel values were normalized by dividing each pixel value in the input image by 255, so that all pixel values ranged between 0 and 1. Scaling the pixel values to this range has several benefits, including improving the numerical stability of the model and facilitates the identification of underlying patterns in the data. Additionally, some activation functions such as the ReLU activation function are designed to work best with inputs in this range.

To ensure that the images in the dataset were compatible with a CNN, they were all resized to a uniform size of $256 \times 256 \times 3$. This standardized the dimensions of the images, making them suitable for processing by the CNN model.

Once the images were resized, the dataset was split into training and testing data in a 75:25 ratio.

To analyse the dataset and train the CNN model, various libraries were used in Python 3, including Keras. These libraries provide a range of tools and functions for building, training, and evaluating neural networks,

4 Results and Discussions

CNN models in general have a lot of configurations and hence the various parameters are manually tuned. Some parameters must be set before the training starts and they are called as hyperparameters. Some of the hyperparameters for the CNN design are epochs, number of dense layers, neurons of each dense layers, dropout etc. In this work, we have varied the number of epochs in the CNN configuration and used ten and twenty epochs. While building the models the number of hidden layers used for training the model were two, four and five. Dropout layers were also introduced in the models. The general CNN architecture usually consists of one convolution layer and one layer of max pool. CNN architecture was also tuned in two ways. Firstly, we considered different models having two convolution layer followed by a max pooling layer and then three convolution layers were applied followed by a max pooling layer.

The results of hyperparameter tuning and tuning of the CNN architecture are tabulated in Tables 1, 2, 3, 4, 5 and 6. These tables contain the number of hidden layers and neurons used in each dense layer, the number of epochs the model was trained and the training and validation accuracies. Table 1 consists of 6 models where the CNN architecture consists of one convolution layer followed by pooling layer. Table 2 also is the result of the CNN architecture along with a dropout layer. Tables 3 and 4 are the models of the CNN architecture with two convolution layers with a pooling layer and Tables 5 and 6 are the models with three convolution layers with a pooling layer.

4.1 The Effect of Number of Epochs

From Table 1 and Table 2, it can be seen that for the conventional CNN, an increase in the number of epochs increases validation accuracy. However, when the dropout layer is added, Model 4 with epoch twenty has the highest accuracy of 98.3% for training and 87.1% for validation.

The same result holds for the other CNN architectures as well. This can be seen in Tables 3, 4, 5 and 6.

4.2 The Effect of Number of Hidden Layers

Various models were created with different number of hidden layers. A manual tuning was done using 2, 4 and 5 hidden layers. For the dataset, it can be seen from all the Tables that the models with 2 hidden layers gave the best accuracy irrespective of the CNN architecture or the inclusion of dropout layers.

Table 1. Using Models with only one Convolution Layer and Pooling Layer.

Model	Number of Hidden Layers	Neurons set used	Number of epochs trained	Training Accuracy	Validation Accuracy
Model 1	2	32, 64	10	0.991	0.890
Model 2	4	32, 64, 128	10	0.603	0.625
Model 3	5	32,64,128	10	0.502	0.496
Model 4	**2**	**32, 64**	**20**	**0.996**	**0.960**
Model 5	4	32, 64, 128	20	0.735	0.636
Model 6	5	32, 64, 128	20	0.992	0.925

4.3 The Effect of Dropout Layers

The dropout layers are generally used to avoid overfitting. Table 2, Table 4, and Table 6 give the results for different models which have a dropout layer. From Tables 1 and 2, it can be observed that the dropout layer in fact reduces the validation accuracy.

Similarly, from Tables 3 and 4 (CNN architecture with 2 convolution layer and a pooling layer) and Tables 5 and 6 (CNN architecture with 3 convolution layers and a pooling layer), it can be observed that when the dropout layers are added, the validation accuracy is reduced.

Table 2. Using Models with one Convolution, Pooling, and Dropout Layers.

Model	Number of Hidden Layers	Neurons set used	Number of epochs trained	Training Accuracy	Validation Accuracy
Model 1	2	32,64	10	0.823	0.738
Model 2	4	32,64,128	10	0.496	0.539
Model 3	5	32,64,128	10	0.597	0.660
Model 4	**2**	**32,64**	**20**	**0.983**	**0.871**
Model 5	4	32,64,128	20	0.752	0.707
Model 6	5	32,64,128	20	0.507	0.441

Table 3. Using Models with 2 Convolution and Pooling Layers.

Model	Number of Hidden Layers	Neurons set used	Number of epochs trained	Train Accuracy	Validation Accuracy
Model 1	2	32, 64	10	0.998	0.917
Model 2	4	32,64,128	10	0.533	0.511
Model 3	6	32,64,128	10	0.480	0.496
Model 4	**2**	**32, 64**	**20**	**1.0**	**0.945**
Model 5	4	32,64,128	20	0.509	0.480
Model 6	6	32,64,128	20	0.507	0.449

Table 4. Using Models with 2 Convolution, Pooling, and Dropout Layers.

Model	Number of Hidden Layers	Neurons set used	Number of epochs trained	Train Accuracy	Validation Accuracy
Model 1	2	32, 64	10	0.948	0.812
Model 2	4	32,64,128	10	0.501	0.496
Model 3	6	32,64,128	10	0.510	0.5
Model 4	**2**	**32, 64**	**20**	**0.998**	**0.964**
Model 5	4	32,64,128	20	0.502	0.503
Model 6	6	32,64,128	20	0.849	0.789

4.4 Changing the CNN Architecture

Models were created using the conventional CNN architecture (one convolution and pooling layers). The details of models studied using this architecture is given in Tables 1 and 2. Similarly the architecture using 2 convolution layer and pooling layers were examined in Tables 3 and 4. The CNN architecture with 3 convolution layers followed by pooling layers were examined in Tables 5 and 6. Among all the models, the model with 2 convolutional layers before 1 layer of pooling, along with dropout layers had the best validation accuracy, as per Model 4 of Table 4.

Table 5. Using Models with 3 Convolution and Pooling Layers

Model	Number of Hidden Layers	Neurons set used	Number of epochs trained	Train Accuracy	Validation Accuracy
Model 1	3	32, 64	10	0.884	0.781
Model 2	6	32,64,128	10	0.491	0.511
Model 3	**3**	**32,64**	**20**	**0.998**	**0.933**
Model 4	6	32, 64,128	20	0.504	0.492

Table 6. Using Models with 3 Convolution, Pooling, and Dropout Layers.

Model	Number of Hidden Layers	Neurons set used	Number of epochs trained	Train Accuracy	Validation Accuracy
Model 1	3	32, 64	10	0.943	0.828
Model 2	6	32,64,128	10	0.504	0.480
Model 3	**3**	**32,64**	**20**	**0.917**	**0.859**
Model 4	6	32, 64,128	20	0.507	0.519

It may be noted that Models 5 and 6, which have 3 convolution layers, were left out of the comparisons on purpose. This was done to make sure that the comparison was fair and meaningful by keeping all models within a maximum of 6 hidden layers.

By sticking to a maximum of 6 hidden layers in each comparison, we concentrate on how different architectural configurations affect the models' performance. This helps us assess the specific impact of design choices like kernel sizes, pooling, and dropout layers more accurately. It eliminates the potential confusion caused by varying layer depths, allowing for a clearer evaluation of the models' overall performance.

Hence, while models 5 and 6 may possess unique characteristics due to their specific architectural configurations, their exclusion from the comparisons helps establish a standardized framework for assessing the influence of other design factors on the models' performance. This deliberate approach enhances the clarity and reliability of the comparison results.

In addition to maintaining a consistent maximum of 6 hidden layers for a proper comparison, Models 5 and 6 were not included in the models with 3 convolution layers due to the potential increase in size and training time.

Models 5 and 6 would have made the models much larger. Adding more layers increases the complexity and size of the models, requiring more memory and computational resources for training and inference.

To keep the comparisons manageable and efficient, the number of hidden layers was limited to 6. This allows for a reasonable training time and ensures a fair evaluation of the models' performance across different architectural choices. By excluding Models 5 and 6, the comparative analysis remains practical and valid without sacrificing efficiency.

4.5 Comparison of the Best Models by Hyperparameter Tuning

In this section, we compare the best performing models from all the hyperparameter tuning models. This comparison is shown in Table 7. It can be seen that the number of epochs is 20 in all the cases and the number of hidden layers is 2. The dropout layers are of no significance. In fact, they reduced the accuracy. Similarly increasing the convolution layers also didn't show any increase in the accuracy.

Table 7. Comparing best models among all the hyperparameter tuning models performed.

Model	Number of Hidden Layers	Neurons set used	Number of epochs trained	Training Accuracy	Validation Accuracy
Model with only one convolution and pooling layers	2	32, 64	20	0.996	0.960
Models with convolution, pooling, and dropout layers	2	32, 64	20	0.983	0.871
Models with 2 convolution and pooling layers	2	32, 64	20	1.0	0.945
Models with 2 convolution, pooling, and dropout layers	**2**	**32, 64**	**20**	**0.998**	**0.964**
Models with 3 convolution and pooling layers	3	32,64	20	0.998	0.933
Models with 3 convolution, pooling, and dropout layers	3	32,64	20	0.917	0.859

5 Conclusion

Detecting driver behaviour can be very important to improve the safety of the vehicles on road. CNN models were used on the dataset which was generated by collecting images from the internet based on the two categories (Safe driving and Phone usage) to determine whether the driver is on a phone.

Various models were successfully tested, and it was found that as the number of epochs increased, the accuracy also increased. Irrespective of the CNN architecture, models with two hidden layers gave the best accuracy. Dropout layers were also added to the CNN architecture, and it was found that the accuracy of the models was better without them. The model with two layers of convolutional before one layer of pooling, along with dropout layers had the best validation accuracy among all the other models.

There are some limitations in this study that do have an impact on the results that the models arrived at. One limitation of this work is the inability to use transfer learning techniques because of limited computational capacity and limited data. The dataset used for this work was small given there was no publicly available dataset regarding the same. Training of the model was done using a dataset created by taking images from the internet websites such as Google. The dataset can be increased by capturing real-time images. The model can be tested in real life scenarios by placing a camera on the dashboard or behind the driving wheel. The model studied is only made to detect the behaviour of driver's phone usage. A model can be made to classify the images that have other types of behaviours like holding a drink while driving, eating etc.

References

1. Road accidents in India - 2021. https://pib.gov.in/PressReleasePage.aspx?PRID=1887097, (Accessed 21 Jan 2023)
2. Gonzalez, A.J., et al.: Detection of driver health condition by monitoring driving behavior through machine learning from observation. Expert Syst. Appli. **199**, 117167 (2022)
3. Koay, H.V., Chuah, J.H., Chow, C.O., Chang, Y.L.: Detecting and recognizing driver distraction through various data modality using machine learning: a review, recent advances, simplified framework, and open challenges (2014–2021). Eng. Appl. Artif. Intell. **115**, 105309 (2022)
4. Kumar, K.K., Kumar, M.D., Samsonu, C. and Krishna, K.V.: Role of convolutional neural networks for any real time image classification, recognition, and analysis (2021)
5. Malik, M., Nandal, R., Maan, U. et al. Enhancement in identification of unsafe driving behaviour by blending machine learning and sensors. Int. J. Syst. Assur. Eng. Manag. (2022)
6. Abadi, M., et al.: TensorFlow: A system for large-scale machine learning. In: OSDI 2016, Savannah, USA, pp. 265–283 (2016)
7. Atiquzzaman, M., Qi, Y. and Fries, R: Real-time detection of drivers' texting and eating behavior based on vehicle dynamics. Trans. Res. Part F: Traffic Psychol. Behav. **58**, 594–604 (2018)
8. Shahverdy, M., Fathy, M., Berangi, R., Sabokrou, M.: Driver behavior detection and classification using convolutional neural networks. Expert Syst. Appl. **149**, 113240 (2020)
9. Mumcuoglu, M.E., et al.: Driving behavior classification using long short term memory networks, In: Proceedings of AEIT International Conference of Electrical and Electronic Technologies for Automotive (AEIT AUTOMOTIVE), pp. 1–6 (2019)

10. Jabbar, R., Al-Khalifa, K., Kharbeche, M., Alhajyaseen, W., Jafari, M., Jiang, S.: Real-time driver drowsiness detection for android application using neural networks techniques. Proc. Comput. Sci. **130**, 400–407 (2018)
11. Chirra, V.R.R., Uyyala, S.R., Kolli, V.K.K.: Deep CNN: a machine learning approach for driver drowsiness detection based on eye state. Rev. d'Intelligence Artif. **33**(6), 461–466 (2019)
12. Niu, Y., Li, Z.: And Fan, Y: Analysis of truck drivers' unsafe driving behaviors using four machine learning methods. Int. J. Ind. Ergon. **86**, 103192 (2021)
13. Cheng, Z.J., Jeng, L.W., Li, K.: Behavioral classification of drivers for driving efficiency related ADAS using artificial neural network. In: IEEE International Conference on Advanced Manufacturing (ICAM), pp. 173–176. Taiwan (2018)
14. Introduction to ANN Set 4 (Network Architectures). https://www.geeksforgeeks.org/introduction-to-ann-set-4-network-architectures/, (Accessed 23 Feb 2023)
15. Deep Learning (CNN) Algorithms. https://www.ecognition.com (Accessed 23 Feb 2023)

Security Defect Identification of Android Applications by Permission Extraction using Machine Learning

Pawan Kumar(✉) and Sukhdip Singh

Department of Computer Science and Engineering, Deenbandu Chhotu Ram University of Science and Technology, Murthal, Sonipat, India
pawankumar.schcse@dcrustm.org

Abstract. The Android platform security is at danger due to malicious applications. Due to the quantity and variety of these applications, traditional solutions are losing their effectiveness, making Android smart phones frequently vulnerable. Features are extracted during static analysis rather than running any code. Static analysis is more effective in general. Features taken from the manifest file can be used in static analysis of an Android application. The proposed model uses static analysis for permission extraction of permission from the application having security defects. A dataset of 1058 applications from the real world, 578 of which are benign and 480 of which are malicious, has been created. The model has been evaluated to ensure the effectiveness of the static analysis method for confirming the security of Android applications using machine learning techniques.

Keywords: Android · Static Analysis · Security Testing · Machine Learning

1 Introduction

Android is the most widely used mobile operating system with a market share of 74% [1]. In addition to such market trends, organizations and individual developers are developing a lot more third-party application. 37 major application categories comprise the Google Play app market [2]. The most enticing features of Android are its open-source nature and broad application marketplace.

Nowadays, Android offers its users a wide range of capabilities with several hundred thousand applications in a wide range of categories. An effective security element to secure system resources and user privacy is provided by the access-control system of the Android permissions system. Applications for Android are more usually attacked by attackers who infect them with harmful malware. In contrast to other operating systems, Android permits the installation of apps from unreliable sources like third-party markets, making it simpler for attackers to integrate and distribute malicious apps. As a result, a large amount of research has looked into approaches for detecting and analyzing Android malware before it has been deployed.

The permission-based security mechanism is typically used by Android to protect user data or prevent access to confidential information by applications. As one of the

R. K. Challa et al. (Eds.): ICAIoT 2023, CCIS 1930, pp. 257–268, 2024.
https://doi.org/10.1007/978-3-031-48781-1_20

most crucial and critical security evaluation procedures on the Android platform, Android app permissions are severely compromised. Since it is nearly impossible to conduct a succession of activities, the permission checking approach is vital for detecting malware without prior approval.

The Java programming language is used to construct most Android applications. An Android program is originally compiled into class files, which are compatible with the JVM and contain Java byte code instructions. These class files are then converted into.dex files that are executable by the Dalvik virtual engine and contain Dalvik byte code instructions. The.dex files are lastly included in an Android APK file for distribution and installation [3]. Unlike conventional Java programs, Android apps don't have a single-entry point. Contrarily, an app consists of one or more components that are listed in its manifest file.

The applications' use of resources, permissions, and other elements validates the results of the functional testing of those aspects. However, analyzing an app's functionality might not be the best way to identify security issues. There are frequently limited tools and methodologies for malware testing, even in situations when developers can test the functional components of these apps. As a result, there is an increasing need for tools that may help developers find and fix security issues in applications prior to being published.

This paper's original contributions can be summarized in the following way:

- A static analysis method is presented for extracting permissions from collection of malicious and benign application's APK files.
- A dataset containing binary features is created.
- This dataset is analyzed to assess the effectiveness of static analysis using machine learning methods.

The structure of the article is as follows: Sect. 2 contains related works; Sect. 3 includes the methods and materials used. The recommended approach is outlined in Sect. 4, Sect. 5 contains implementation, Sect. 6 of the article discusses the results, and Sect. 7 has a conclusion.

2 Related Work

The literature has discussed several Android malware testing initiatives. The papers that are most pertinent to malware testing are listed in the section that follows. This section goes over relevant research in analyzing security vulnerabilities in Android app development.

Through the development of static analysis tools, various efforts have been made to look for security vulnerabilities in Android applications. For instance, Crylogger [4] finds Android app misuses cryptography APIs; whereas FixDroid [5] checks a developer's code for common security problems and offers suggestions.

DroidSpan [6] presented an innovative classification method for Android apps that is based on behavioral summary for Android applications that depicts the distribution of sensitive access from light-weight summarizing.

The JOWMDroid [7] suggests an estimating method that makes use of an entropy approach to choose the best characteristics, reducing the necessity for a standard mechanism that picks all traits to tell apart between benign and malicious malware. This is especially crucial because APK files have numerous characteristics that would make any regular method cumbersome and unusable.

The authors of [8] describe a malware detection method that uses a hybrid analytic feature set, which consists of API calls, permissions, and system calls. To understand the relationship between static and dynamic features, they employ the tree augmented naive Bayes model.

Static and dynamic analysis are combined in the SAMADroid [9] technique on both distant and local hosts. They detect malicious applications with great accuracy by combining NB, RF, and SVM. Additionally, it has the benefits of low energy usage and increased storage effectiveness. While the StormDroid [10] method sequences sensitive API calls and permissions directly in the source code, the former extracts dynamic and static features from these calls.

The CuckooDroid [11] framework makes it possible for Cuckoo Sandbox to function thanks to a misuse detector that is frequently used to identify recognized malware and categorize android malware by hybrid analysis techniques. The suggested model takes use of the low rate of false positives along with the ability of anomaly detection leveraging SVM and a linear classifier.

The API and permissions needed to use the static analysis capabilities are extracted by the authors of [12] from the source code. They also consider the overall amount of time the system needs to extract the dynamic analysis features. Based on this, a detection accuracy of 93.33% was obtained.

After reviewing the above-mentioned articles, it has been concluded that there is a need for more efficient permission extraction mechanisms for android applications. This work focuses explicitly on permission extraction from Android application's APK files for identifying security vulnerabilities that would help developers to develop more secure applications.

3 Preliminary Methods

This section introduces the preliminary methods used in the article:

3.1 Static Code Analysis

Static code analysis is a method for identifying security vulnerabilities by examining at the source code [13]. This approach avoids using an Android emulator or actual Android smartphone to launch the application. Code obfuscation and dynamic code loading are two significant drawbacks of this strategy. Static analysis has the advantages of low computation costs, short computation times, and low resource consumption.

There are two main methods for discovering security vulnerabilities: the first method is signature-based and tests an application to see whether it meets a set of rules or instructions before classifying it as malware; the second method uses machine learning algorithms to find malicious behavior. To train the model and forecast a novel or unidentified malware, features taken from known malware are used.

3.2 Machine Learning

Machine learning is a technique that plays a significant role in data science. Using machine learning approaches, algorithms are taught to produce classifications or predictions and to identify the Android applications that contain security vulnerabilities [14]. A branch of computer science and artificial intelligence called machine learning is concerned with using data and algorithms to mimic human learning processes and continuously increase accuracy. Following are the machine learning algorithms used in this article for malware testing:

K-Nearest Neighbor (KNN). It is a supervised machine learning algorithm. The technique allows for the resolution of both classification and regression problem statements [15]. The letter "K" stands for the quantity of nearest neighbors to a new unidentified variable that needs to be forecasted or categorized.

After collecting all the prior data, a new data point is categorized using the KNN algorithm built on relationship. This implies that new data can be consistently and rapidly categorized using the KNN approach.

Support Vector Machine (SVM). One of the extremely famous supervised learning algorithms is SVM, which may be used to solve problems involving classification and regression. However, Machine Learning Classification problems are where it is most frequently used. The SVM method's objective is to identify the ideal boundary, or decision line, for categorizing the n-dimensional space, allowing us to quickly classify fresh data points in the future [16].

The name of this greatest decision boundary is a hyperplane. SVM generates the hyperplane by selecting the best points and vectors. These intense cases are referred to as support vectors, which is the name of the algorithm known as an SVM.

Logistic Regression (LR). LR is a categorization method that employs supervised learning to forecast the chance of a target variable [17]. There are only two valid classes since the objective or related variable is dichotomous. LR predicts the result of a dependent categorical variable. There must consequently be a discrete or categorical value as a result. It provides probabilistic values in the 0 to 1 range rather than an exact number between 0 and 1. It can be either Yes or No, 0 or 1, true or false, etc.

Naive Bayes (NB). NB algorithm is a categorization algorithm that applies the Bayes theorem with the strong presumption that each predictor is independent of the others [18]. In other words, it is assumed that the existence of a feature does not depend on the existence of any other features within the same class.

The fundamental aim of Bayesian classification is to ascertain the posterior probabilities, or P (L | features), which is the likelihood of a label given some observable features. We may express this quantitatively as follows with the use of the Bayes theorem:

$$\mathrm{P}(L|features) = \frac{P(L) * P(features|L)}{P(features)} \tag{1}$$

Here,

- P(*L*|*features*) is the posterior probability of class.
- *P*(*L*) is the prior probability of class.
- *P*(*features*|*L*) is the likelihood which is the probability of predictor given class.
- *P*(*features*) is the prior probability of predictor.

4 Experimental Setup

The proposed methodology is divided into three steps. Figure 1 depicts the overall process flowchart. The first of these is the detection of security vulnerabilities in Android applications. The second is the generation of a binary dataset by permission extraction from malicious and benign Android application APK files using AXMLPrinter [19]. The third step entails using machine learning methods to evaluate the proposed model's performance. The detailed steps are as follows:

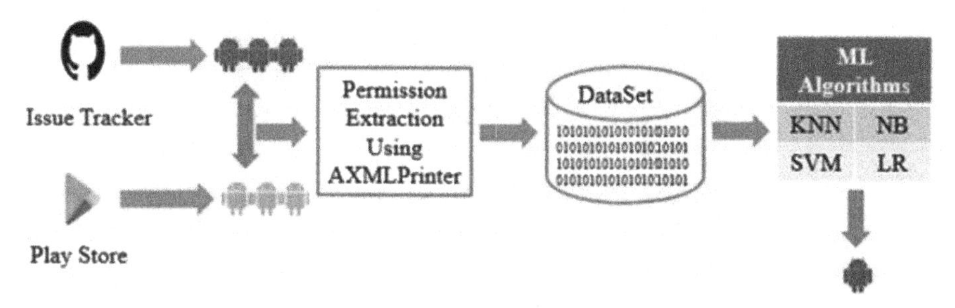

Fig. 1. Proposed permission based malicious applications detection framework.

Step 1. Android applications with security vulnerabilities have been identified using a key-word-based search method on the GitHub issue tracker [20]. A manual scan has been done for security vulnerabilities that users and developers have reported. A list of 648 applications with security-related concerns was generated by this approach and 544 open-source applications has been shortlisted that have at least one security-related issue after looking for the keywords only 518 applications have been filtered in the issue trackers. After considering the applications that had security problems that the developer could verify, fix, or describe, only kept 480 of them. These apps' APK files were acquired from Apkpure.com. To verify the applications' cleanliness, 578 clean apps have been downloaded from Google Play and ran a Virus Total scan on them. 37 key application categories make up the Google Play application catalogue and collected samples from 32 of these categories. 578 clean and 480 infected APK files from 1058 real-world applications were used for the feature extraction process using static analysis.

Step 2. The AndroidManifest.xml file is extracted from APK files using AXMLPrinter [19] and examined. The metadata in this file includes information about the main activity, component construction, and application permission requests. Based on these metadata, we extract the various components to the appropriate Java objects in the subsequent data flow analysis, concluding the results and identifying the vulnerabilities as defined in Android reference. A dex.class (byte code file) is created once the binary code (APK file) has been decompiled. The dataset contains 1058 rows for real-world applications and 277 columns for Android permissions extracted as defined in android reference.

Step 3. Four machine learning algorithms—KNN, NB, SVM and LR— are applied on the dataset to examine the performance of the proposed model for malware prediction using accuracy, precision, recall, and F1-measure metrics.

5 Results and discussion

```
INTERNET                   170
READ_PHONE_STATE           166
ACCESS_NETWORK_STATE       144
WRITE_EXTERNAL_STORAGE     119
ACCESS_WIFI_STATE          119
READ_SMS                   103
RECEIVE_BOOT_COMPLETED      89
WRITE_SMS                   87
ACCESS_COARSE_LOCATION      71
CHANGE_WIFI_STATE           69
dtype: int64
```

Fig. 2. Top 10 malicious permissions used by Android applications.

The top 10 permissions used by malicious applications in the dataset are illustrated in Fig. 2. The most popular of these permissions is the INTERNET, which is used in 170 applications. The second frequently used permission by malicious applications is READ_PHONE_STATE. 166 malicious applications use READ_PHONE_STATE. CHANGE_WIFI_STATE, which has been used by 69 applications, is the least-used of the top 10 malicious permissions retrieved from the provided dataset as shown in Fig. 2.

Figure 3 displays the top 10 benign permissions utilized by Android apps. A total of 88 benign applications uses the permission INTERNET, which is the most popular of them all. The second most common benign permission that applications employ is WRITE_EXTERNAL_STORAGE. WRITE_EXTERNAL_STORAGE permission is utilized by 67 applications. The number of applications using the READ_CONTACTS permission, as shown in Fig. 3, is the lowest of the top 10 benign permissions gleaned from Android applications.

Accuracy is one of the most often used evaluation measures in malware research because it demonstrates how well a model has performed overall. The abbreviations TP and TN, respectively, indicate for the number of correctly categorized permissions used by malicious and benign applications, in the confusion matrix. FP and FN represent the total number of permissions that were wrongly categorized.

$$\text{Accuracy} = \frac{\text{TP} + \text{TN}}{\text{TP} + \text{FP} + \text{TN} + \text{FN}} \tag{2}$$

```
INTERNET                   88
WRITE_EXTERNAL_STORAGE     67
ACCESS_NETWORK_STATE       50
WAKE_LOCK                  34
RECEIVE_BOOT_COMPLETED     27
ACCESS_WIFI_STATE          26
READ_PHONE_STATE           25
VIBRATE                    17
READ_EXTERNAL_STORAGE      15
READ_CONTACTS              14
dtype: int64
```

Fig. 3. Top 10 benign permissions used by Android applications.

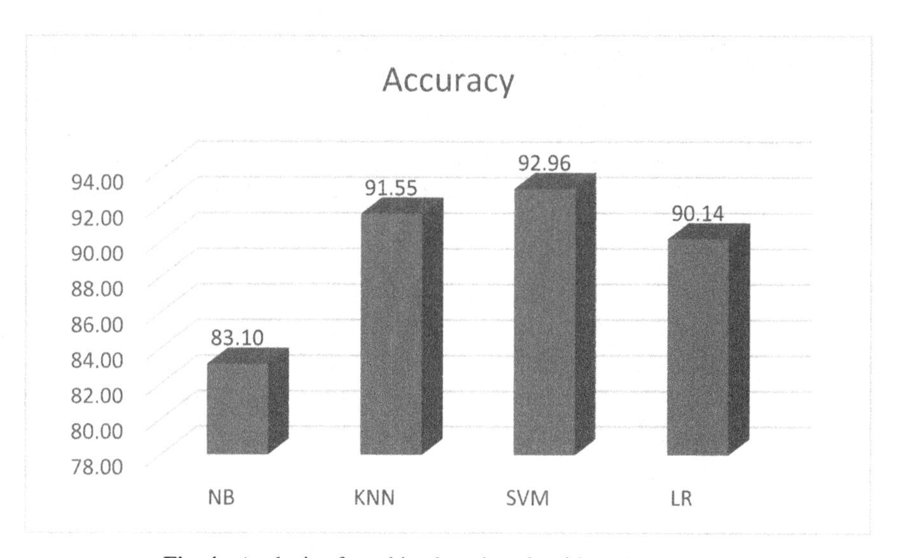

Fig. 4. Analysis of machine learning algorithms (Accuracy).

SVM is the machine learning algorithm with the highest accuracy (92.96%) as shown in Fig. 4. The SVM algorithm surpasses other machine learning algorithms in terms of accuracy. KNN and LR algorithms have 91.55% and 90.14% accuracy while NB has the lowest accuracy (83.10%) among all the machine learning algorithms.

Precision is determined by dividing the total number of true positives by the sum of true positives and false positives. The formula is as follows:

$$\text{Precision} = \frac{\text{TP}}{\text{TP} + \text{FP}} \tag{3}$$

NB is the machine learning algorithm with the highest precision (92.86%) as shown in Fig. 5. The NB algorithm surpasses other machine learning algorithms in terms of

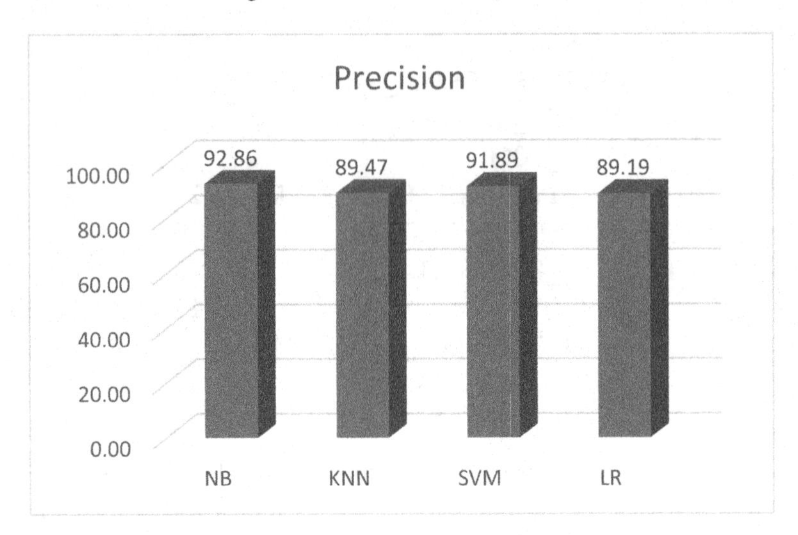

Fig. 5. Analysis of machine learning algorithms (Precision).

precision. KNN and SVM algorithms have 89.47% and 91.89% precision while SVM has the lowest precision (89.19%) among all the machine learning algorithms.

Recall is a measure that evaluates the fraction of true positive predictions among all possible positive predictions.

$$\text{Recall} = \frac{\text{TP}}{\text{TP} + \text{FN}} \tag{4}$$

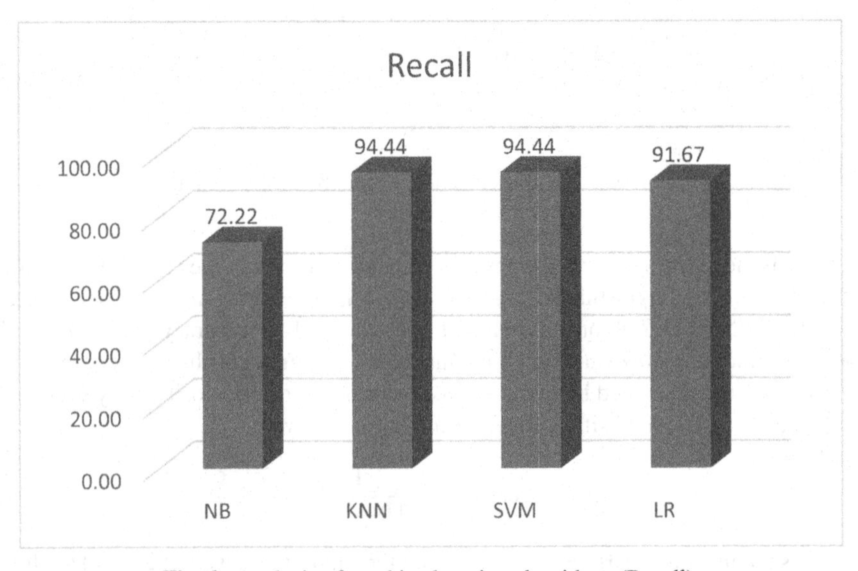

Fig. 6. Analysis of machine learning algorithms (Recall).

SVM and KNN are the machine learning algorithms with the highest recall (94.44%) as shown in Fig. 6. The SVM and KNN algorithms surpass other machine learning algorithms in terms of recall. LR algorithm has 91.67% recall while NB has the lowest recall (72.22%) among all the machine learning algorithms.

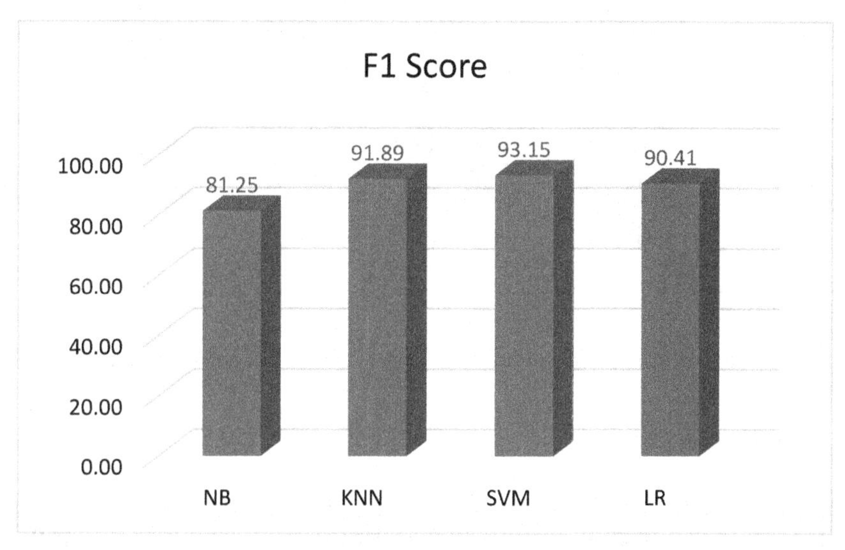

Fig. 7. Analysis of machine learning algorithms (F1-Score).

The F1-score offers a means to combine recall and precision into a single measure that accounts for both measures. It represents the harmonic average of these two measures and has a value ranging from 0 to 100%.

$$\text{F1} - \text{score} = 2 * \frac{\text{Precision} * \text{Recall}}{\text{Precision} + \text{Recall}} \tag{5}$$

SVM is the machine learning algorithm with the highest F1-score (93.15%) as shown in Fig. 7. The SVM algorithm surpasses other machine learning algorithms in terms of F1-score. LR and KNN algorithms have 90.41% and 91.89% F1-score while NB has the lowest F1-score (81.25%) among all the machine learning algorithms.

As a result, SVM is the machine learning method with the highest accuracy (92.96%). The SVM approach surpasses other machine learning algorithms in terms of the other three metrics (accuracy, recall and F1-score), even though the KNN and SVM perform best in terms of recall (94.44%). NB has the best precision (92.86%) and the lowest accuracy (83.10%), recall (72.22%), and F1-score (81.25%) of all the algorithms.

The findings of the model show that the SVM algorithm has attained promising performance. The SVM algorithm, which attains 92.96% accuracy is 0.15% better than the standard KNN algorithm and has the highest prediction accuracy.

Additionally, the proposed model received the highest F1-score and precision as shown in Fig. 8. Now, the proposed model can successfully learn the unbalanced data as compared to the baseline models (MUDFLOW [21], MalPat [22], LinRegDroid [23]).

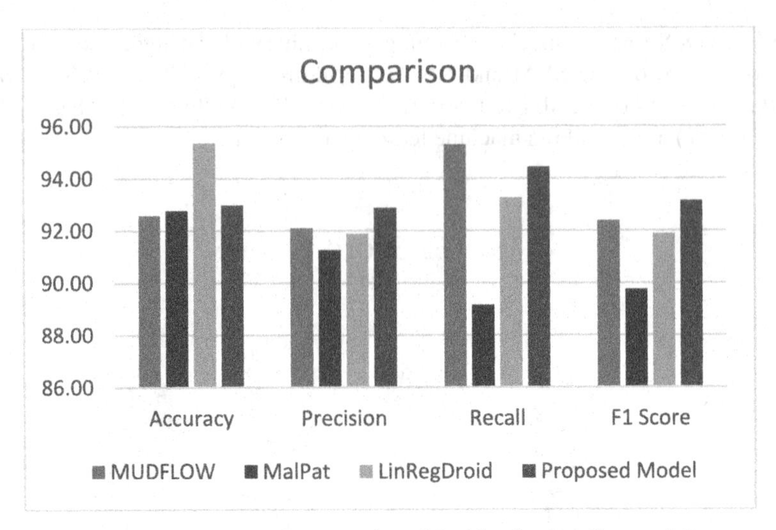

Fig. 8. Comparison of proposed model with other baseline models.

6 Conclusion

The proposed model demonstrated the permission extraction from APK files of malicious and benign android applications to make a binary dataset is done using AXMLPrinter [19]. In-depth use of baksmali [24] can efficiently produce extremely readable and semantically complete byte code files. The binary dataset has been used for binary classification using machine learning algorithms (NB, KNN, LR and SVM). The trained model then uses four evaluation indicators to test the datasets.

SVM employs a hyperplane to optimize the margin of separation between classes and is measured along a line normal to the hyperplane. This is because the dataset utilized in this article is linearly separable. The results show that the SVM algorithm performs better in terms of accuracy than any other algorithms. The proposed model is superior in terms of malware analysis when compared to the other baseline models as shown in Fig. 8. For assisting developers in avoiding security vulnerabilities, the proposed model has stronger directing implications.

References

1. Global mobile OS market share 2012–2022 | Statista. https://www.statista.com/statistics/272698/global-market-share-held-by-mobile-operating-systems-since-2009/ (Accessed 28 Sep 2022)
2. Android Apps on Google Play. https://play.google.com/store/apps (Accessed 28 Sep 2022)
3. Sharma, T., Rattan, D.: Malicious application detection in android — a systematic literature review. Comput. Sci. Rev. **40**, 100373 (2021). https://doi.org/10.1016/j.cosrev.2021.100373
4. Piccolboni, L., Di Guglielmo, G., Carloni, L.P., Sethumadhavan, S.: CRYLOGGER: detecting crypto misuses dynamically. In: 2021 IEEE Symposium on Security and Privacy (SP), vol. 2021, pp. 1972–1989 (May 2021). https://doi.org/10.1109/SP40001.2021.00010

5. Nguyen, D.C., Wermke, D., Acar, Y., Backes, M., Weir, C., Fahl, S.: A stitch in time: supporting android developers in writingSecure Code. In: Proceedings of the 2017 ACM SIGSAC Conference on Computer and Communications Security, pp. 1065–1077 (Oct 2017). https://doi.org/10.1145/3133956.3133977
6. Cai, H.: Assessing and improving malware detection sustainability through app evolution studies. ACM Trans. Softw. Eng. Methodol. **29**(2), 1–28 (2020). https://doi.org/10.1145/3371924
7. Cai, L., Li, Y., Xiong, Z.: JOWMDroid: android malware detection based on feature weighting with joint optimization of weight-mapping and classifier parameters. Comput. Secur. **100**, 102086 (2021). https://doi.org/10.1016/j.cose.2020.102086
8. Surendran, R., Thomas, T., Emmanuel, S.: A TAN based hybrid model for android malware detection. J. Inf. Secur. Appl. **54**, 102483 (2020). https://doi.org/10.1016/J.JISA.2020.102483
9. Arshad, S., Shah, M.A., Wahid, A., Mehmood, A., Song, H., Yu, H.: SAMADroid: a novel 3-level hybrid malware detection model for android operating system. IEEE Access **6**, 4321–4339 (2018). https://doi.org/10.1109/ACCESS.2018.2792941
10. Chen, S., Xue, M., Tang, Z., Xu, L., Zhu, H.: StormDroid: a streaminglized machine learning-based system for detecting android malware. In: Proceedings of the 11th ACM on Asia Conference on Computer and Communications Security, pp. 377–388 (May 2016). https://doi.org/10.1145/2897845.2897860
11. Wang, X., Yang, Y., Zeng, Y., Tang, C., Shi, J., Xu, K.: A novel hybrid mobile malware detection system integrating anomaly detection with misuse detection. In: Proceedings of the 6th International Workshop on Mobile Cloud Computing and Services - MCS 2015, pp. 15–22 (Sep 2015). https://doi.org/10.1145/2802130.2802132
12. Liu, Y., Zhang, Y., Li, H., Chen, X.: A hybrid malware detecting scheme for mobile Android applications. In: 2016 IEEE International Conference on Consumer Electronics (ICCE), pp. 155–156 (Jan. 2016). https://doi.org/10.1109/ICCE.2016.7430561
13. Senanayake, J., Kalutarage, H., Al-Kadri, M.O., Petrovski, A., Piras, L.: Android source code vulnerability detection: a systematic literature review. ACM Comput. Surv. (2022). https://doi.org/10.1145/3556974
14. Keyvanpour, M.R., Barani Shirzad, M., Heydarian, F.: Android malware detection applying feature selection techniques and machine learning. Multimed. Tools Appl., 1–15 (2022). https://doi.org/10.1007/s11042-022-13767-2
15. Guo, G., Wang, H., Bell, D., Bi, Y., Greer, K.: KNN model-based approach in classification. LNCS(LNAI, LNB), vol. 2888, pp. 986–996. Springer, Cham (2003). https://doi.org/10.1007/978-3-540-39964-3_62/COVER
16. Cortes, C., Vapnik, V.: Support-vector networks. Mach. Learn. **20**(3), 273–297 (1995). https://doi.org/10.1007/BF00994018
17. Zou, X., Hu, Y., Tian, Z., Shen, K.: Logistic regression model optimization and case analysis. In: 2019 IEEE 7th International Conference on Computer Science and Network Technology (ICCSNT), pp. 135–139 (Oct 2019). https://doi.org/10.1109/ICCSNT47585.2019.8962457
18. Webb, G.I.: Naïve Bayes. In: Sammut, C., Webb, G.I. (eds.) Encyclopedia of Machine Learning, pp. 713–714. Springer US, Boston, MA (2010). https://doi.org/10.1007/978-0-387-30164-8_576
19. Home - XML Printer. https://www.xmlprinter.com/ (Accessed 29 Sep 2022)
20. GitHub Issues. https://github.com/issues?q=is%3Aopen+is%3Aissue+android+security (Accessed 28 Sep 2022)
21. Avdiienko, V., et al.: Mining apps for abnormal usage of sensitive data. In: Proceedings of International Conference on Software Engineering, vol. 1, pp. 426–436 (Aug 2015). https://doi.org/10.1109/ICSE.2015.61

22. Guanhong, T., Zibin, Z., Ziying, G., Lyu, M.R.: MalPat: mining patterns of malicious and benign android apps via permission-related APIs. IEEE Trans. Reliab. **67**(1), 355–369 (2018). https://jglobal.jst.go.jp/en/detail?JGLOBAL_ID=201802287387115284, (Accessed Mar. 03 2023)
23. Sahin, D.O., Akleylek, S., Kilic, E.: LinRegDroid: detection of android malware using multiple linear regression models-based classifiers. IEEE Access **10**, 14246–14259 (2022). https://doi.org/10.1109/ACCESS.2022.3146363
24. baksmali. https://github.com/JesusFreke/smali (Accessed 29 Sep 2022)

Analysis of Employee Attrition using Statistical and Machine Learning Approaches

Akash Kumar Singh(✉) and Prateek Thakral

Department of Computer Science and Engineering and Information Technology, Jaypee University of Information Technology, Waknaghat, Solan, Himachal Pradesh, India
akash.singh2761@gmail.com

Abstract. Attrition means exit of employees from an organization due to any reason. The exit may be voluntary or involuntary in nature. Increased rate of employee attrition leads to loss of an organization's resources in terms of time, cost, customer satisfaction as well as talent acquisition team's efforts. The pattern of statistical approaches used in this paper to study the causes of attrition provides insights for the decision makers of organizations about the possible constraints due to which employees may think of leaving. This paper analyses an employee dataset and proposes a machine learning based model with an F1 Score of 0.99 and accuracy of 98.63% that can be used and deployed by the companies on their datasets to predict attrition. The timely prediction of attrition using this method will help the companies to prevent the possible attrition of employees and will help in retaining them for good.

Keywords: Attrition · Machine Learning · Prediction · Support Vector Machine · Decision Tree · Random Forest Classifier · Logistic Regression

1 Introduction

Employee Attrition refers to the gradual reduction in the workforce of an organization when employees leave, either voluntarily or involuntarily.

When employees leave, it takes time to find their replacement. The incoming replacement has to be then groomed according to the structure and environment of the organization. All these activities consume significant amount of time and in turn, the projects may get delayed. Employee attrition may either be in the form of voluntary turnover or in the form of downsizing. Irrespective of the form, it negatively affects the critical relationships between the consumer-oriented employees and their consumers, which degrades customer outcomes and financial performance of the organizations [1]. Also, recruitment in firms is to a great extent done with the help of recruitment consultants. So, the expenditure of recruitment increases as attrition increases. This may prove to be problematic as increasing costs may pressurize the earnings margins of a company, especially in the present time of economic slowdown.

Every company has some employees who possess expertise as well as a very significant experience in their respective areas of work that particularly is done within a

R. K. Challa et al. (Eds.): ICAIoT 2023, CCIS 1930, pp. 269–279, 2024.
https://doi.org/10.1007/978-3-031-48781-1_21

company. If a company loses such employees, it may be very hard to find optimum replacement. This may result into a compromising situation in terms of end-product quality. "Take our 20 best people away, and I will tell you that Microsoft will become an unimportant company," Bill Gates once said.

Considering the above stated factors, it becomes very clear that attrition at an increased rate is not at all desirable for any organization and every organization should take measures in order to curb any possible attrition. This paper analyses an employee dataset and models the attrition prediction using four machine learning algorithms out of which Random Forest Classifier is found to be the best with an F-1 score of 0.99, accuracy of 98.63% and AUC (Area under the Curve) of 0.9577.

The rest of the paper is organised in such a way that Section 2 presents previous works done to predict and analyse employee attrition using machine learning and other data science techniques. Section 3 describes the proposed methodology of our study which includes data pre-processing, analysis of the data as well as examination of machine learning algorithms like Logistic Regression, Decision Tree, Random Forest and Support Vector Machine. Sections 4 presents the results of the evaluation of the machine learning models and Section 5 presents the Conclusion of this study.

2 Related Work

Since attrition is a rising concern for business houses, there have been several studies based on its predictability using machine learning as well as other new generation computational techniques.

Khalifa et al. [2] employed dozens of machine learning algorithms using MATLAB to predict attrition and measured the performances in terms of accuracy as well as average prediction speed and average training time for the dataset of 1471 records. Logistic Regression was found to be the best model for prediction with 87.78% accuracy. The said model had the average prediction speed of 30180 observations per second and average training time of 7.01 seconds.

Seelam et al. [3], out of Logistic Regression, KNN, Naive Bayes and Random Forest, found Random Forest to be the best algorithm with 86.95% accuracy when applied on the dataset of around 1500 records. In terms of AUC with respect to ROC graph, Logistic Regression was found to be the best with area of 0.84.

Sekaran and Shanmugam [4] used explainable AI techniques for the purpose. Using LightGBM, a gradient boosting framework, a model was trained upon the dataset, which got an AUC score of 0.837 on test data. Local Interpretable Model-Agnostic Explainer (LIME) and SHapley Additive exPlainer (SHAP), both of which are powerful Explainable AI models, were deployed for evaluating the factors responsible for employee attrition.

Shankar et al. [5] used advanced techniques of the Machine Learning domain like Random Forest Classifier, Gradient Boosting Classifier, Neural Networks Classifier and TabNet to address the issue of attrition. Among all the nodels, MLP performed the best and had an accuracy of 88%.

Habous et al. [6] used Logistic Regression, Bernoulli Naive Bayes, Multimonial Naive Bayes, Decision Tree, Gaussian Naive Bayes and Random Forest. Best prediction of attrition was done using Random Forest model with 86.4% accuracy. Chakraborty et al. [7] tested five different models including Gradient Booster, SVM, K-Nearest Neighbors and found Random Forest model to be the best with 90.20% accuracy.

Patro et al. [8], alongwith prediction of attrition, also suggested an approach that might be useful for the company to retain its employees at the minimum cost. Implementations of 40 different models were done using AutoML algorithm of H2O Machine Learning platform. Stacked ensemble of all models gave the best F1 score of 0.6086 and thus, was used as the main reference model.

Mehta and Modi [9] got the best result with ensemble technique Gradient Boosting with an accuracy of 95.05%.

Bannaka et al. [10] proposed models for prediction of employee attrition as well as for employee retention prediction. Attrition was best predicted by Random Forest model with an accuracy of 89.70%. Retention was best predicted by XGBoost model with an accuracy of 90.44%.

Joseph et al. [11] out of 6 different machine learning models, found Random Forest Classifier to be the best with 86% accuracy and F1 score of 0.8599. Yahia et al. [12] used deep data driven predictive approaches alongwith machine learning approaches on a dataset of 1470 samples and another dataset of 15000 samples. Voting Classifier algorithm was found to be the best with an accuracy of 93% on dataset of 1470 samples and 98% on dataset of 15000 samples.

Alduayj and Rajpoot [13] used machine learning algorithms, adaptive synthetic approach as well as manual undersampling of data. The best performance was of 96.7% accuracy using KNN algorithm with Synthetic Balanced data.

Mhatre et al. [14] used big data and machine learning approaches for predicting attrition and found XGBoost algorithm to be the best with an accuracy of 96%. The said algorithm had a runtime of 1.76 seconds.

Bhartiya et al. [15] predicted attrition using Random Forest Classifier with an accuracy of 83.3%. Jain and Nayyar [16] used XGBoost for the purpose and achieved an accuracy of 89%. Hebbar et al. [17] used Random Forest with an accuracy of 90% for prediction. Ray and Sanyal [18] used Adaptive Probabilistic Estimation model to analyse and predict attrition of employees.

3 Proposed Methodology

The implementation of the Machine Learning algorithms was done in Python language using Google Colaboratory. Libraries like matplotlib, seaborn, pandas, numpy, scikitlearn, etc were used for various different purposes.

3.1 Dataset

The employee attrition dataset was obtained from Kaggle Machine Learning Repository [19].

The dataset contained five (5) files of .csv format –

- employee_survey_data.csv
- general_data.csv

- manager_survey_data.csv
- in_time.csv
- out_time.csv

Apart from these 5 files, there was a Data Description File named data_dictionary.xlsx describing relevant details about the above mentioned files. It provided the meanings as well as levels of all the variables in the above files.

Two files – in_time.csv and out_time.csv were excluded from being utilised for the proposed analysis of employee attrition.

The other three files were merged on Employee ID using the Pandas Merge command in Python and the resultant was used as the final dataset for analysis. The merged dataset consisted of 4410 rows (or, records of 4410 different employees) and 29 columns (or, 29 variables).

3.2 Phases of analysis

The analysis was carried out in three phases – Data Pre-processing, Statistical Analysis, and Modelling & Classification.

Data Pre-processing. Variable EmployeeID had distinct values for each record. Variables EmployeeCount, Over18 and StandardHours had same values for each record, i.e. 1, Y and 8 respectively. Hence, all these four columns were dropped. The dataset was then checked for presence of NULL values and it was found that columns 'NumCompaniesWorked', 'TotalWorkingYears', 'EnvironmentSatisfaction', 'JobSatisfaction', 'WorkLifeBalance' had 19, 9, 25, 20 and 38 NULL values respectively. Mode was used to fill the NULL values in 'EnvironmentSatisfaction', 'JobSatisfaction' and 'WorkLifeBalance'. Similarly, Median was used for 'NumCompaniesWorked' and Mean was used for 'TotalWorkingYears'. After the pre-processing, the dataset was of 4410 rows and 25 columns. Out of 4410 records, 711 had Attrition value 'Yes' which means 711 records were of those employees who had left the company and the remaining 3699 records were those of currently serving the company. This shows that there is an attrition of 16.12%.

Statistical Analysis. Various variables were plotted against the counts of attrition to determine the parameters of concern for the company. Thus, the parameters that needed to be addressed right away were identified for the decision makers.

Modelling and Classification. A Heat Map was plotted to check the correlation between variables (shown in Fig. 1). There was no such correlation that needed to be addressed.

Some Machine Learning approaches do not handle the categorical variables. So, Label Encoding of the dataset was done by using LabelEncoder from sklearn.preprocessing. The dependent variable 'Attrition' was separated from the independent variables. Then, splitting of dataset into Training Data and Testing Data was done. Testing Data accounted for 30% of the total dataset. Stratification was used to ensure equal proportions of the classes of 'Attrition' variable were present in both, training and test data.

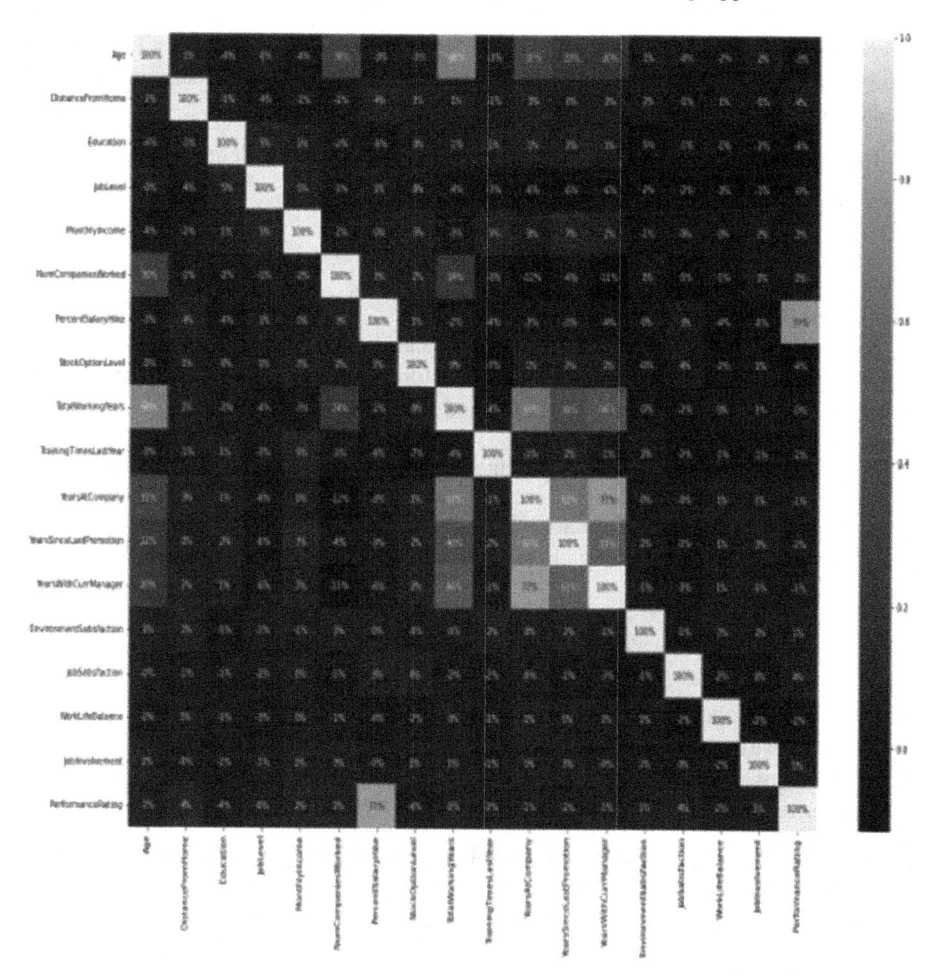

Fig. 1. Heatmap

Using the training data, models were then fitted using the following Machine Learning algorithms –

Logistic Regression (LR). Logistic Regression is used to measure the relationship between a categorical dependent variable and one or more independent variables. Its equation is generated as follows –

$$\log\left(\frac{\mathrm{P(x)}}{1-\mathrm{P(x)}}\right)=\mathrm{mx}+\mathrm{c} \tag{1}$$

$$\mathrm{P(x)}=\frac{\mathrm{e^{x}}}{(1-\mathrm{e^{mx+c}})} \tag{2}$$

This equation uses Maximum Likelihood Estimation (1–2).

Decision Tree (DT). Information Gain and Gini Index are the two popular criteria that can be used for building decision tree, which is a powerful tool for classification and prediction.

Entropy is a common way to measure the impurity. Higher the entropy, higher is the impurity. Entropy is calculated as in Eq. (3).

$$\text{Entropy} = \sum -p_x \log p_x, \tag{3}$$

where p_x is the probability of the class x of the set

Entropy is used to calculate Information Gain which in turn helps to know how significant a given attribute of the feature vectors is. Information Gain is the key to determine the ordering of attributes in the nodes of a decision tree.

$$\text{Information Gain} = \text{entropy of parent node} - \text{average entropy of child nodes} \tag{4}$$

Attribute with highest Information Gain is selected for splitting first.

When a variable is randomly chosen, Gini Index gives the degree or probability of it being wrongly classified.

$$\text{Gini} = 1 - \sum (p_i)^2 \tag{5}$$

where p_i is the probability of an object being classified to a particular class

While constructing a decision tree, the feature with the lowest Gini Index is preferred to be chosen as the root node. Gini Index is less computationally intensive and that's why preferred over Information Gain.

Random Forest (RF). Random Forest that can be used for both, classification as well as regression problems, works on the principle of Bagging. Bagging can be used to overcome the problem of overfitting in decision trees.

Bagging works in three steps:

i. Main dataset is broken into multiple subsets. These subsets can have fraction of rows and columns from main dataset.
ii. Classifier is built for each of the subsets.
iii. Results of all the classifiers are combined and label is assigned to the object. Thus, Random Forest is a powerful algorithm as it combines the result of multiple classifiers.

Support Vector Machine (SVM). It is an algorithm which plots the data points in a graph and finds an optimal plane which can separate two classes with high accuracy. Support Vectors are basically the coordinates of individual data points in the dataset which lie on the positive and negative hyperplanes. A line, or plane in case points are plotted in multi-dimensions, which separates two classes very well is called Hyperplane. Margin is the distance between extreme data points of both the classes. Hyperplane with highest margin is selected.

4 Result

Each machine learning model was first evaluated with the help of testing data on the basis of –

- Accuracy
- Confusion Matrix (from sklearn.metrics)
- Area Under the Curve (AUC)

The Confusion Matrices of all the models have been displayed in Fig. 2. The evaluation has been summarised in Table 1.

Table 1. Evaluation Summary

	Accuracy	Confusion Matrix				Area Under the Curve
		TN (True Negative)	FP (False Positive)	FN (False Negative)	TP (True Positive)	
LR	0.8405	1094	16	195	18	0.5350
DT	0.9312	1101	9	82	131	0.8034
RF	0.9863	1110	0	18	195	0.9577
SVM	0.8390	1110	0	213	0	0.5

In case of confusion matrices, correct predictions, i.e. True Positives and True Negatives depict the correct classifications by the model and need to be maximized. False Positives and False Negatives depict the misclassifications and should be as much less as possible. As per the observations from Fig. 2, Random Forest is the best performer. Also, Random Forest algorithm got the best accuracy of 98.63%.

Based on the confusion matrices and by using the classification report that belongs to sklearn.metrics, further evaluation was done based on the following parameters –

$$\text{Precision} = \frac{\text{TP}}{\text{FP} + \text{TP}} \tag{6}$$

$$\text{Recall (or, Sensitivity)} = \frac{\text{TP}}{\text{FN} + \text{TP}} \tag{7}$$

$$\text{F1 Score} = \frac{2 * (\text{Precision} * \text{Recall})}{\text{Precision} + \text{Recall}} \tag{8}$$

Support – Number of actual occurences of the class in the dataset. It doesn't vary between models. For class '0' i.e. for 'Employees who have not left', the support was 1110 and for class '1' i.e. for 'Attrition', support was 213.

Following observations were made on the basis of the above parameters as shown in Table 2.

The parameters like Precision, Recall and F1 Score are of very high importance considering the skewed nature of the target variable. The target variable 'Attrition' is

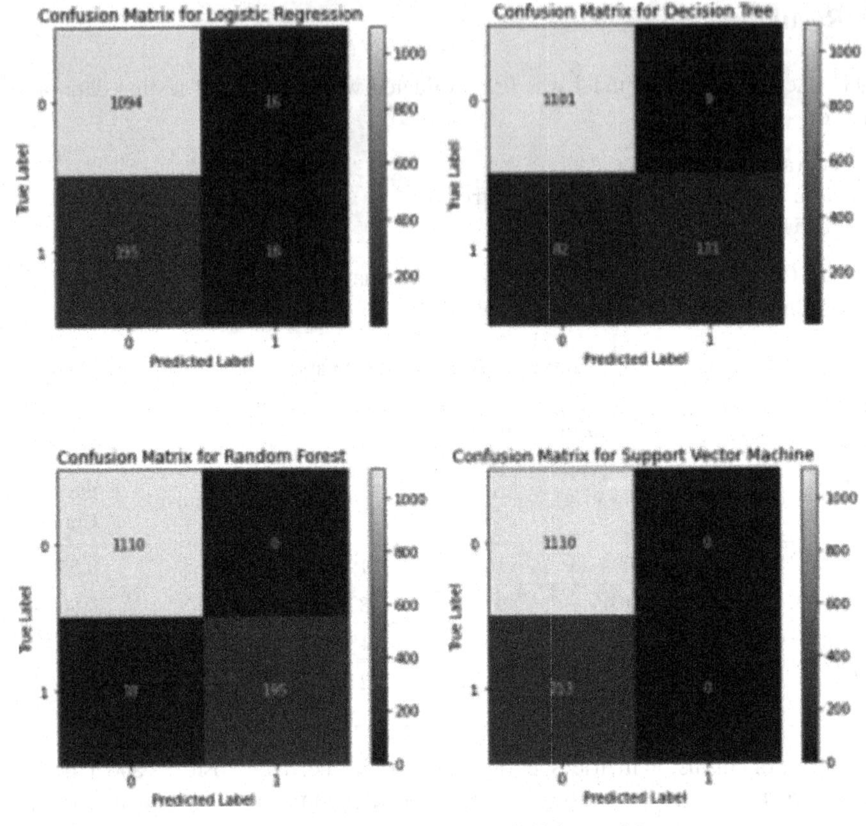

Fig. 2. Confusion Matrices of all the models

Table 2. Observations based on Classification Report

	Precision	Recall	F1 Score
LR	0.80	0.84	0.79
DT	0.93	0.93	0.93
RF	0.99	0.99	0.99
SVM	0.70	0.84	0.77

has large difference of between the counts of 'Yes (1)' and 'No (0)'. Only 16.12% of the total records were of people who left the company, i.e. 'Yes (1)'. In such cases, the evaluation parameters Precision, Recall and F1 Score (which is a combination of Precision and Recall) must be preferred in place of Accuracy. However, in this paper we have considered Accuracy as one of the evaluation parameters to better describe the functioning of machine learning models.

As per the findings tabulated in Table 2, Random Forest algorithm got the best results. It achieved highest Precision, Recall and F1 Score of 0.99.

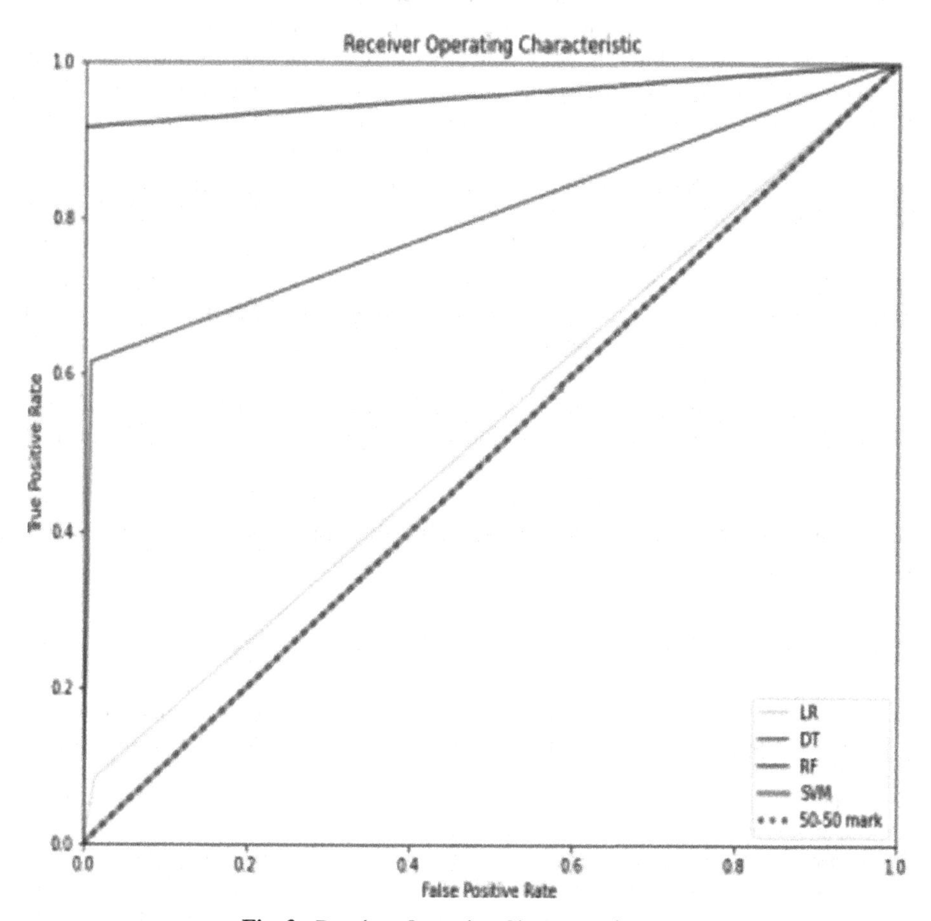

Fig. 3. Receiver Operating Characteristic Curve

The ROC Curve (Receiver Operating Characteristic Curve) plotted between True Positive Rate (TPR) and False Positive Rate (FPR) for all the models has been shown in Fig. 3. The ROC Curve provides an easy way to distinguish and compare the performances of multiple machine learning models at a time. It depicts the Area Under the Curve (AUC). Higher the AUC, better is the performance of the model. Random Forest algorithm got the highest AUC score of 0.9577.

5 Conclusion

Attrition is a major problem for all the corporates globally. During the period of FY 2021 – FY 2022, the IT industry specifically faced huge attrition related issues that were mainly because of high growth in the workforce demand in this particular sector following the Covid related restrictions and a paradigm shift to digitization.

The Global Economic Slowdown is creating complex situations such as rise in input costs, due to which margins of earnings are negatively affected in many sectors. Thus, organizations are looking forward to cost reduction and a great contributor to this can be reduction in recruitment cost.

It is of utmost importance, especially during such unprecedented time, for the employers to keep a check on their workforce requirements and have a predictive analysis of the attrition in the organization and subsequently take appropriate measures to curb any possible attrition. This definitely will help the organizations in cost reductions, especially on the front of recruitment expenditure.

The analysis and the pattern of recommendations made in this paper may be of great help to the decision makers in the organizations who are looking forward to curb attrition. By analysing and comparing the outcomes of each model, the Random Forest model with an F1 score of 0.99, accuracy of 98.63% and AUC of 0.9577 was found to be the best for analysis of employee attrition. Such a model can be deployed on the company specific manpower dataset to obtain predictive insights related to attrition.

References

1. Subramony, M., Holtom, B.C.: The long-term influence of service employee attrition on customer outcomes and pbrofits. J. Service Res. 15(4) 460–473 (2012). https://doi.org/10.1177/1094670512452792
2. Khalifa, N., Alnasheet, M., Kadhem, H.: Evaluating machine learning algorithms to detect employees' attrition. In: 3rd International Conference on Artificial Intelligence, Robotics and Control (AIRC), Cairo, Egypt (2022). https://doi.org/10.1109/AIRC56195.2022.9836981
3. Seelam, S.R., Kumar, K.H., Supritha, M.S., Gnaneswar, G., Reddy, V.V.M.: Comparative study of predictive models to estimate employee attrition. In: Seventh International Conference on Communication and Electronics Systems (ICCES), Coimbatore, India (2022). https://doi.org/10.1109/ICCES54183.2022.9835964
4. Sekaran, K., Shanmugam, S.: Interpreting the factors of employee attrition using explainable AI. In: International Conference on Decision Aid Sciences and Applications (DASA), Chiangrai, Thailand (2022). https://doi.org/10.1109/DASA54658.2022.9765067
5. Shankar, R.S., Piyadarshini, V., Neelima, P., Raminaidu, C.H.: Analyzing attrition and performance of an employee using machine learning techniques. In: Fifth International Conference on Electronics, Communication and Aerospace Technology (ICECA), Coimbatore, India (2021). https://doi.org/10.1109/ICECA52323.2021.9676102
6. Habous, A., Nfaoui, E.H., Oubenaalla, Y.: Predicting employee attrition using supervised learning classification models. In: Fifth International Conference On Intelligent Computing in Data Science (ICDS), Fez, Morocco (2021). https://doi.org/10.1109/ICDS53782.2021.9626761

7. Chakraborty, R., Mridha, K., Shaw, R.N., Ghosh, A.: Study and prediction analysis of the employee turnover using machine learning approaches. In: IEEE 4th International Conference on Computing, Power and Communication Technologies (GUCON) University of Malaya, Kuala Lumpur, Malaysia (2021). https://doi.org/10.1109/GUCON50781.2021.9573759
8. Patro, A.C., Zaidi, S.A., Dixit, A., Dixit, M.: A novel approach to improve employee retention using machine learning. In: 10th IEEE International Conference on Communication Systems and Network Technologies, Bhopal, India (2021). https://doi.org/10.1109/CSNT51715.2021.9509601
9. Mehta, V., Modi, S.: Employee attrition system using tree based ensemble method. In: Second International Conference on Communication, Computing and Industry 4.0 (C214), Bangalore, India (2021). https://doi.org/10.1109/C2I454156.2021.9689398
10. Bannaka, B.M.D.E., Dhanasekara, D.M.H.S.G., Sheena, M.K., Karunasena, A., Pemadasa, N.: Machine learning approach for predicting career suitability, career progression and attrition of IT graduates. In: 21st International Conference on Advances in ICT for Emerging Regions: 042 – 048, Colombo, Sri Lanka (2021). https://doi.org/10.1109/ICter53630.2021.9774825
11. Joseph, R., Udupa, S., Jangale, S., Kotkar, K., Pawar, P.: employee attrition using machine learning and depression analysis. In: Fifth International Conference on Intelligent Computing and Control Systems, Madurai, India (2021). https://doi.org/10.1109/ICICCS51141.2021.9432259
12. Yahia, N.B., Hlel, J., Palacios, R.C.: From big data to deep data to support people analytics for employee attrition prediction. IEEE Access **9**, 60447 - 60458 (2021). https://doi.org/10.1109/ACCESS.2021.3074559
13. Alduayj, S.S., Rajpoot, K.: Predicting employee attrition using machine learning. In: 13th International Conference on Innovations in Information Technology (IIT), Al Ain, United Arab Emirates (2018). https://doi.org/10.1109/INNOVATIONS.2018.8605976
14. Mhatre, A., Mahalingam, A., Narayanan, M., Nair, A., Jaju, S.: Predicting employee attrition along with identifying high risk employees using big data and machine learning. In: 2nd International Conference on Advances in Computing, Communication Control and Networking (ICACCCN), Greater Noida, India (2020). https://doi.org/10.1109/ICACCCN51052.2020.9362933
15. Bhartiya, N., Jannu, S., Shukla, P., Chapaneri, R.: Employee attrition prediction using classification models. In: 5th International Conference for Convergence in Technology (I2CT) Pune, India (2019). https://doi.org/10.1109/I2CT45611.2019.9033784
16. Jain, R., Nayyar, A.: Predicting employee attrition using XGBoost machine learning approach. In: International Conference on System Modeling & Advancement in Research Trends, Moradabad, India (2018). https://doi.org/10.1109/SYSMART.2018.8746940
17. Hebbar, A.R., Patil, S., Rajeshwari, S.B., Saqquaf, S.S.M.: Comparison of machine learning techniques to predict the attrition rate of the employees. In: 3rd IEEE International Conference on Recent Trends in Electronics, Information & Communication Technology, Bangalore, India (2018). https://doi.org/10.1109/RTEICT42901.2018.9012243
18. Ray, A.N., Sanyal, J.: Machine learning based attrition prediction. In: Global Conference for Advancement in Technology (GCAT) Bangalore, India (2019). https://doi.org/10.1109/GCAT47503.2019.8978285
19. Kaggle Dataset Link. https://www.kaggle.com/datasets/vjchoudhary7/hr-analytics-casestudy, (Accessed 20 Jan 2023)

A Time Series Analysis-Based Stock Price Prediction Framework Using Artificial Intelligence

Harmanjeet Singh(✉) and Manisha Malhotra

University Institute of Computing, Chandigarh University, Sahibzada Ajit Singh Nagar (Mohali), Punjab, India
{harmanjeetsingh.uic,manisha.mca}@cumail.in

Abstract. Forecasting stock prices have recently emerged as an essential component of the economic realm. Stock price forecasting is regarded as a challenging endeavor due to the volatility and noise of stock market activity. In many stock price prediction scenarios, the Facebook Prophet, LightGBM and ARIMAX models have been demonstrated to be competitive versus other models. This research presents an architecture based on a time series model, such as Facebook Prophet, Light Gradient Boost Machine (GBM), and Autoregressive Integrated Moving Average with Explanatory Variable (ARIMAX) to accurately predict stock prices. Experiments with multiple potential outcomes are conducted to evaluate the suggested framework using the stock price data set. The model was trained on ADANI stock price data over the previous fourteen years using Facebook Prophet, LightGBM and ARIMAX and evaluated using the Root Mean Square Error metric (RMSE).

Keywords: Time series · Stock prediction · ARIMAX · Facebook Prophet · LightGBM

1 Introduction

In the past, stand-alone deep learning techniques were used to achieve the goal of predicting stock prices. But, employing the deep learning neural network in conjunction with state-of-the-art methodologies in the execution of iterative tasks delivers a more significant degree of success. In the present era of the internet, where ML (machine learning) and DL (deep learning) approaches are armed with the rigorous capability for study and forecasts regarding stock prices [1]. There are existing businesses and organizations operating within the financial sector that offer their services to commercial businesses, retail establishments, and financial institutions worldwide. This industry is comprised of institutions such as banks, financial businesses, insurance firms, and real estate groups, as mentioned in [2]. Traders can engage in transactions involving currencies, equities, and derivatives thanks to the virtual trading platforms made available by agents of financial markets. The actions of financial institutions profoundly influence the economic

R. K. Challa et al. (Eds.): ICAIoT 2023, CCIS 1930, pp. 280–289, 2024.
https://doi.org/10.1007/978-3-031-48781-1_22

landscape. Because of the stock market, buying and selling publicly traded companies' shares can occur on or off specialized stock exchanges. Due to this market, even people who start with very few financial resources have the potential to amass a significant fortune. There is much interest in getting an accurate estimate of stock values in both the academic and corporate worlds, and for a good reason. Profits can be increased through stock price forecasting by academics and business specialists. A stock market is where stock transactions, trades, and distributions occur. Stocks must be moved about for any of these activities to occur. It has long been utilized by multinational corporations as one of their primary sources of finance [3]. Not only does the issuing of stocks bring in a considerable quantity of money, but it also fosters a greater concentration of funds, contributing to the consolidation of the organic structure of corporate capital. Predicting the price of a share of stock, although the stock market is notorious for its high levels of noise and its dynamic behaviors, is a highly challenging endeavor.

On the other hand, predicting stock prices requires extensive research and development time. Prices of stocks shift for several interrelated reasons, including geopolitical unrest, political maneuvering, shifts in economic conditions, shifts in market trends, advances in technology, and, perhaps most significantly, the attitudes of investors towards a particular stock. Time series or Deep learning is a new method that has recently come into existence [4], and is one of the numerous methods developed to anticipate the close price of a stock [5]. In this respect, the Long Short-Term Memory (LSTM) network, which intended to remember the temporal pattern for extended durations, is commonly employed for tasks involving time-series analysis [6, 4].

1.1 Contribution and Organization of Paper

In this work, a review of previous research on predicting stock prices using techniques based on artificial intelligence and hybrid models was done. The vast bulk of the methods that were proposed throughout the many publications was founded on the deep learning framework and hybrid outcomes for various stock market indexes. This study discusses the significant stock of a firm that uses AI-based time series approaches and the function of time series models with hyper-parameter tuning of features to get optimal outcomes.

The remaining parts of the article are organized as described below. In Sect. 2, detailed research methods are presented. This section includes the most current publication in stock price forecasting using AI-based time series models and methods. The architectures of the models that are being offered are discussed in Sect. 3. Section 4 presents the experiment research using data sets collected from the NSE website. Before moving on to the data pre-processing and model creation phase, this section widens the investigation of the technical indicators involved. In addition, this part provides a complete discussion of the model description, evaluation parameters, and outcomes after the section. The study comes to a close with a discussion of potential future options in Sect. 5.

2 Literature Review

In [7], Principal component analysis (PCA) was used to reduce the number of dimensions before the LSTM model was used to predict the stock's close price for the next trading day. This was done because technical indicators have a lot of dimensions. This was done before applying the model to forecast the stock's close price for the following trading day. Compared to the LSTM standard model, it was revealed that the quantity of redundant data dropped, and the model achieved greater accuracy in its predictions. This was a significant finding. In a later step, an LSTM model without PCA and an LSTM model with PCA were compared to see how accurate and consistent the predictions were. This was done in order to determine whether or not the PCA-equipped LSTM model produced more accurate results.

In [2], ML and DL algorithms were applied to unstructured data gathered from social media and structured data in the form of international data and technical indicators to forecast the value of stocks. Most of this stock price projection was based on hybrid data acquired in areas where English is not the predominant language. Both LSTM and GA were utilized to determine the parameters for the stock market indexes. Based on evaluation criteria such as Mean Absolute Percentage Error (MAPE) and RMSE, these parameters were compared with those chosen by other models such as Back Propagation Neural Network Genetic Algorithm (BPNNGA), Least Squares Support Vector Regression Genetic Algorithm (LSSVRGA), Random Forest Genetic Algorithm (RFGA), and Extreme Gradient Boost Genetic Algorithm (XGBoostGA). By attaining MAPE values that varied from those of the other four models by a margin of less than five percent, the Long Short-Term Memory employed Genetic Algorithm (LSTMSGA) model produced exceptionally accurate prediction results.

In [3], a hybrid strategy that is both time and resource efficient, has been proposed to reduce the risk associated with investing during Black Swan occurrences. Experiments were carried out using the S\&P 500, the BSE Sensex, and the Nifty 50 as test subjects. It was claimed that the model achieved an accuracy of 86 percent during "black swan" events, which are unpredicted events that happen worldwide, such as a recession, a long disease, or war. The model was evaluated based on the accuracy parameter, and it was claimed that it achieved an accuracy of 80 percent for single step ahead prediction.

In [8], hybrid models of Bi-directional Long Short-Term Memory (Bi-LSTM) and Gated Recurrent Unit (GRU) were suggested, where first the individual performances of the Bi-LSTM model, the GRU model, and the standard neural model were investigated and assessed. Then a comparison was held between the individual models and the hybrid models. Model accuracy was determined using Mean Squared Error (MSE), Root Mean Squared Error (RMSE), Mean Absolute Error (MAE), Mean Squared Log Error (MSLE), and Mean Absolute Percentage (MAPE) parameters.

In [9], it was recommended that the execution of the framework occur in two steps: selecting stocks for the portfolio and forecasting stock prices. Both of these phases would take place after the framework had been designed. Selecting stocks for a portfolio begins with constructing a mean-variance model using the stocks' historical data. It is followed by the determination of the sharp ratio. If the ratio is greater than one, the operation will continue to utilize Synthetic Minority Over-sampling (SMOTE) techniques. If this is not the case, a different group of stocks will be chosen.

In [10], LSTM and CNN-based hybrid model was presented, followed by an Attention block. Unfortunately, the model ran into the issue of over-fitting, which necessitated a change in the model's architecture in the form of an adjustment to the amount of neurons present in each layer and the addition of a drop-out layer after each of the layers.

It is essential to emphasise that accurate prediction relies heavily on effective attribute extraction, which is why this topic warrants special attention.

To efficiently learn the dynamics for share price from a data-set, [11] offers a hybrid DL framework combine with CNNs model and LSTM model. The first step taken into consideration due to a time series data set containing stock prices was the extraction of various time scales from the data set using the CNN layer. The original prices were blended at a later date using time scale extraction. It was also thought about how to use multiple LSTMs to understand temporal dependencies for different time-scale features. While attempting to anticipate the close price of a stock, all of the features that have been gathered by LSTMs are integrated using linked neuron networks. Still, the above strategy doesn't fully understand how the interdependence of stock prices that are very long doesn't work.

In [12], it was possible to do a comparative examination of twenty-five distinct ensembles' worth of regressors and classifiers. The study's results showed that stacking and mixing ensembles worked better than bagging and boosting. While evaluating the models, the RMSE parameter was considered, and it was reported that stacking obtained 0.0001–0.001 and blending achieved 0.002–0.01, both of which were deemed to be superior to other ensemble strategies.

In [13], the authors described hybrid deep learning models capable of forecasting and/or analyzing time series data-sets. These models were made using multi-scale features, and the authors used many pipelines or networks to pull out the multi-scale features. This resulted in a large and complex model that was not considered favorable for training or testing and was also irrelevant to insight patterns.

A new hybrid approach introduced in [14], LSTM auto-encoder and stacked LSTM network comprise the first and second layers of the LSTM model's fundamental structure, respectively. An auto-encoder method was observed to be quite beneficial for noise reduction in inputted data, which results in negligible mistakes in the case of a basic long short term memory model and a standard multi-layer perceptron. This finding was made after it was discovered that an auto-encoder method had shown quite beneficial results. The conclusion was reached when more errors were produced when the parameter had a high value.

Ensemble model of LSTM networks for Intraday trading forecasting was proposed in [15], in which a series of LSTM network models are used individually before being joined in series to evaluate overall performance. The network inputs are taken from a broad variety of technical analysis indicators. The results showed that using Ensemble LSTM significantly improved over typical ensemble weighting procedures. For example, a model with an equally weighted technique, or one with an ideal single model, may be evaluated based on its power to forecast the price of 22 stocks out of 44 US big-cap businesses. Additionally, the outcomes of the ensemble LSTM model are superior to those of standalone models like the lasso and ridge logistic regression. Although surpassing several different benchmarks, the recommended solution has a significantly higher overall cost in terms of computation. The training duration of the ensemble model for each stock took roughly one hundred and fifty minutes, whereas the training period of lasso for the same job just took a few seconds.

Multi-filter neural networks such as recurrent and convolutional networks were proposed in [16], to extract features from time series data. These hybrid neural networks outperformed single-structure neural networks when put in comparison with state of art parameters. Although it is an inevitable fact that for short term duration learning, CNNs models are excellent whereas Recurrent Neural Networks (RNN) are considered to be excellent in long term duration. However, in [17] networks were observed to struggle in comprehend the extremely long stock price data-sets.

Despite this, the time series that we have provided considers all of the problems that have been brought up to correctly predict stock prices. ARIMAX, Facebook Prophet, and LightGBM are the three components that make it up.

3 Architecture of Proposed Models

In this research, time series models were deployed on stock named as Adani Port to forecast stock price. Initially, the dataset was preprocessed. Further, three windows were designed, one for every three days, one for every seven days, and one for every thirty days, with features corresponding to "high," "low," "volume," "turnover," and "trades." For each time window, average of attributes was calculated. Then, for each time window, standard deviation was calculated. After that, the dataset was split into two subsets and the mean value of each feature was put in the empty cells of each subset. The data collected before 2019 served as our training data set, while the data collected after 2019 served as our test data set. Because stock prices are highly impacted by noise and disturbance from the outside, in order to stabilize this noise, we include a list of exogenous features. These features include high mean, low mean, high standard deviation, low standard deviation, volume mean, volume standard deviation, turnover mean, and turnover standard deviation in durations of a day, day of the week, a week, and a month.

After this, the proposed model is trained with AutoARIMA, FBProbhet and LightGBM. The following steps were carried out in suggested framework's stock prediction process:

a) **Data collection:** Stock data is gathered from yahoo finance from the period of Nov. 2007 to April. 2021 and supplied into the pre-processing step during this stage.
b) **Data Pre-processing:** In this phase, involves transforming raw data into something that can be analysed.
c) **Data Modeling:** a stock prediction model is built and the converted data are input into it. To properly anticipate the closing price of a stock, data analysis is carried out in this stage.
d) **Performance Evaluation:** It is here that the model's output is compared against actual output in order to determine whether or not the model is accurate. Model improvement or data preparation enhancement is needed if this model does not meet its stated goals. Otherwise, inform the user of the findings.
e) **Result:** Once the model has been shown, the user may see the findings and predictions.

4 Experiment Study

Figure 1 shows the five critical features from data-set that can be use in prediction of stock movement. These features are stock's open price, close price, high price, low price and last close price.

4.1 Dataset

This analysis makes use of the historical data on the stock that Yahoo Finance generated. Figure 1 illustrates an example of this data collection, which has a variety of attributes such as open (the stock's current open price), close (its current close price), high (the stock's highest price in a day), low (the stock's lowest price in day), close (the stock's closing price in a day), and last (last close price of a stock).

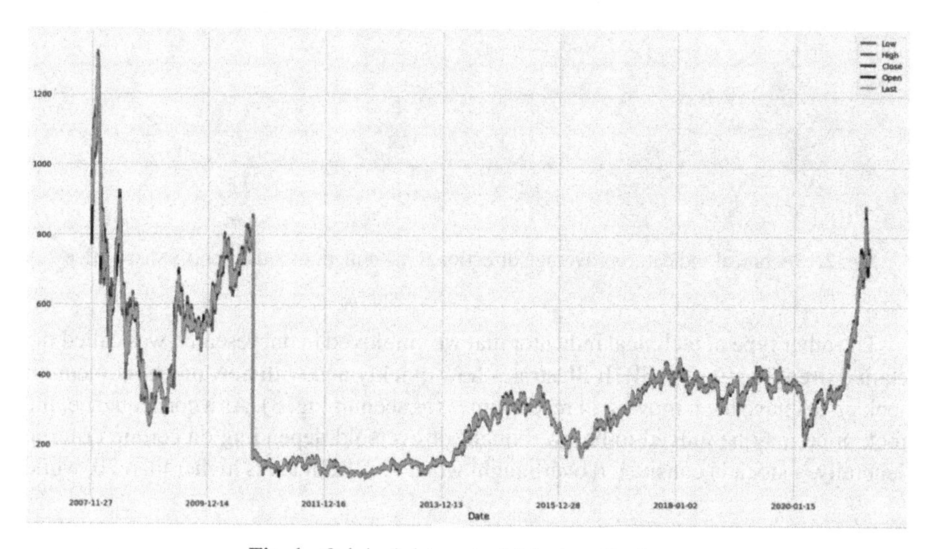

Fig. 1. Original data-set of Adani port's share.

The direction of the stock movement is reflected in the adjusted characteristics to provide an accurate depiction of a company's value after considering its operational operations. This data set provides information on the price movement of Adani stock's throughout the given period. Figure 1 depicts Adani's share price close index.

One set of data is used for training, while the other is utilized for testing, for two sets of data. The train data set accounts for 70 percent of the source data set, whereas the test data set comprises 30 percent. Data ranging from November 27, 2007 (which is a Tuesday) through April 19, 2017 (which is a Wednesday) was used to train the models. The models are then assessed using data ranging from Thursday, April 20, 2017, until Thursday, April 30, 2021.

4.2 Technical Indicators

For the stock price of Adani Port, an average directional movement was determined where ADX < = 25 is considered not in trends, ADX = > 25 to ADX < = 50 is considered in trends and ADX = > 50 is considered in high trends. It performs well in determining the direction of a trend, which is useful for market analysts and investors (Fig. 2).

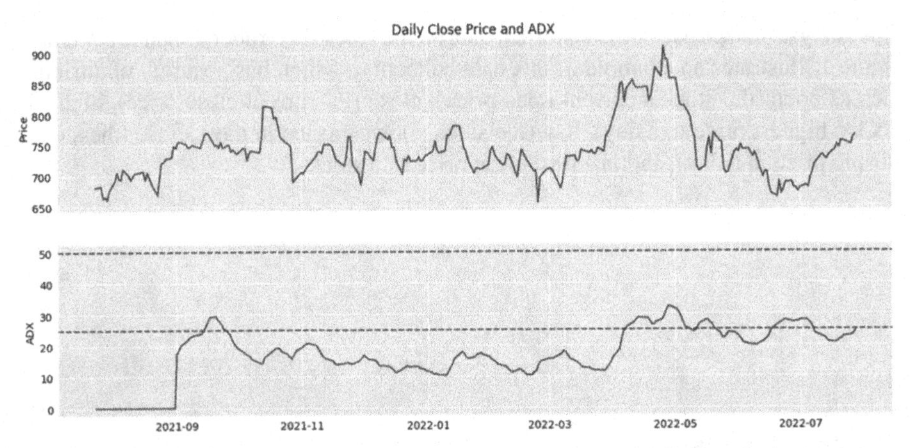

Fig. 2. Technical indicator as average directional movement of Adani port's share price

The other type of technical indicator that we employed in our research was called the relative strength index (RSI). It illustrates how quickly and with how much momentum stock prices have been moving in recent times (as seen in Fig. 3). As a consequence, the stock price may be judged to be overbought or oversold depending on certain criteria. Generally, a stock is considered overbought when its RSI index is higher than 70, while

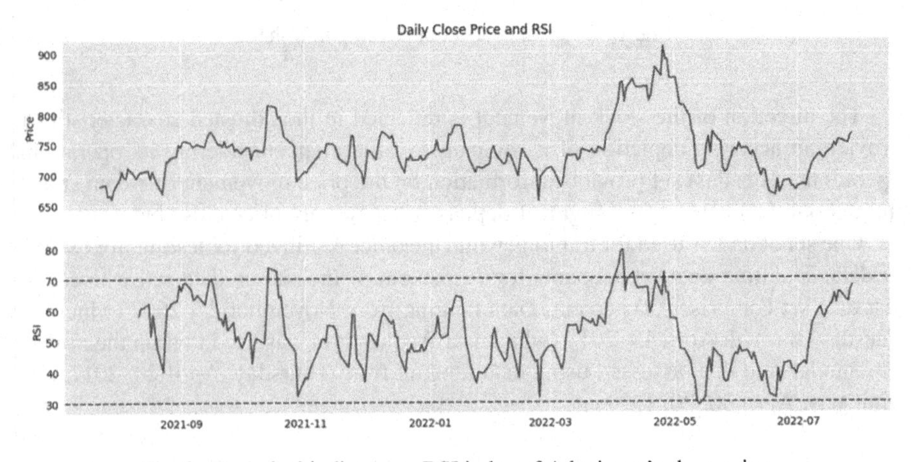

Fig. 3. Technical indicator as RSI index of Adani port's share price

a stock is considered to be oversold when its RSI index is lower than 30. Further information from the RSI indicator suggests that stock prices may be getting ready to make a trend reversal or corrective retreat. This index, in general, assists the investor in making decisions on the optimal moment to purchase or sell shares. Figure 3 illustrates Adani port's daily close price as well as the RSI index over the specified period.

4.3 Data Pre-processing and Model Design

In order to create the model of the time series, we followed the process that is mentioned below:

Following the separation of the data set into a train set of data before the year 2019 and a test set of data after the year 2019, based on the close price column, we processed the training and test data in such a way that every next day's close price is dependent on the different windows size such as 3-days, 7-days, and 30-days the previous values of stock's open price, close price, high price, low price, and last close price, which are then marked as features. This was accomplished Following the completion of the process involving the preparation of the data, we computed the mean and standard deviation of each characteristic in order to arrive at a point where we could validate the presence of stationary data. In order to create our models, we relied on the library resources provided by the Prophet, PMDARIMA, and LightGBM. In order to develop the AUTO ARIMA model, we used the VWAP (volume average weighted price) of the train dataset in conjunction with some exogenous characteristics. The model is constructed with the loss parameter set to "root mean squared error," and it also has the "mean absolute error" setting.

In order to develop the Facebook Prophet model and LightGBM, date and close price columns from training data were taken into consideration with VWAP columns.

4.4 Result and Discussion

In order to successfully implement all of the time series models, the frameworks TensorFlow 2.8.0 and Keras 2.8.0 were utilized. The researchers used a 64-bit Windows 10, an Intel i5-8265U central processing unit (CPU), 16 gigabytes of random-access memory (RAM), and 64-bit Windows to train the models. In the instance of Auto ARIMA, it took around 210 ms, generating an RMSE of training data of 27.12, followed by an MAE of training data of 10.61. In the instance of LightGBM, it took around 213 ms, giving an RMSE of 89.98 for the training data, followed by an MAE of 44.76 for the training data. In Facebook prophet, it took around 202 ms, generating an RMSE of training data of 29.72, and MAE of training data of 16.26.

The trained model is given test data so that the "Loss," "Mean Squared Error," and "Mean Absolute Error" parameters can be used to measure how well it can predict outcomes. This is achieved by utilizing the mean squared error library, which is imported from the sklearn package. The model generated the results that are displayed in Table 1.

The graph of the same is shown in Fig. 4 where the ARIMAX forecast line gives better prediction results than other models.

Table 1. Evaluation matrices of prediction result on training data in time series models.

Stock Name	Model's evaluation of training records		
	Model	RMSE	MAE
Adani Port	ARIMAX	27.12	10.61
	LightGBM	89.98	44.76
	FBProphet	29.72	16.26

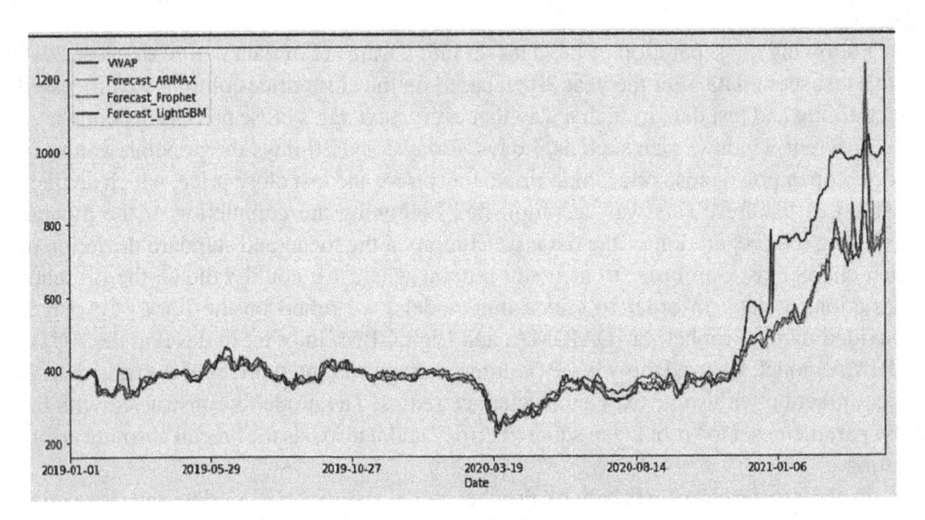

Fig. 4. Predicted price of Adani port's share.

5 Conclusion

Computerized financial market price forecasting is attracting investors. Consequently, ARIMAX, FBProphet, and LightGBM time series models may predict stock prices. In this study, we invented a time series-based stock market forecasting approach. The suggested model calculated ADANI Ports' pricing with the lowest RMSE and MAE. Our goal is to use classical, ensemble, and time series methods together to make more reliable future work models.

In future research, the suggested technique may be applied to stock market data from different nations, and comparisons can be made with XGBoost, SVM, and ANN. The SVM and ANN models, on the other hand, can be built using several different things, such as investor mood, interest rates, the country's political situation, news, and currency rates. In addition, we will assess the accuracy of our proposed model's predictions for various nonlinear and non-stationary time series, including exchange rates, crude oil prices, wind speed, temperature, precipitation, earthquakes, and tourist arrivals.

References

1. Nti, I.K., Adekoya, A.F., Weyori, B.A.: A systematic review of fundamental and technical analysis of stock market predictions. Artif. Intell. Rev. **53**(4), 3007–3057 (2019)
2. Lin, Y.L., Lai, C.J., Pai, P.F.: Using deep learning techniques in forecasting stock markets by hybrid data with multilingual sentiment analysis. Electronics **11**(21), 3513 (2022)
3. Bhanja, S., Das, A.: A black swan event-based hybrid model for Indian Stock Markets' trends prediction. Innov. Syst. Softw. Eng. **bhanja2022black**, 1–5 (2022). https://doi.org/10.1007/s11334-021-00428-0
4. Jing, N., Wu, Z., Wang, H.: A hybrid model integrating deep learning with investor sentiment analysis for stock price prediction. Expert Syst. Appl. **178**, 115019 (2021)
5. Sarvesh, S., Sidharth, R.V., Vaishnav, V., Thangakumar, J., Sathyalakshmi, S.: A hybrid model for stock price prediction using machine learning techniques with CNN. In: 5th International Conference on Information Systems and Computer Networks (ISCON) (2021)
6. Kumar, D., Meghwani, S.S., Thakur, M.: Proximal support vector machine-based hybrid prediction models for trend forecasting in financial markets. J. Comput. Sci. **17**, 1–13 (2016)
7. Zheng, X., Xiong, N.: Stock price prediction based on PCA-LSTM model. In: 5th International Conference on Mathematics and Statistics (2022)
8. Md. Ebtidaul Karim, Md. Foysal, Sunanda Das,: Stock Price Prediction Using Bi-LSTM and GRU-Based Hybrid Deep Learning Approach. In: Ashish Khanna, Deepak Gupta, Vineet Kansal, Giancarlo Fortino, Aboul Ella Hassanien, (ed.) Proceedings of Third Doctoral Symposium on Computational Intelligence: DoSCI 2022, pp. 701–711. Springer Nature Singapore, Singapore (2023). https://doi.org/10.1007/978-981-19-3148-2_60
9. Padhi, D., Padhy, D.N., Bhoi, A.K., Shafi, J., Yesuf, S.: An intelligent fusion model with portfolio selection and machine learning for stock market prediction. Comput. Intell. Neurosci. **06**, 18 (2022)
10. Jiang, W.: Applications of deep learning in stock market prediction: re-cent progress. Expert Syst. Appl. **184**, 115537 (2021)
11. Hao, Y., Gao, Q.: Predicting the trend of stock market index using the hybrid neural network based on multiple time scale feature learning. Appl. Sci. **10**(11), 3961 (2020)
12. Nti, I., Adekoya, A., Weyori, B.: A comprehensive evaluation of ensemble learning for stock-market prediction. J. Big Data **7**, 3 (2020)
13. Eapen, J., Bein, D., Verma, A.: Novel deep learning model with CNN and bi-directional LSTM for improved stock market index prediction. In: IEEE 9th Annual Computing and Communication Workshop and Conference (CCWC) (2019)
14. Al-Thelaya, K.A., El-Alfy, E.S.M., Mohammed, S.: Forecasting of bahrain stock market with deep learning: methodology and case study. In: 8th International Conference on Modeling Simulation and Applied Optimization (ICMSAO), pp. 1–5. Manama, Bahrain (2019)
15. Borovkova, S., Tsiamas, I.: An ensemble of LSTM neural networks for high-frequency stock market classification. J. Forecast. **38**(6), 600–619 (2019)
16. Long, W., Lu, Z., Cui, L.: Deep learning-based feature engineering for stock price movement prediction. Knowl.-Based Syst. **164**, 163–173 (2019)
17. Kamara, A.F., Chen, E., Liu, Q., Pan, Z: Combining contextual neural networks for time series classification. Neurocomputing **384**, 57–66 (2020).

Towards Better English to Bharti Braille Neural Machine Translation Through Improved Name Entity Translation

Nisheeth Joshi[1,2](✉) and Pragya Katyayan[1,2]

[1] Department of Computer Science, Banasthali Vidyapith, Tonk, Rajasthan, India
nisheeth.joshi@rediffmail.com

[2] Centre for Artificial Intelligence, Banasthali Vidyapith, Tonk, Rajasthan, India

Abstract. In this paper, the expansion of English to Bharti Braille neural machine translation system has been carried out. It is shown how a baseline Neural Machine Translation (NMT) model can be improved by adding some linguistic knowledge to it. This was done for five language pairs where English sentences were translated into five Indian languages and then subsequently to corresponding Bharti Braille. This has been demonstrated by adding a sub-module for translating name entities. The approach shows promising results across language pair and improvement in the quality of NMT outputs. The least improvement was observed in English-Tamil language pair with 8.79% and the most improvement was observed in English-Hindi language pair with 15.01%.

Keywords: Neural Machine Translation · Name Entity Recognition · Linguistic Embellishments

1 Introduction

Braille is a writing system used by people who are visually impaired. It is a system of raised dots that can be read by touch. Each character, or cell, is made up of six dots arranged in a rectangular grid, with up to two dots raised in any given position. The Braille code assigns a unique combination of dots to represent each letter of the alphabet, punctuation marks, and other symbols. It can also be used to represent music and mathematical notation. The Braille system was invented by Louis Braille in 1824 and is widely used around the world as a primary means of literacy for the visually impaired.

Machine translation (MT) is the use of software to automatically translate text from one natural language (such as English) to another (such as Spanish). The goal of machine translation is to produce a high-quality, fluent translation that is equivalent in meaning to the original text. There are several different types of machine translation, including rule-based machine translation, statistical machine translation, and neural machine translation.

Rule-based machine translation (RBMT) uses a set of pre-defined rules to translate text, while statistical machine translation (SMT) uses statistical models trained on

R. K. Challa et al. (Eds.): ICAIoT 2023, CCIS 1930, pp. 290–300, 2024.
https://doi.org/10.1007/978-3-031-48781-1_23

large bilingual corpora (texts in two languages) to generate translations. Neural machine translation (NMT) is the most recent type of machine translation and uses deep neural networks to generate translations.

Machine translation is widely used in a variety of applications, such as online translation services, localization of software and websites, and multilingual information retrieval. However, it's important to note that machine translation is not always perfect and may produce translations that are not entirely accurate or idiomatic.

Braille machine translation is a technology used to convert written text into Braille, a writing system used by people who are visually impaired. The translated text can be read using a Braille display, which is a device that uses a series of raised dots to represent letters and other characters. The process of converting text to Braille is called Braille transcription. There are several software and online tools that can be used for Braille machine translation.

In this paper we discuss the discussed the working of our machine translation system which incorporates NMT approach. Through experiments we show the improved our proposed approach over the baseline NMT system.

2 Literature Review

Abualkishik and Omar developed a system for translating Quran verses into Braille [1]. The system translated various deviations as well. It was found that the system could only work well with only one variation. The authors also made a partially successful attempt to transcribe the different versions into Braille [2]. They further extended the work and were able to transcribe three versions of Quran verses into Braille. Al-Salman proposed a low cost solution of translating Arabic text into Braille and vice versa [3]. The system developed was rule based system where rules were used for translation. Through this system the visually impaired people we able to use the internet and communicate with the world. Azam et al. worked on developing a transfer based machine translation system for Bangla to Braille [4]. They evaluated their system on a literary text where Bangla poem was translated into Braille and was provided to a visually impaired people for validation. It was found that the system was able to mostly translate all the text into Braille. This was one of the first attempts to bridge the knowledge barrier that the visually impaired people face in their daily lives.

Blenkhorn proposed a technique to transcribe braille to text using a rule-based mechanism [5]. The approach developed was lightweight and could be configured for other languages as well. In their paper, they demonstrated an algorithm for converted English text into braille. Cleave worked on developed a system which showed appropriate translation which at that time was being done by humans using some electro-mechanical machines [7]. Blenkhorn also proposed a scheme of translating text into braille. They showed the translation of text using a dictionary-based mechanism which was able to translated characters in several languages into braille. This helped braille experts to quickly change the braille mappings without having any knowledge of programming. This approach was very effective and popular among this community. Fahiem worked on developing a translation system which translated Urdu text into braille. Urdu is a context sensitive language, as it has different forms of glyphs for an alphabet, depending on the position of that character in the word. Thus, system developed was also context sensitive [9]. They used an optical character recognition system to recognize braille code for Urdu. They then developed a mechanism which then translated the braille text into Urdu which was context sensitive. George et al. developed a device which helped visually impaired people to read English text. This used a simple transliteration technique which simply transliterated a character into braille [10]. The letter were first recognized and then were sent for mapping which were then converted into respective braille encodings and were shown on the user interface.

Gotoh et al. developed a system named BrailleMUSE which translated music into braille. This computer-based system was the first of this kind [11]. The system took music from the internet and translated into braille. Hossain et al. highlighted the lack of tools for the visually disabled people for studying in Bangladesh [12]. It was highlighted that due to this reason a large number of visually impaired people were deprived of basic education. They developed transfer-rules and thus developed a transfer-based machine translation system which translated Bangla to braille. This system used a automata-based approach where DFA was developed for translation of text. The output of the system was given to the visually impaired people who confirmed that the system was able to properly translate the text from Bangla to braille.

Hossain et al. worked on developing a machine translation system using the same approach. They also developed rules using DFA and regular expressions [13]. They then used a statistical approach which did the job of translating text to braille fairly well. In their study, it was reflected that regular expressions play a very important role in representing language which can be used by machines [14]. They validated their claims though several experiments and concluded that use of DFA and regular expressions were very effective in translation text from Bangla to braille. They primarily employed elimination methods for structure and state and achieved satisfactory results.

Iain and Pasquale worked on development of a device which could back translate braille text into natural language [15]. First, the device performed scanning of the page which has braille text in it. Then, only the braille text was converted into natural language. This further bridged the barrier between people who could see and people who were visually impaired. Ingham worked on developing braille books using two mechanisms (i.) presses which used zinc plates for embossing; and (ii.) by taking help of volunteers [16]. This proved to be a low-cost solution of producing books in braille. They further showed techniques which were cost effective in input generation and streamlined the entire working. Li and Yan developed a SVM classifier which extracted braille text from images [20]. They reduced this problem into a simple image processing problem where the image was taken as an input though camera, was preprocessed, segmented and finally braille encoding extraction. This approach was simple and very effective in extracting braille text. One of the reasons could be braille being a fixed language in nature. Li et al. proposed a simple character recognition system which extracted braille text using Haar features and SVM classification [21].

Mahbub-Ul-Islam et al. felt the need of providing books in braille so that visually impaired people could study [22]. They worked on developing several approaches which transcribed Bangla text into braille. Finally, they so worked on developing a model based on DFA for translation of Bangla text into braille. Nian-Feng and Li-rong showed the constraints of general public in understanding braille and accordingly work with it [23]. Shimomura et al. developed a translation system from Japanese to braille. It was developed in C and Python [25]. They first performed morphological analysis of Japanese using MeCab engine. Then, they used TensorFlow to develop their neural machine translation system. The NMT model was trained on 1 lac Japanese lexicons.

Wang et al. worked on developing a new method of translating Chinese to braille. Since in Chinese one has to first perform word segmentation, they worked on developed a corpus-based machine translation system for translating Chinese text into braille [30]. They employed statistical machine learning for this task which used context mining of the corpus and identified the tone marking. Zhang et al. identified the need for translation of literary text into braille so that visually impaired and understand their culture [31]. As braille is a cryptic language, its translation using computer-based mechanism was needed. Although several points were highlighted by the authors but they missed out of explaining the hardware implementation or the design of the algorithm.

In Indian Context Das et al. explored the possibility of computerization of two braille equipment which were being used in production of text in braille [8]. A machine translation system was developed which translated text from English to braille. To test the performance of the system, it was provided to a blind boys' academy in West Bengal, India for testing and evaluation. Jariwala and Patel worked on developing system for translation of Gujarati text into braille [17]. They highlighted the need for a translation tool which would help visually impaired people living in Gujarat. They developed a simple rule-based machine translation system of translating text in English, Hindi and Gujarati to braille [18]. This system helped in translation text into text and finally embossing into a book. Lahiri et al. showed the working of their system "Sparsha" which helped the visually impaired people in reading as well as writing text in braille using computers [19]. They explained their approach of translation of text to and from Bharti braille.

Salah and Ram studied several transliteration systems used in transcribing text into braille and then worked on developing a translation system for Malayalam to braille [24]. Shreekanth and Udayshankara worked on developing Hindi braille corpus as their were none available at that time. They worked on developing a Bharati braille-bank database [26]. They portrayed the contradictions in several research findings and concluded that this all was happening due to a lack of common encodings. Singh and Bhatia developed a transliteration system for transcribing English and Hindi text to braille [27]. They primarily used a dictionary-based approach for this study. Their approach was simple but yet effective as it correctly converted English and Hindi text into braille [28]. They splitted the input sentence into words and then extracted characters and finally performed lookup-up in the dictionary. Vyas and Virparia studied several techniques for transcribing braille characters. He also looked at their properties, extracted features from them [29]. They portrayed the characteristics of Gujarati and problems in transcribing and recognizing braille characters in Gujarati.

3 Experimental Setup

In order to develop a machine translation system. We first collected English monolingual corpus and translated it into five Indian languages viz Bengali, Gujarati, Hindi, Marathi and Tamil. Table 1 shows the 11 constructs in English which were specifically focused.

Table 1. Characteristics of English Monolingual Corpus.

S. No.	Construct
i.	Simple Sentence
ii.	Infinitive Sentence
iii.	Gerund Sentence
iv.	Participle Sentence
v.	Appositional Sentence
vi.	Initial Adverb Sentence
vii	Coordinate Sentence
viii.	Copula Sentence
ix.	Wh-Sentence
x.	Relative Sentence
xi.	Discourse Construct

One lac English sentences were collected and then got them translated by human annotators. The translated sentences were then vetted by two human annotators.

Once the vetting process was competed, this corpus was used for training NMT models. Two NMT models- a baseline NMT model and another name entity aware NMT model, were trained.

4 Proposed System

For both the models, some common steps are followed which are the de-facto methods in current state-of-the-art NMT systems. These were:

4.1 Source Text Rewriting

The source English text which has its default structure as subject-verb-object is converted to the default structure of Indian languages i.e., subject-object-verb. For this, the English sentences were first parsed using Stanford CoreNLP library [32]. It was our first choice as it is the most popular and well-maintained library for computational processing of English.

A transfer grammar which converted English syntax into the syntax of Indian languages, was constructed. For this, 843 handcrafted rules were developed. A snapshot of these rules is shown in Fig. 1.

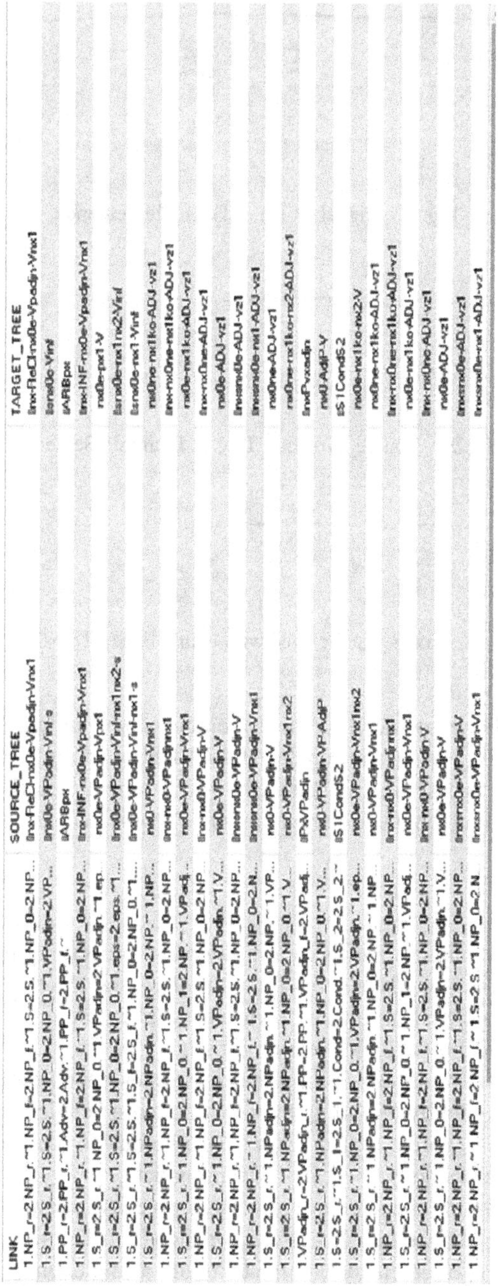

LINK	SOURCE_TREE	TARGET_TREE
1.NP_r=2.NP_r.~1.NP_f=2.NP_f.~1.S=2.S.~1.NP_0=2.NP...	Inx-ReCl-nx0e-Vpadjn-Vnx1	Inx-ReCl-nx0e-Vpadjn-Vnx1
1.S_r=2.S_r.~1.S=2.S.~1.NP_0=2.NP_0.~1.VPadjn=2.VP...	Inx0e-VPadjn-Vinf-s	Isnx0e-Vinf
1.PP_r=2.PP_r.~1.Adv=2.Adv.~1.PP_f=2.PP_f.~	IARBpx	IARBpx
1.NP_r=2.NP_r.~1.NP_f=2.NP_f.~1.S=2.S.~1.NP_0=2.NP...	Inx-INF-nx0e-Vpadjn-Vnx1	Inx-INF-nx0e-Vpadjn-Vnx1
1.S_r=2.S_r.~1.NP_0=2.NP_0.~1.VPadjn=2.VPadjn.~1.ep...	nx0e-VPadjn-Vpx1	nx0e-px1-V
1.S_r=2.S_r.~1.S=2.S.~1.NP_0=2.NP_0.~1.eps=2.eps.~1...	Inx0e-VPadjn-Vinf-nx1nx2-s	Isnx0e-nx1nx2-Vinf
1.S_r=2.S_r.~1.S=2.S.~1.S_f=2.S_f.~1.NP_0=2.NP_0.~1....	Inx0e-VPadjn-Vinf-nx1-s	Isnx0e-nx1-Vinf
1.S_r=2.S_r.~1.NPadjn=2.NPadjn.~1.NP_0=2.NP.~1.NP...	nx0-VPadjn-Vnx1	nx0ne-nx1ko-ADJ-vz1
1.NP_r=2.NP_r.~1.NP_f=2.NP_f.~1.S=2.S.~1.NP_0=2.NP...	Inx-nx0-VPadjnnx1	Inx-nx0ne-nx1ko-ADJ-vz1
1.S_r=2.S_r.~1.NP_0=2.NP_0.~1.NP_1=2.NP.~1.VPadj...	nx0e-VPadjn-Vnx1	nx0e-nx1ko-ADJ-vz1
1.NP_r=2.NP_r.~1.NP_f=2.NP_f.~1.S=2.S.~1.NP_0=2.NP...	Inx-nx0-VPadjn-V	Inx-nx0ne-ADJ-vz1
1.S_r=2.S_r.~1.NP_0=2.NP_0.~1.VPadjn=2.VPadjn.~1.V...	nx0e-VPadjn-V	nx0e-ADJ-vz1
1.NP_r=2.NP_r.~1.NP_f=2.NP_f.~1.S=2.S.~1.NP_0=2.NP...	Inxsnx0e-VPadjn-V	Inxsnx0e-ADJ-vz1
1.NP_r=2.NP_r.~1.NP_f=2.NP_f.~1.S=2.S.~1.NP_0=2.N...	Inxsnx0e-VPadjn-Vnx1	Inxsnx0e-nx1-ADJ-vz1
1.S_r=2.S_r.~1.NPadjn=2.NPadjn.~1.NP_0=2.NP.~1.VP...	nx0-VPadjn-V	nx0ne-ADJ-vz1
1.S_r=2.S_r.~1.NPadjn=2.NPadjn.~1.NP_0=2.NP_0.~1.V...	nx0-VPadjn-Vnx1nx2	nx0ne-nx1ko-nx2-ADJ-vz1
1.VPadjn_r=2.VPadjn_r.~1.PP=2.PP.~1.VPadjn_f=2.VPadj...	IPxVPadjn	InxPvxadjn
1.S_r=2.S_r.~1.NPadjn=2.NPadjn.~1.NP_0=2.NP_0.~1.V...	nx0-VPadjn-VP-AdjP	nx0-AdjP-V
1.S=2.S_r.~1.S_1=2.S_1.~1.Cond=2.Cond.~1.S_2=2.S_2.~	IS1CondS2	IS1CondS2
1.S_r=2.S_r.~1.NP_0=2.NP_0.~1.VPadjn=2.VPadjn.~1.ep...	nx0e-VPadjn-Vnx1nx2	nx0e-nx1ko-nx2-V
1.S_r=2.S_r.~1.NPadjn=2.NPadjn.~1.NP_0=2.NP.~1.NP...	nx0-VPadjn-Vnx1	nx0ne-nx1ko-ADJ-vz1
1.NP_r=2.NP_r.~1.NP_f=2.NP_f.~1.S=2.S.~1.NP_0=2.NP...	Inx-nx0-VPadjnnx1	Inx-nx0ne-nx1ko-ADJ-vz1
1.S_r=2.S_r.~1.NP_0=2.NP_0.~1.NP_1=2.NP.~1.VPadj...	nx0e-VPadjn-Vnx1	nx0e-nx1ko-ADJ-vz1
1.NP_r=2.NP_r.~1.NP_f=2.NP_f.~1.S=2.S.~1.NP_0=2.NP...	Inx-nx0-VPadjn-V	Inx-nx0ne-ADJ-vz1
1.S_r=2.S_r.~1.NP_0=2.NP_0.~1.VPadjn=2.VPadjn.~1.V...	nx0e-VPadjn-V	nx0e-ADJ-vz1
1.NP_r=2.NP_r.~1.NP_f=2.NP_f.~1.S=2.S.~1.NP_0=2.NP...	Inxsnx0e-VPadjn-V	Inxsnx0e-ADJ-vz1
1.NP_r=2.NP_r.~1.NP_f=2.NP_f.~1.S=2.S.~1.NP_0=2.N	Inxsnx0e-VPadjn-Vnx1	Inxsnx0e-nx1-ADJ-vz1

Fig. 1. Transfer rules using by the transfer engine

Thus, the overall working of this phase was to first parse a English sentence and then by applying the transfer rules convert the English structure into corresponding Indian

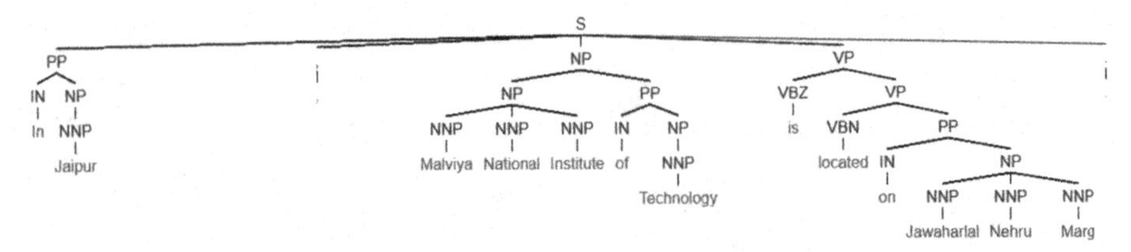

Fig. 2. English Parse of Example Sentence.

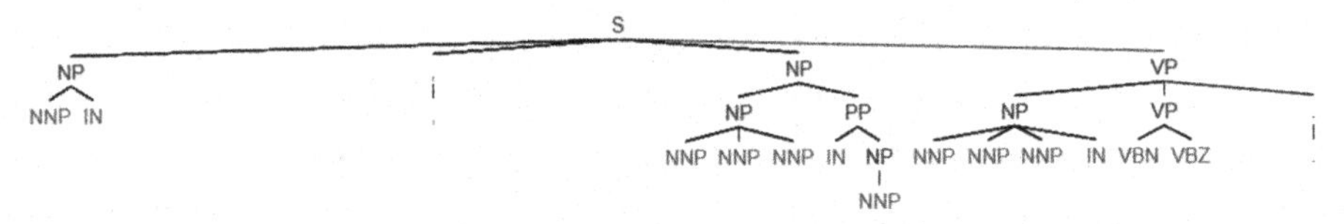

Fig. 3. Source Syntactic Tree of Example Sentence.

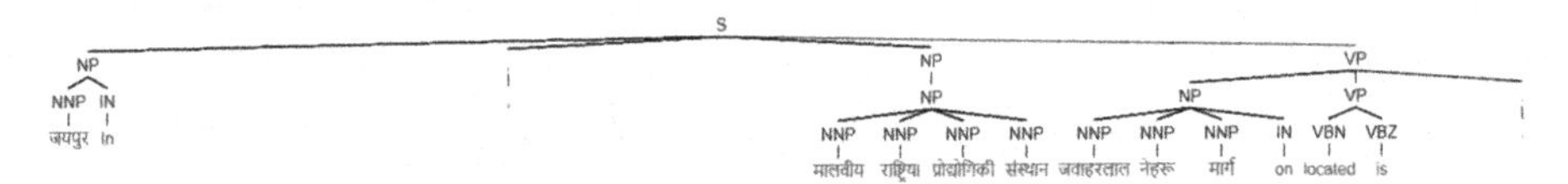

Fig. 4. Source Syntactic Tree with Name Entity Translated text.

language structure. For example, if we gave a sentence like, "In Jaipur, Malviya National Institute of Technology is located on Jawaharlal Nehru Marg." Then using Stanford parser, we will get a parse tree as shown in Fig. 2 which would then be transferred (converted) to its Indian language (target language) equivalent as shown in Fig. 3.

4.2 Sub Wording

Sub wording is a technique used in natural language processing to divide words into smaller units, called sub-words or sub-word units. The goal of sub wording is to capture the meaning of a word by breaking it down into smaller, more basic units that can be more easily understood by machine learning models.

Sub-word models are quite helpful in machine translation as they improve the performance of these model by providing them with a more detailed understanding of the meaning of individual words. It also helps to handle out-of-vocabulary (OOV) words, which are words that are not present in the training data.

Thus, in this example the source words were processed with sub-wording using a popular sub-wording method known as byte pair encoding. Through this we got the most frequent pair of bytes (or characters) in the text with a single, unused byte.

The above two process were applied on the entire one lac English sentence corpus. Once this process was competed, and a baseline NMT model was trained using PyTorch library.

Another system was also trained which augmented the transfer tree by first recognizing the name entities in the corpus and then translating or transliterating them in the target Indian language. For example, if the English text was to be translated into Hindi, then at first, the system would recognize the name entities in the input sentence. In the example they would be "Jaipur", "Malviya National Institute of Technology" and "Jawaharlal Nehru Marg" recognized as location, organization and location respectively.

Table 2. Layered Output of the Input Source Sentence

Input Sentence	**In Jaipur, Malviya National Institute of Technology is located on Jawaharlal Nehru Marg**
English Parse in LISP Notation	[S [PP [IN In] [NP [NNP Jaipur]]] [, ,] [NP [NP [NNP Malviya] [NNP National] [NNP Institute]] [PP [IN of] [NP [NNP Technology]]]]] [VP [VBZ is] [VP [VBN located] [PP [IN on] [NP [NNP Jawaharlal] [NNP Nehru] [NNP Marg]]]]]] [. .]]
LISP Notation of Syntax Transferred Tree	[S [NP [NNP] [IN]] [, ,] [NP [NP [NNP] [NNP] [NNP] [NNP]]] [VP [NP [NNP] [NNP] [NNP] IN]] [VP [VBN] [VBZ]] [. .]]
NE Translated Parse Tree	[S [NP [NNP जयपुर] [IN In]] [, ,] [NP [NP [NNP मालवीय] [NNP राष्ट्रिया] [NNP प्रोद्योगिकी] [NNP संस्थान]]] [VP [NP [NNP जवाहरलाल] [NNP नेहरू] [NNP मार्ग] [IN on]] [VP [VBN located] [VBZ is]] [. .]]

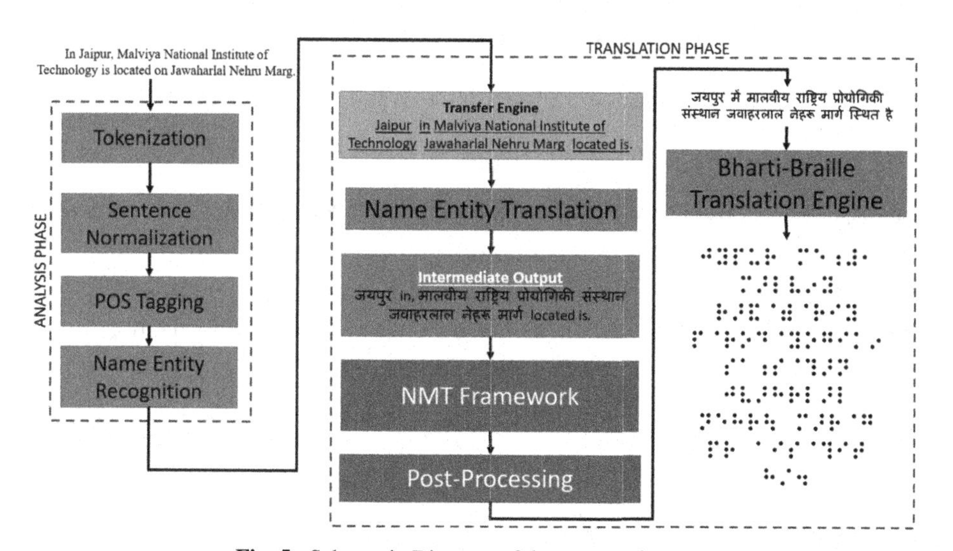

Fig. 5. Schematic Diagram of the proposed system

Next, the system would translate these entities using a separate knowledge base that was constructed for organization names and location names. In case if these are not found in the knowledgebase then the same would get translated. As for the other name entities (the ones which are not tagged as organization or location), they would also get transliterated into the target language. Thus, in our example, the target name entities would be "जयपुर", "मालवीय राष्ट्रिय प्रोद्योगिकी संस्थान" and "जवाहरलाल नेहरू मार्ग".

These are replaced by the English name entities in the parse tree. Thus, the resultant parse tree has a mix of English text and Hindi translated text. This is shown in Fig. 4.

Next the augmented (NE translated) sentence is sent for sub-wording and then finally is provided to PyTorch library for NMT model for training. Figure 5 shows the schematic diagram of the entire working of the system. The working of the entire systems is also enumerated in Table 2.

Next, the text translated for the NMT systems was sent the Bharti Braille translation engine which would then translate it into the corresponding Bharti Braille encoding.

5 Evaluation

Both the MT systems were tested on one thousand sentences (for each language pair). BLEU evaluation metric was used to compare the results of the two systems. In all cases the name entity induced NMT models performed better than baseline NMT models. The result of this study is shown in Table 3. These are system level scores where the individual translation scores were added and were divided by one thousand. The same is shown in Eq. (1). Here, BLEU-Score$_i$ is the individual BLEU score for each sentence which are then added for all n sentences and are then divided by total number (n) of sentences.

$$System - level\ BLEU\ Score = \frac{\sum_{i=1}^{n} BLEU - Score_i}{n} \tag{1}$$

Table 3. Evaluation results of NMT models.

Language Pair	Baseline Models	Name Entity Induced NMT Models	Improvement
English-Bengali	0.4653	0.5570	9.17%
English-Gujarati	0.5879	0.7124	12.45%
English-Hindi	0.4982	0.6483	15.01%
English-Marathi	0.5219	0.6591	13.72%
English-Tamil	0.4211	0.5090	8.79%

6 Conclusion

In this paper, the development of a name entity augment NMT system was shown. The improvement of the NMT system was shown by adding a name entity translation sub-module. The proper treatment of name entities in the translation process, improves the performance of the system. This is evident through the comparison of results of baseline NMT system with name entity induced NMT system where in all the language pairs there has been some improvement in the results. The performance improvement ranged from 8% to 15% where English-Tamil has shown the least improvement with a gain of 8.79% and English-Hindi showed the most improvement with a gain of 15.01%.

As an extension to this work, authors would also like to observe the change in performance of the NMT models when multi-word expressions are handled in the same manner. This would further strengthen our hypothesis that by adding linguistic knowledge, the performance of the vanilla NMT models can be improved.

Acknowledgements. This work is supported by the funding received from SERB, GoI through grant number CRG/2020/004246 for project entitled, "Development of English to Bharti Braille Machine Assisted Translation System".

References

1. Abualkishik, A.M., Omar, K.: Quranic braille system. Int. J. Hum. Social Sci. **3**(4), 313–319 (2009)
2. Abualkishik, A.M., Omar, K.: Quran vibrations in Braille code. In: 2009 International Conference on Electrical Engineering and Informatics, vol. 1, pp. 12–17. IEEE (Aug 2009)
3. Al-Salman, A.S.: A bi-directional bi-lingual translation Braille-text system. J. King Saud Univ.-Comput. Inform. Sci. **20**, 13–29 (2008)
4. Azam, S., Gani, M.O., Khan, A.I., Mahbub-ul-Islam, F.M., Hossain, S.A.: Architecture and implementation of Bangla text to braille translation. In: 2nd International Conference on Computer Processing of Bangla (2011)
5. Blenkhorn, P.: A system for converting braille into print. IEEE Trans. Rehabil. Eng. **3**(2), 215–221 (1995)
6. Blenkhorn, P.: A system for converting print into Braille. IEEE Trans. Rehabil. Eng. **5**(2), 121–129 (1997)
7. Cleave, J.P.: A programme for Braille transcription. In: Information Theory, pp. 184–194. Academic Press, Inc. (1956)
8. Das, P.K., Das, R., Chaudhuri, A.: A computerised Braille transcriptor for the visually handicapped. In: Proceedings of the First Regional Conference, IEEE Engineering in Medicine and Biology Society and 14th Conference of the Biomedical Engineering Society of India. An International Meet, pp. 3–7. IEEE (Feb 1995)
9. Fahiem, M.A.: A deterministic turing machine for context sensitive translation of Braille codes to Urdu text. In: International Workshop on Combinatorial Image Analysis, pp. 342–351. Springer, Berlin, Heidelberg (Apr 2008)
10. George, J., George, V., Deepan, A., Valson, J.N.: Optical character recognition based handheld device for printed text to braille conversion. In: 2011 3rd International Conference on Machine Learning and Computing, ICMLC 2011, pp. 172–175
11. Gotoh, T., Minamikawa-Tachino, R., Tamura, N.: A web-based Braille translation for digital music scores. In: Proceedings of the 10th International ACM SIGACCESS Conference on Computers and Accessibility, pp. 259–260 (Oct 2008)
12. Hossain, S.A., Mahbub-ul-Islam, F.M., Azam, S., Khan, A.I.: Bangla Braille Adaptation. In: Technical Challenges and Design Issues in Bangla Language Processing, pp. 16–34. IGI Global (2013)
13. Hossain, S.A., Biswas, L.A., and Hossain, M.I.: Regular expression for Bangla-2-braille machine Translator. In: International Conference on Advances in Engineering and Technology, pp. 54–57 (Mar 2014)
14. Hossain, S.A., Biswas, L.A., Hossain, M.I.: Analysis of Bangla-2-braille machine translator. In: 2014 17th International Conference on Computer and Information Technology (ICCIT), pp. 300–304. IEEE (Dec 2014)

15. Iain, M., Pasquale, A.: A Portable device for the translation of braille to literary text. In: Proceedings of ACM Conference on Assistive Technologies (ASSETS), pp. 231–232 (2006)
16. Ingham, K.R.: Braille, the language, its machine translation and display. IEEE Trans. Man-Mach. Syst. **10**(4), 96–100 (1969)
17. Jariwala, N.B., Patel, B.: Conversion of Gujarati text into Braille: a review. Int. J. Innov. Adv. Comput. Sci. **4**(1), 59–64 (2015)
18. Jariwala, N.B., Patel, B: Transliteration of digital Gujarati text into printable Braille. In: 2015 Fifth International Conference on Communication Systems and Network Technologies, pp. 572–577. IEEE (Apr 2015)
19. Lahiri, A., Chattopadhyay, S.J., Basu, A.: Sparsha: A comprehensive Indian language toolset for the blind. In: Proceedings of the 7th International ACM SIGACCESS Conference on Computers and Accessibility, pp. 114–120 (Oct 2005)
20. Li, J., Yan, X.: Optical braille character recognition with support-vector machine classifier. In: 2010 International Conference on Computer Application and System Modeling (ICCASM 2010), vol. 12, pp. V12–219. IEEE (2010)
21. Li, J., Yan, X., Zhang, D.: Optical braille recognition with haar wavelet features and support-vector machine. In: 2010 International Conference on Computer, Mechatronics, Control and Electronic Engineering, vol. 5, pp. 64–67. IEEE (Aug 2010)
22. Mahbub-ul-Islam, F.M., Khan, A.I., Gani, M.O., Azam, S., Hossain, S.A.: Computational model for Bangla text to braille translation. In: 2nd International Conference on Computer Processing of Bangla (2011)
23. Nian-feng, L., Li-rong, W.: A kind of Braille paper automatic marking system. In: 2011 International Conference on Mechatronic Science, Electric Engineering and Computer (MEC), pp. 664–667. IEEE (Aug 2011)
24. Salah, C., Ram, A.R.: A review paper on Malayalam text to braille transliteration. Int. J. Curr. Eng. Technol. **5**(4) (2015)
25. Shimomura, Y., Kawabe, H., Nambo, H., Seto, S.: Braille translation system using neural machine translation technology I - code conversion. In: Jiuping, Xu., Ahmed, S.E., Cooke, F.L., Duca, G. (eds.) Proceedings of the Thirteenth International Conference on Management Science and Engineering Management: Volume 1, pp. 335–345. Springer International Publishing, Cham (2020). https://doi.org/10.1007/978-3-030-21248-3_25
26. Shreekanth, T., Udayashankara, V.: A new research resource for optical recognition of embossed and hand-punched Hindi Devanagari braille characters: Bharati braille bank. Int. J. Image. Graph. Signal Process. **7**(6), 19 (2015)
27. Singh, M., Bhatia, P.: Automated conversion of English and Hindi text to Braille representation. Int. J. Comput. Appl. **4**(6), 25–29 (2010)
28. Singh, M., Bhatia, P.G.: Enabling the Disabled with Translation of Source Text to Braille (Doctoral dissertation) (2010)
29. Vyas, H.A., Virparia, D.P.V.: Optical Gujarati braille recognition: a review. Int. J. Emerg. Technol. Appl. Eng. Technol. Sci. (2014)
30. Wang, X., Yang, Y., Liu, H., Qian, Y.: Chinese-Braille translation based on Braille corpus. Int. J. Adv. Pervasive Ubiquit. Comput. (IJAPUC) **8**(2), 56–63 (2016)
31. Zhang, X., Ortega-Sanchez, C., Murray, I.: Hardware-based text-to-braille translator. In: Proceedings of the 8th International ACM SIGACCESS Conference on Computers and Accessibility, pp. 229–230 (Oct 2006)
32. Manning, C.D., Surdeanu, M., Bauer, J., Finkel, J.R., Bethard, S., McClosky, D.: The stanford CoreNLP natural language processing toolkit. In: Proceedings of the 52nd Annual Meeting of the Association for Computational Linguistics: System Demonstrations, pp. 55–60 (2014)

Real-Time AI-Enabled Cyber-Physical System Based Cattle Disease Detection System

K. S. Balamurugan[1(✉)], R. Rajalakshmi[2], Chinmaya Kumar Pradhan[3], and Khalim Amjad Meerja[4]

[1] Karpaga Vinayaga College of Engineering and Technology, Chengalpattu, Tamilnadu, India
profksbala@gmail.com

[2] Vignan's Nirula Institute of Technology and Science for Women, Guntur, Andhra Pradesh, India

[3] Vignan's Lara Institute of Technology and Science, Guntur, Andhra Pradesh, India

[4] Koneru Lakshmaiah Education Foundation, Guntur, Andhra Pradesh, India
kmeerja@kluniversity.in

Abstract. Cattle farming is the second most thing in the farmer's life after agriculture. In India, milk is a part of our daily lives. Nowadays, due to increasing cattle farming, monitoring the cattle behaviors, health status and predicting the calving time, estrus periods are essential to know the status of the cows which leads to various health issues for people who are consuming that diseased animal's milk and other products of that milk. Camera-based surveillance system for monitoring cattle is expensive and complex. The proposed AI-Enabled Cyber-physical system (CPS)-based monitoring system has reduced the complexity of the system by predicting the abnormal behavior of the cow in advance by using wearable sensors module. IoT-based sensor data is evaluated by a Weight factor distributed algorithm for finding the activities of cows and a Convolution Neural Network (CNN) is used for image analysis with trained datasets. The performance of AI-Enabled CPS-based systems shows high accuracy, and less complexity when compared to other SVM and CNN.

Keywords: Convolutional Neural Network (CNN) · Artificial Intelligence (AI) · Sensors · Cyber-physical system (CPS)

1 Introduction

According to UNDP predictions, the world population will reach 9.5 billion by 2050, and the requirements for animal products globally (e.g., milk, dairy products, meat) will increase by 70%. Today, farmers are facing problems with infrastructure, connectivity, animal monitoring, and diseases. There are a variety of skin issues that cattle face, some of which are manageable while others are more challenging. In Ethiopia, ringworm, bovine papillomatosis (warts), and lumpy skin disease (LSD) are common skin conditions [1]. The most well-known and frequently used algorithm is CNN. The primary advantage of CNN over its forerunners is that it recognizes the pertinent features automatically and

R. K. Challa et al. (Eds.): ICAIoT 2023, CCIS 1930, pp. 301–313, 2024.
https://doi.org/10.1007/978-3-031-48781-1_24

without human oversight. CNN has extensively used in a variety of fields, such as face recognition, voice processing, computer vision, etc. Like a traditional neural network, the structure of CNNs was inspired by the neurons in human and animal brains. More specifically, the visual cortex of a cat's brain is made up of a convoluted pattern of cells, and the CNN simulates this pattern [2]. The objective measurement of animal stress using digital technology is one of the goals of several developments in smart livestock monitoring. Biometrics and artificial intelligence (AI) are also used to evaluate the impact of these developments on the welfare and production of livestock [3].

Cattle undergo stress when they are unable to respond behaviorally or physiologically to physical or environmental stressors. These difficulties disrupt homeostasis, and an adaptive response is triggered in an effort to bring everything back into balance, making a preliminary determination of deviant behaviors in cattle is one of the key challenges in the management of animals kept in group housing, The main obstacle to achieve optimal reproductive performance is failing to recognize gestures in the proper and appropriate manner. In today's intensive dairy production, it is extremely important to accurately predict lameness activity. Due to the freedom of movement of cattle, wireless communication is a significant problem in cow monitoring systems. A very difficult problem is the lack of data about endemic diseases and the reliability of available data [4].

To overcome the above-mentioned issues and challenges, a well-developed system for monitoring the cattle and predicting the disease, and calving time in advance is essential. Section 2 focuses on the related work about cattle physiological and behavioural responses to be monitored. The proposed system is described in Sect. 3 regarding sensors and the CNN algorithm. The experimental setup and results are discussed in Sect. 4. In Sect. 5, the Conclusion & Future Scope are summarized from the experimental studies.

2 Related Work

Much research is going on monitoring cattle farming using different technologies such as IoT, and AI-based systems. Controlling and automating farming operations, the Internet of Things and artificial intelligence are making a significant contribution to modern agriculture. To anticipate future challenges in farming practices, the data generated by various sensors must be managed and analyzed using machine learning and deep learning-based methods [5]. Guo, et al. created an automated platform for dairy cattle temperature measurement and monitoring. The platform classified and identified dairy cattle images using the YOLO V3-tiny deep learning algorithm (you only look once, YOLO). There were three layers of YOLO V3-tiny identification in the system: (i) the body of a dairy cow; (ii) an individual number (also known as an ID); (iii) An identification thermal image of the eye socket [6].

Chaudhry, et al. concluded that a necessary technological advancement is livestock health monitoring. ML-based predictive technology can assist in predicting theearly detection of cattle diseases, and precision cattle farming is a useful tool for veterinarians to monitor the crop's health [7]. Anita z, et al. evaluated two distinct methods for detecting rumination using various machine learning algorithms and epoch lengths. The findings indicate that an accelerometer ear tag can detect rumination with high accuracy (98.4%)

[8]. Ezann, et al. suggested that the rapid growth of these fields is made possible by the convergence of AI methods like machine learning, expert systems, and analytical technologies with the collection of massive and intricate data [9]. Han, et al. stated that after noise reduction, data are used to develop a deep- learning- based LSTM model for cattle state dynamics. This model can predict the cattle's state change in the following cycle. When the predicted results are compared to the actual results, the model's accuracy and effectiveness are shown [10].

Veerasak, et al. suggested a method for locating foot and mouth disease (FMD) outbreaks that have wreaked havoc on Thailand's cattle industry for the past ten years. For authorities to develop a strategy for preventing FMD outbreaks, it is essential to have a prediction of outbreaks based on relevant risk factors that has a high prediction accuracy [11]. Neethirajan, et al. suggested a hands-on, realistic, and practical approach to affective state recognition is provided by AI technologies. This makes it easier for ethologists and animal handlers to understand why animals behave the way they do and how to improve their welfare and productivity [12]. Bezawit, et al. explained the numerous factors that influence the potential economic benefits of livestock farming. The prevalence of disease in livestock is one of these factors. The economic benefits of livestock farming have been impacted significantly by this condition [13]. Bao, et al. implemented the proposed model to make animal farming easier, a methodical application that uses smart algorithms, data processing, and basic devices for collecting data must be developed [14]. Singh, et al. examined by monitoring the health of cattle is essential for ensuring their health for milk production. Real-time monitoring and prediction are two applications for which digital technologies have attracted a lot of attention. Lower-quality milk production is the result of inadequate health monitoring [15]. Balamurugan et al. examined the performance of the self-powered multisensory wireless network with a lora RF module [16]. Pavlovic et al. reported on a method for developing algorithms that use two feature selection techniques to systematically reduce dimensionality to classify key cattle states. These are applied to knowledge- specific and generic time series extracted from raw accelerometer data and are based on Mutual Information and Backward Feature Elimination. After that, the extracted features are used to train Hidden Markov Model-based classification models [17].

Sabrina et al. studied animal behavior, the estrus cycle, the behavioral examination of grazing animals, and the observation of animal posture response. These observations describe how animals were observed by attaching accelerometer sensors to their necks or legs. Able to classify grazing, resting, and walking as the three states. Because it uses a single sensor that records animals' positions, the results aren't very precise, especially in the grazing and resting states [18]. From the studies of cattle care management systems, it is clear that they are unable to meet real-time applications. So, a novel emerging system is required for real-time.

3 Proposed System

Artificial Intelligent enabled CPS system for cattle disease detection system has IoT based sensor analyzing system, Camera surveillance using CNN algorithm. Keeping the complexity and efficiency in mind, we came up with a solution for the cattle care system using AI and CNN. The fundamental terms of an expert system are discussed in this section, as are the functionalities of the components. The formulation of CNN architecture and its application to train the model is further developed during this process. In the proposed system first, a wearable 3-axis accelerometer along with neck and leg sensor modules are connected to every animal for continuous monitoring. Sensors such as temperature sensors, motion sensors, accelerometers, and various others are used for temperature, motion monitoring, and movement. The data recorded will be analyzed and trained with help of a weight factor algorithm.

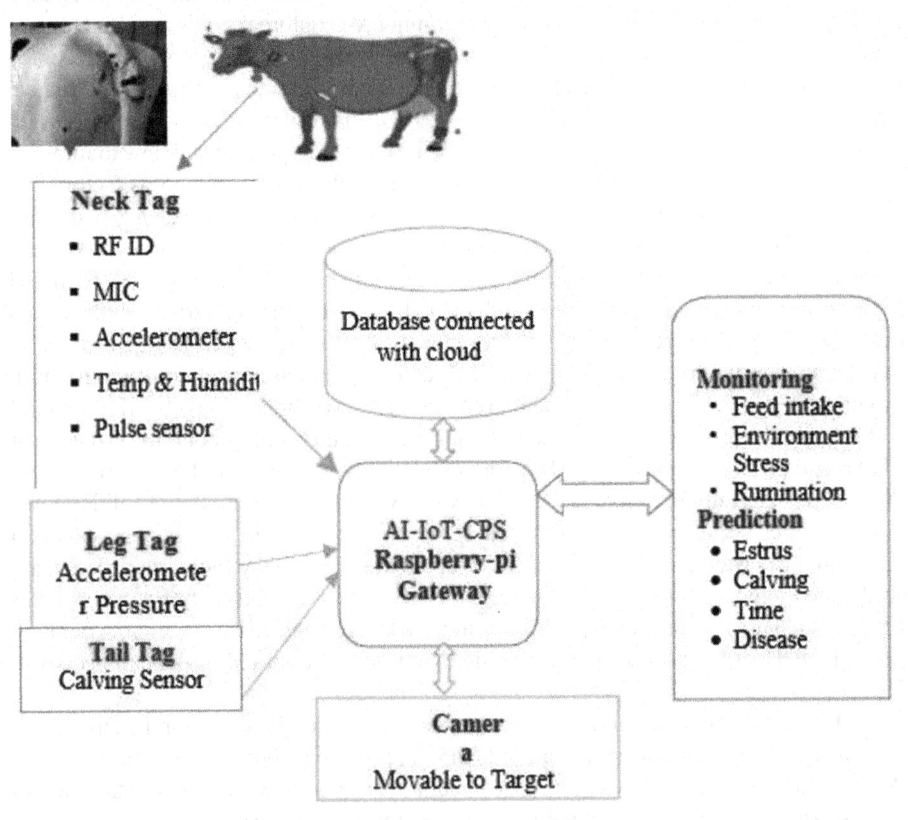

Fig. 1. Proposed System Architecture.

CNN for human healthcare results, Weight factor algorithm is triggered which one early studied and published in [16]. Then by using the image capture facility pictures of the mouth and breast images are captured. By using the CNN the images are analyzed and disease is detected by comparing them with the trained datasets. This suggested procedure helps to reduce the complexity of systems that have the image capture camera monitor for every cattle. The cattle which are observed abnormally will only undergo image capture and simultaneously for diagnosis. Monitoring every moment of cattle leading to the detection of the infected or diseased animal easily and immediately thereafter isolating the infected animal from other healthy animals is much-needed and important.

The fundamental terms of an expert system are discussed in this section, as are the functionalities of the components. The mathematical formulation of CNN architecture and its application to the training model is further developed.

3.1 System Architecture

The primary purpose of an intelligent computer program, also known as an expert system, is to be able to imitate the thought process of human specialists in order to resolve complex problems in any field. The goal of the Animal Disease Detection System (ADDS) is to quickly and precisely identify the kind of illness from the image's observed characteristics. As shown in Fig. 1, a group of Sensors, Database Storage, CNN layers, an Explanation unit and a Train and Data Set make up a typical expert system of any kind. The data for a specific domain use case and the structured and unstructured knowledge of accumulated experiences are stored in the knowledge base, typically by the domain expert. The user query and the data in the knowledge base are correlated, analyzed, and justified by the CNN layers to produce the result. Because the explanation of rules corresponds to the method of knowing about problems, the conventional rule-based expert system has its own advantages. However, there are unavoidable drawbacks to relying solely on the rules to justify: The methods of reasoning ought to be identical. It is unable to self-learn, is not predictive, and even lacks effective treatment options when inaccessible information is present in known data.

3.2 Sensors Module

The symptoms and clinical signs of a number of cattle diseases and how they affect cattle behavior are discussed. In this section, after taking into account those diseases, the conditions are mapped to the sensors that are relevant for analyzing the behavioral changes and health that cattle reveal about that disease. There are a variety of sensors available for specific measurement of cow activity. The microphone, temperature sensor, and accelerometer (pedometer,) etc are the most frequently used sensors. The first illustration shows a general framework for a cattle health monitoring system that includes an accelerometer module, electrode module, humidity, temperature, and so on. The wireless system is generally categorized on the basis of technique, data, algorithms and performance. It consists of sensor type, sensor location, measurement type and the alerts given by the sensor for the occurrence of an event. In earlier days, the health of cattle has been monitored visually by veterinarians.

3-Axis Accelerometer Sensor: With so many behavioral indicators, there is a chance to improve outcomes, output, and well-being by being able to identify and swiftly address animal health problems. However, gathering data on quantifiable animal activity on pasture through direct observation or video surveillance is labour and time- intensive, and the presence of a watcher can interfere with typical behavioural patterns. It is challenging to consistently observe animal behavior in a pastoral system, especially when there are many animals spread out over a large area. For the technology to be trusted and for users to use it, the relationship between sensor analysis and observed behavior must be proven. Ninety-seven percent of 66 pertinent studies used 3- dimensional (x, y, and z-axis) accelerometer sensors to measure acceleration values within three orthogonal spatial axes that captured the motion dynamics of the animal. Although collar-mounted accelerometers detect the right-left, front-back, and up-down directions, ear-mounted triaxial accelerometers in sheep detect the accelerometer datasets from the x-axis, y-axis, and z-axis corresponding to the directions of up-down, right-left and correspondingly.

3.3 Database

A database is a collection of data that has been organized to make it simple to manage and update. Any type of data can be stored, maintained, and accessed using databases. They gather data on individuals, locations, or objects. It is gathered in one location so that it can be seen and examined. To analyze Cattle care behaviour and predict disease, Standard datasets are not available, so a number of images are collected from different farming, trained, classified and the dataset is prepared.

3.4 Convolutional Neural Network (CNN)

CNN is a Deep Learning method that can take in an input image, and give various elements and objects in the image importance (learnable weights and biases), and be able to distinguish between them. CNN requires substantially less pre-processing than other classification techniques. CNN architecture was influenced by how the Visual Cortex is organised and is similar to the connectivity network of neurons in the human brain. Only in this constrained area of the visual field, known as the Receptive Field, do individual neurons react to stimuli. Some of these fields intersect to cover total area visibility. Selection of a model that can analyze visual imagery and the feature extraction of various diseases, one category of the neural network is CNN. They are frequently utilized in computer vision applications due to their reputation for extracting important image features. When compared to the conventional multilayer perceptron model for processing vectors, CNN employs a distinct architecture. CNN have better computational capabilities than SVM-based image classification algorithms, which only do a limited amount of feature engineering. They have a clever way of looking at images, zooming in and out on pixels that are next to each other in small areas, and then they put those readings into a pooling layer. In essence, there is an output along with the previously mentioned convolutional layer, pooling layer, and regular ANN after that (fully connected).

As shown in Fig. 2, illustrate the CNN architecture. CNN employs a variety of filters to identify distinct edges, which are then aggregated to discover a variance in the classification. Even though fully connected feed-forward neural networks can be trained to classify vectors, using this architecture for images is impractical. Due to the large image input sizes, using a multilayer perceptron module will result in exponential weight increases. For instance, the first hidden layer of an image measuring 250 x 250 pixels would contain four neurons. When compared to the size of the input image, the total weight used between the first hidden layer and the inputlayer would be 250*250*4, which is four times more. Memory requirements for this module are enormous. CNN learn to recognize edges in images in addition to reducing number of weights. By reducing the sparsity and increasing the generality, the image will be convolved into a smaller matrix during the iterative convolution process. On convolution, there are only 25 learnable parameters required for filters of size 5*5. It accomplishes this by resolving the evident vanishing or exploding gradient issues in the multilayer perceptron module.

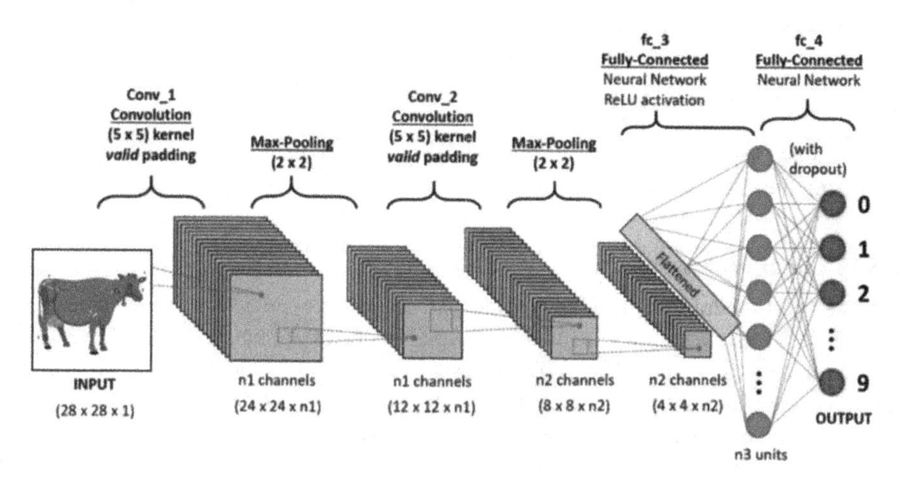

Fig. 2. CNN Architecture.

The output of convolutional layer will be of size (T-g+1) if we have a T-T image convoluted with the filter of size g-g. We must sum up the contributions from previous layers to calculate the pre-nonlinearity input to X_{ij} in our layer:

$$x_{ij}^{L} = \sum\nolimits_{a=0}^{f-1} \sum\nolimits_{b=0}^{f-1} W_{ab} \cdot Y_{(i+a)(j+b)}^{f-1} \tag{1}$$

The CNN non-linearity is given equation no (2)–(6).

$$y\_ij^\wedge l = \sigma x \tag{2}$$

$$V_{dw} = \beta * V_{dw} + (1 - \beta) * dw2 \tag{3}$$

$$V_{bw} = \beta * V_{bw} + (1 - \beta) * db2 \tag{4}$$

$$\mathbf{W} = \mathbf{W} - a * \left(\frac{dw}{\sqrt{Vdw}} \right) \tag{5}$$

$$\boldsymbol{b} = \boldsymbol{b} - a * \left(\frac{db}{\sqrt{Vdw}} \right) \tag{6}$$

where, Vdw is small change in weight, Vbw is a small change in biases, β is momentum and W is the value of weight, α is the learning rate and b is the bias value. The gradients error of weight obtained by partial derivatives with total error is denoted by dw. The gradients error of biases obtained by partial derivatives with total error is denoted by db.

In order to incorporate location-invariant features into neural networks, max-pool layers are introduced. As they move through the output layer, they take the maximum value from the convolutions. For instance, if we take an input layer of size TT and a max-pool kernel of size k, the max-pool layers will produce an output of size T/k*T/k because the max function will reduce each k block to a single value.

3.5 Steps Involved in ADDS Model

The Animal Disease Detection System (ADDS) is used to predict the illness of cattle by analyzing the captured image.

3.5.1 Initial Processing

The predictor module, the training module, and the data pre- processing module make up the ADDES model. At first, the model uses images of sick animals that users have taken with their devices as input. The RGB images are turned into vectors by the preprocessing module, where they are subjected to RGB vector normalization and resized to the desired size in accordance with the designed CNN architecture.

3.5.2 Learning Process of ADDS

Here, discussed how the proposed convolutional architecture came to be. It needs to build our network in such a way thatprecise results are obtained in the most optimal manner as the network's depth significantly increases the memory requirement. Ideal CNN architecture for image classification is based on this concept. Convolutional architecture chose ReLu over theconventional sigmoidor tanh to overcome the previous limitation. When compared to the more conventional activation units, ReLu is extremely quick. If the derivatives computed are less than or equal to 0, they are given the value 0; otherwise, they are given the value 1. The use of ReLu in deep neural networks has demonstrated that the training process produces moderately sparse results. Max pooling layers and convolution layersspan four levels in our model. The suggested system performs a 2D convolution with a kernel size of 3*3 and a stride of convolution 1 after providing an initial input size of 230*230*3. The suggested system has sixteen of these filters and normalizes the layer with a batch normalization layer. The ReLu function, whose output is max pooled using a stride of convolution 2, is then used to activate the end layer. With 32 filters, repeat the same procedure twice, with a 50% dropout in the fourth convolutional layer. Perform the image smoothly to produce an output size of 4608 vectors by providing the multilayer perceptron layer with an input of 12*12*32. The hidden layer of the multilayer perception layer is 512 vectors in size and has a 20% dropout. Softmax cross-entropy is used to activate the size 13 vector that makes up the final output layer. The process flow diagrams depict the various convolution layers and fully connected layers of the model. The flow of execution of the process in convolution and fully connected layers are shown in Fig. 3 and 4.

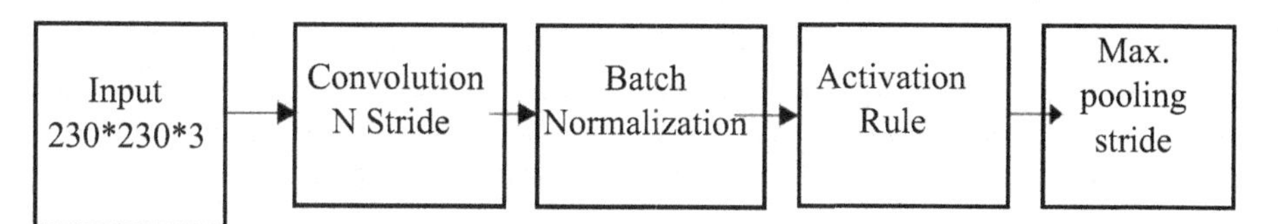

Fig. 3. Image Input to a Convolution layer.

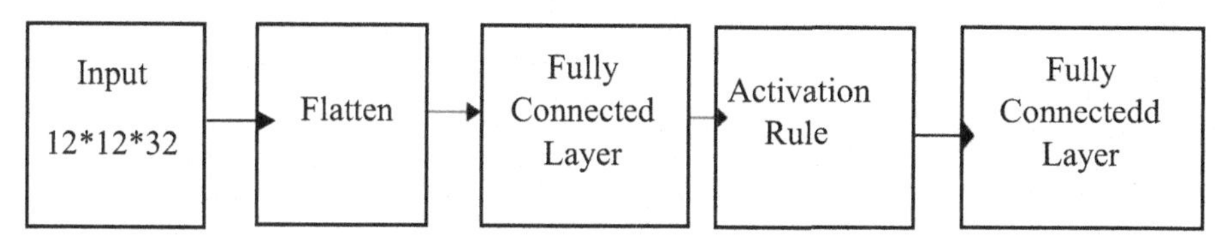

Fig. 4. Image Processing at Fully Connected layer.

4 Experimental Setup

Figure 5 demonstrates that proposed model accurately predicts the cow activities. As a result, accurate predictions can be made and a lot of money and time can be saved without the need for a veterinary specialist, which could ensure a strong improvement in the diagnosis of livestock diseases. The prearranged network, which uses segmentation and feature extraction to classify the diseases, is used to carry out the experimental strategy in this section. The training set for the 15 classes of diseases in experiment consisted of a total of 25,970 images. Each RGB image has a size of 230 × 230 × 3, and the samples have an average of 50 points. The data sequences are fed into the CNN for pattern recognition after the features are extracted from the samples. With a test accuracy of 98.05%, were able to achieve a satisfactory eigen variance for the spatial visualization of data. Sensors play a key role in the data analytics and the monitoring at every instance is possible. The various features which are to be monitored are measured by using the sensors. The movements of the cattle such as feeding, rumination, temperature and motion, leg movements and various important required data are collected by sensor modules.

Fig. 5. Test the disease prediction of the cattle.

By keeping the actual values in mind, the recorded values are compared and analyzed for further proceedings. The recorded values only decide whether the further evaluation is to be done or not. The aim of each experiment is to improve the accuracy of the tests. The test results from various algorithm run using different samples.

Table 1. Characteristics of Sensors.

Sensors	Measuring Range	Accuracy	Sensitivity
Accelerometer	1g to 250 g	+ 1%	300 mV/g
Temperature Sensor	−2100° to 1760 °C	+ 0.1^{0}	31.5 mV/C^{0}
Humidity Sensor	0 to 78% RH	0.5 to 5% RH	25 to 100%RH
Pulse Oximeter	95 to 100%	80 to 90 %	80 to 100%
Respiration Sensor	156% to 226%	0.39	95.8 to 100%

The values within the range limits for each cattle parameter were reached from measurements taken within 24 h: ranging from a minimum body temperature of 37.9 °C to a maximum of 39 °C, a heart rate of 80 bpm to 158 bpm, humidity of 2% to 98%, and general cattle health of 78% to 96%. Body temperature was 38.20 °C heart rate was 120 beats per minute, humidity was 76%, and general cattle health was 88% for the measurements taken on a farm with 15 cattle. Table 1 provides the specifications of the sensors used in the proposed system which are more required such as measuring range, accuracy, sensitivity and logging details.

Table 2. Predict the Health Condition by sensors value.

Temperature (°C)	Pulse (bpm)	Humidity	Tri-Accelerometer data			Health
			X	Y	Z	
38.5	49	46	933	1060	892	84%
38.5	75	97	629	968	641	83%
38.4	80	68	969	906	96	80%
38.05	56	47	272	18	676	92%
39	71	39	1073	1094	576	86%
38.05	59	88	426	475	988	88%
37.9	62	92	844	203	880	88%
38	77	92	398	976	294	87%
37.9	69	54	454	1092	629	96%

By the SVM classification algorithm, different rate results are shown in Table 2. This effective CNN is able to extract valid patterns from a real-time dataset without losing any important information. The majority of the experiments show that our algorithm performs well when visualized. SVM-based classification helps us distinguish our algorithm's output from other ones. While varying the noise, CNN-based classification reduces loss more than other image classification algorithms, as shown in Table 3.

Table 3. Output readings of different diseases using CNN.

Disease	Training Samples	Test Samples	Accuracy
Anthrax	50	25	98.89%
Black Quarter	50	25	100%
Rabies	50	25	99.25%
Blue Tongue	50	25	98.85%
Tetanus	50	25	97.70%

Along with value of IoT-CPS system which one mentioned in Table 2 and Image processing data which one mentioned in Table 3 are considered as a input of different Machine learning algorithms. Also, calculated computing time, computation cost and accuracy. By comparing the output reading values of different algorithms with respect to CNN gives the highly accurate output with minimum loss deviation, shown in below Fig. 6.

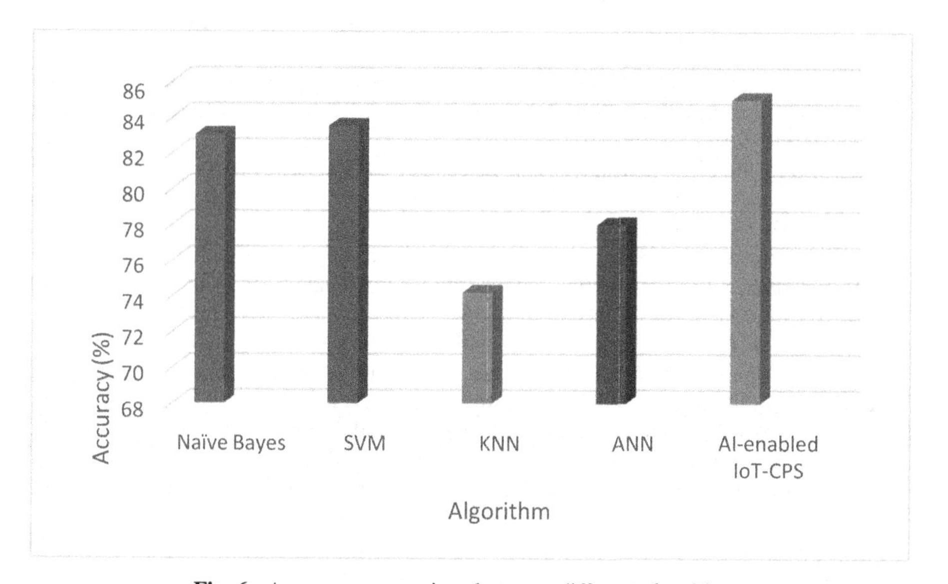

Fig. 6. Accuracy comparison between different algorithms.

5 Conclusion

Artificial Intelligence enabled IoT-CPS system is developed for monitoring the cattle and identifying the disease in advance if any cattle are affected. The proposed system has two modules, IoT-based CPS is used to sense the cattle temperature, pulse, leg, neck, and tail movement with the help of respective sensors. By analyzing the sensed data, proposed to weigh factors, find the cavity, ensure time exactly, and also predict the abnormal behaviour of cows. If IoT-CPS is triggered, then the proposed AI-CNN systemanalyzes the captured mouth, tongue, and breast image along with the sensed data, predicting breast cancer, brucellosis, and Lumpy virus in advance. Compared to the existing emerging technologies like SVM, and CNN, the proposed AI-CPS-IoT systemprovides 85% accuracy, less complexity, and low-cost system.

References

1. Laura, G., Federica, S., Sabrina, C., Matteo, F.: Bovine papillomatosis hiding a zoonotic infection epitheliotropic viruses in bovine skin lesions. Multidisciplinary Digital Publishing Institute (2020). https://doi.org/10.3390/pathogens9070583
2. Seetham, E., Onuodu, F.E.: Object detection using convolution neural network transfer learning. Int. J. Innov. Res. Eng. Multidisc. Phys. Sci. (2022)
3. Fuentes, S., Viejo, C.G., Tongson, E., Dunshea, F.R.: The livestock farming digital transformation: implementation of new and emerging technologies using Artificial Intelligence. Animal Health Res. Rev. **23**, 1–13 (2022). https://doi.org/10.1017/S1466252321000177
4. Sharma, B., Koundal, D.: Cattle health monitoring system using wireless sensor network: a survey from innovation perspective. IET Wirel. Sensor Syst. **8**(4), 143–151 (2018). https://doi.org/10.1049/iet-wss.2017.0060
5. Subeesh, A., Mehta, C.R.: Automation and digitization of agriculture using artificial intelligence and internet of things. Artif. Intell. Agric. **5**, 278–291 (2021). https://doi.org/10.1016/j.aiia.2021.11.004
6. Guo, S.-S., et al.: Development ofan automated body temperature detection platform for face recognition in cattle with YOLO V3-tiny deep learning and infrared thermal imaging. Appl. Sci. **12**, 4036 (2022). https://doi.org/10.3390/app12084036
7. Singh, D., Singh, R., Gehlot, A., Akram, S.V., Priyadarshi, N., Twala, B.: An imperative role of digitalization in monitoring cattle health for sustainability. Electronics **11**(17), 2702 (2022). https://doi.org/10.3390/app12084036
8. Anita, Z., et al.: Detection of rumination in cattle using an accelerometer ear-tag: A comparison of analytical methods and individual animal and generic models. Comput. Electron. Agric. **192**, 106595 (2022). https://doi.org/10.1016/j.compag.2021.106595
9. Ezanno, P., et al.: Research perspectives on animal health in the era of artificial intelligence. Vet. Res. **52**, 1–15 (2021). https://doi.org/10.1186/s13567-021-00902-4
10. Han, X., Lin, Z., Clark, C., Vucetic, B., Lomax, S.: AI based digital twin model for cattle caring. Sensors **22**(19), 7118 (2022). https://doi.org/10.3390/s22197118
11. Punyapornwithaya, V., Klaharn, K., Arjkumpa, O., Sansamur, C.: Exploring the predictive capability of machine learning models in identifying foot and mouth disease outbreak occurrences in cattle farms in an endemic setting of Thailand. Prev. Veter. Med. **207**, 105706 (2022). https://doi.org/10.1016/j.prevetmed.2022.105706
12. Neethirajan, S.: Affective state recognition in livestock—artificial intelligence approaches. Animals **12**(6), 759 (2022). https://doi.org/10.3390/ani12060759

13. Lake, B., Getahun, F., Teshome, F.T.: Application of artificial intelligence algorithm in image processing for cattle disease diagnosis. J. Intell. Learn. Syst. Appl. **14**(04), 71–88 (2022). https://doi.org/10.4236/jilsa.2022.144006
14. Bao, J., Xie, Q.: Artificial intelligence in animal farming: a systematic literature review. J. Clean. Prod. **331**, 129956 (2022). https://doi.org/10.1016/j.jclepro.2021.129956
15. Singh, D., Singh, R., Gehlot, A., Akram, S.V., Priyadarshi, N., Twala, B.: An imperative role of digitalization in monitoring cattle health for sustainability. Electronics **11**(17), 2702 (2022). https://doi.org/10.3390/electronics11172702
16. Balamurugan, K.S., Sivakami, A.: LoRa-IoT based self-powered multi- sensors wireless network for next generation integrated farming. Int. J. Sci. Technol. Res. **8**, 1527–1533 (2019)
17. Pavlovic, D., et al.: Behavioural classification of cattle using neck-mounted accelerometer-equipped collars. Sensors **22**(6), 2323 (2022). https://doi.org/10.3390/s22062323
18. Mac, S.E., Lomax, S., Clark, C.E.: Dairy cow and calf behavior and productivity when maintained together on a pasture-based system. Animal Biosci. **36**(2), 322 (2023). https://doi.org/10.5713/ab.22.0135

Unlocking the Power of AI: A Real-Time Translation of Sign Language to Text

Yashmita, Shahina Bano, Aastha Chaudhary, Binwant Kaur, S. R. N. Reddy, and Rishika Anand(✉)

Computer Science and Engineering, IGDTUW, Delhi 110006, India
{yashmita027btcse19,shahina032btcse19,aastha040btcse19, binwant030btcse19,srnreddy,rishika003phd19}@igdtuw.ac.in

Abstract. Sign language is an important means of communication for deaf and mute individuals all around the globe. It provides them with the opportunity to interact and socialize with the rest of society without feeling left out. But a large majority of the population does not know Sign Language. This paper proposes a Sign Language Recognition system based on a deep learning model to recognize American sign language alphabets (A-Z). The model uses MobileNet V1 pre-trained model to detect hand signs from the self-captured dataset. Using the Hand tracking module for data collection, the problem of lack of racial diversity was solved. The predicted labels are presented in the form of text. The complete implementation of the system is done on NVIDIA® Jetson Nano™. The model achieved a training accuracy of 99.74%. The system detects American Sign Language gestures in real time and translates them, making communication with the hearing impaired easier for everyone.

Keywords: SLR · Sign Language Translator · Deep Learning · Jetson Nano · Real-Time Sign Detection

1 Introduction

Sign language is a crucial means of communication for the deaf and mute community. Despite its long-standing usage, non-signers often struggle to comprehend the language, leading to a communication gap between the two communities. Effective communication is vital for an individual to progress in life, and sign language plays a significant role in facilitating this interaction. Sign language is made up of manual and non-manual restrictions such as hand orientation and motion, which are executed in 3D space. It differs from spoken language in terms of its structure, as sign language sentences have special configurations and movements. Different countries have their own official sign languages, each with its own set of rules and structural differences from spoken language. The most commonly used sign languages are ASL (American sign language) and BSL (British sign language) but none of them is universal.

The paper introduces a novel approach to ASL recognition using a combination of hand-tracking and deep learning. The novelty of the approach lies in the use of a hand-tracking module that captures images of sign language gestures with the hand skeleton

R. K. Challa et al. (Eds.): ICAIoT 2023, CCIS 1930, pp. 314–330, 2024.
https://doi.org/10.1007/978-3-031-48781-1_25

and joints visible. This enables the model to learn the key points of the hand for each particular sign, making it more accurate in recognizing the signs.

The hand-tracking module allows for a more detailed understanding of the hand's movement and positioning, which is crucial in recognizing sign language gestures. Traditional approaches to sign language recognition have relied on image processing techniques that do not take into account the hand's actual position and movement, leading to lower accuracy rates.

Another novelty of the paper is the use of the MobileNet model, which has 27 convolutional layers, making it a deeper model than commonly used counterparts like VGG16. This depth allows for better feature extraction and higher accuracy rates in recognizing sign language gestures.

The implementation of the system is performed on Jetson Nano, which acts as the core processor for sign language recognition. Jetson Nano is a small and powerful computer that provides real-time processing capabilities, making it ideal for sign language recognition applications.

Overall, the novelty in this paper lies in the combination of hand-tracking and deep learning techniques for sign language recognition, as well as the use of a deeper model and real-time processing capabilities for implementation. This approach has the potential to significantly improve sign language recognition accuracy and enhance communication accessibility for the hearing-impaired community.

1.1 Need for Sign Language

Languages are the elementary means through which one individual can communicate with others. [2] With more than 466 million people with hearing and speech impairments around the world, the bridge between signers and non-signers is quite wide. Some studies have also shown that many children with autism spectrum disorder have demonstrated better quality communication when using sign language.

Individuals need to communicate well to express themselves and live a better life. Those having hearing and speech impairments find it difficult to communicate with others and thus can feel left out. Individuals facing an early onset of hearing loss might face limited job opportunities. Hearing loss can cause a drop in confidence and self-esteem.

To provide the necessary education to deaf and mute individuals, there arises a need of building a technology that can help bridge the gap with the rest of society. Various researchers have been going on to develop a system that can translate sign language to spoken languages and vice-versa. This will not only help signers to express themselves to the rest of society but also help non-signers to communicate with the signers as well. All of this can be done with the help of a translator through which the gestures of the sign language can be represented in speech as well as text formats.

2 Existing Work

There is a significant amount of research done in Sign Language Recognition (SLR) using several vision-based and sensor-based deep learning approaches. Some early works on this topic use Hidden Markov Model [3] for Hand Gesture Recognition. The authors of

[4] present an algorithm that can be used to extract and classify 2-Dimensional motion within an image sequence basis motion trajectory. In [5], a standard CNN model is used with three groups of layers having an accuracy of 70%. In [6], the model is developed using a pre-trained model GoogLeNet and training is done on ASL Finger Spelling Dataset.

2.1 Jetson Nano

Jetson Nano has been chosen after analyzing many embedded systems as given in Table 1 based on their specs because it strikes a compromise between quick translation and constrained memory and processing capabilities. It is a portable server with a quad-core ARM®-based CPU and NVIDIA Maxwell architecture GPU. It is a reasonably priced system with high CPU performance rates and is easily accessible when required. The Jetson Nano is a gadget made exclusively for the Internet of Things that is about the size of a smartphone. As long as the network is accessible, it can be installed anywhere and whenever. Jetson Nano is cost-effective and compatible with several languages. It is user-friendly since it is relatively simple to use.

Table 1. Comparison of different Embedded Systems

Equipment		Nvidia Jetson Nano [7]	Arduino UNO [8]	Raspberry Pi 4 [9]	Verdin iMX8M [10]	Intel Nuc [8]
CPU	Model	ARM A57	ATmega328P	ARM Cortex-A72 64-bit	4 Arm Cortex-A53	Intel i7, i5 or i3 processor
	Cores	4	8	4	64	4–8
	Freq	1.43 GHz	16 MHz	1.5 GHz	1.8 GHz	3–5 GHz
GPU	Model	128-core Maxwell	-	Broadcom Video Core VI (32-bit)	OpenGL® ES 2.0	Intel Arc GPU
	Power	5W-10W	-	2.56W-7.30W	3.4–6.5 W	225 W
RAM		4GB	2 KB	1GB, 2GB, 4GB, or 8GB	2GB -8GB LPDDR4	4GB
Ports		4x USB 3.0 ports	1 USB port (Serial over USB communication only)	2 USB 3.0 ports	2x USB 2.0 OTG controllers	2x USB 3.0 ports

(continued)

Table 1. *(continued)*

Equipment	Nvidia Jetson Nano [7]	Arduino UNO [8]	Raspberry Pi 4 [9]	Verdin iMX8M [10]	Intel Nuc [8]
Camera	2x MIPI CSI-2 DPHY lanes	OV7670 Camera Module	MIPI CSI port	1x MIPI CSI (4-lane) with PHY	VIVOTEK* FD8169A
Applications	Smart traffic control system	Door control systems, light dimmers, IoT devices, etc	IoT based smart mirror	Aerospace, Automotive, Defense Systems, etc	Build Windows mini-PCs, create headless Linux media server
Connectivity	Gigabit Ethernet, Wireless networking adapter	Wi-Fi, Ethernet, SIM	Gigabit Ethernet, RJ45 cable	Wi-Fi, Bluetooth, Ethernet ISC, SPI, QSPI, UART, PWM, GPIO, JTAG	Wi-Fi, Ethernet, Port: 3 USB 3.0, 1 USB 2.0, 1 USB C, HDMI output
Cost	$99	$27.95	$55	$67.75	$649
OS	Linux4Tegra	No official support	Ubuntu Mate, Snappy Ubuntu Core, etc	Toradex Embedded Linux, Windows Embedded, INTEGRITY, QNX	Windows 10, Ubuntu
Flash Memory	microSD card	32KB	8GB eMMC	16GB eMMC	64GB
Languages	C/C + +, Python, Java, JavaScript, Go and Rust	C/C + +	Python	C/C + +	Any supported programming language by OS

2.2 Sign Language

As seen from Table 2, most of the research is done using ASL. The VGG16 pre-trained model achieves the highest accuracy of 98.84%. Some researchers developed their own models, leading to a greater accuracy of 99.64%. The most common algorithm used in the previous works is Convolutional Neural Networks (CNN).

Table 2. Study of existing work on sign language recognition

Ref	Sign language	Algorithm	Model	Dataset	Sample size	Simulated/ Deployed	Accuracy
[12]	ASL	CNN	GoogLeNet	ASL Finger spelling Dataset from University of Surreys CVSSP	65,000	Deployed	> 70%
[13]	ASL	CNN	VGG16 Net	Self-Captured	43,120	Simulated	98.84%
[14]	Indian Sign Language	CNN, SVM	-	Self-Captured	36,000	Deployed	99.64%
[15]	American, Indian, Italian, and Turkey Sign Language	Random Forest, SVM, KNN, Decision Tree, ANN, Naïve Bayes	-	Multiple datasets including existing and self captured	156000, 4972, 12856, and 4124	Simulated	99.29%
[16]	Bengali Sign Language	CNN	-	Self-captured	1000	Simulated	98.20%
[17]	ASL	K-convex hull, ANN	-	Self-captured	1850	Deployed	94.32%
[18]	Turkish Sign Language	LR, KNN, RF, ANN	-	Self-captured	87000	Simulated	98.97%
[19]	ASL	ANN	-	Self-captured	520	Deployed	96.15%

3 Proposed Work

The suggested architecture makes use of the NVIDIA Jetson Nano kit as its main system for SLR. ASL has been used instead of the Thai and Hong Kong sign languages that other researchers had used. A simple system has been developed that can be used by everyone. A pre-trained model is used to actively train the model to reach a greater accuracy, and the dataset has been both collected and developed with the greatest variety.

3.1 Goals

To proceed with the research, a few goals have been defined which will be the outline covering the objectives of sign language to text conversion.

i. The very first goal is to collect a dataset of various images which will be used at the time of training and testing of the algorithm. Data Collection and creation were done using the Hand tracking module from CVZone and OpenCV library using the concept of 21 hand marks of the cropped image of the hand to detect a hand gesture [20].
ii. The second goal is to make the necessary installation of packages and libraries in the hardware i.e., Jetson Nano. Since the hardware has 2GB of RAM, the number of required packages installed was kept as minimum as possible. The packages included OpenCV, MediaPipe, Matplotlib, TensorFlow, python, PyCharm, etc. For integrating a camera with the model, some additional packages were also installed.
iii. Next step included the training of the deep learning-based model. It was done using the collected dataset of 24 classes of images. To do that, epoch and batch size was defined in the model that could be adjusted as per the requirements. The more the epoch, the fitter the model. Batch size is defined by how many images are being sent in one pass to the model for training.
iv. For testing purposes, a camera was integrated with the Jetson Nano so that users can project hand gestures in front of it. The camera will detect hand gestures and show the output on the screen in text format. In this manner, real-time sign language detection will be done.

3.2 Sign Language to Text

This section describes how the hand gestures are converted into text format and displayed on the screen. After training the model on the collected dataset, the trained model was saved and can be easily imported anytime. So, the classifier model is imported from the saved file and the labels are defined for each letter.

The hand gestures are captured in real-time. With the help of the video capture function and the hand detector function, the hand of the user is located. The image is cropped to fit the hand landmarks [20]. Images are resized to ensure that every image is of the same size and modified to have a white background, and renamed as 'imgWhite'. The resulting image is used for the final sign detection. After resizing and changing the dimensions of the image, it is fed to the imported model to classify or detect the sign in real time. That is how the real-time sign gestures are predicted and the predicted label is printed on the screen for each alphabet.

3.3 Architecture

The architecture in Fig. 1 shows all the steps followed in this project. Firstly, the ASL dataset was created by capturing images from various angles and shuffling them for better training. The dataset contains 2400 images belonging to 24 classes. The next step was preprocessing the dataset, which included resizing the images to a fixed height and width of 224x224 to ensure all images have the same dimensions. Then, an image with a white background was produced and labeled as 'imgWhite' for consistency. This approach also aids in enhancing the visibility of the hand key points utilized for a particular sign gesture. A pre-trained CNN model was used in training with the new dataset. A camera was integrated with Jetson Nano to capture images in real-time and provide input to the trained classifier. The classifier predicts the alphabet from the input image and provides a text output. Thus, translating the sign language to text. All the steps were performed on NVIDIA Jetson Nano.

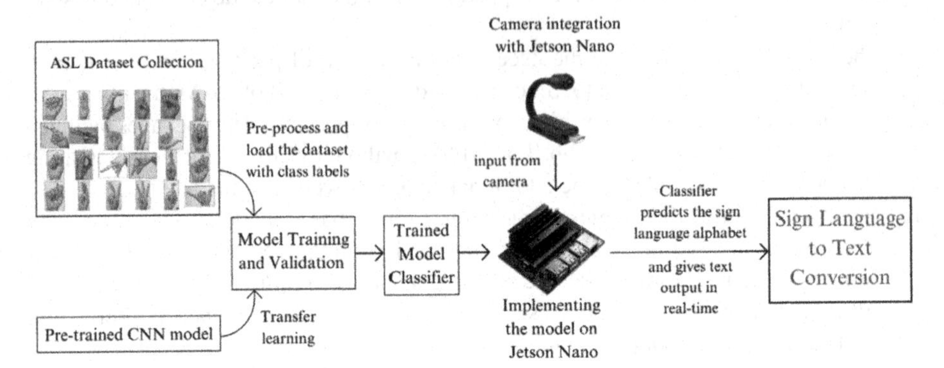

Fig. 1. Proposed Architecture

3.4 Workflow

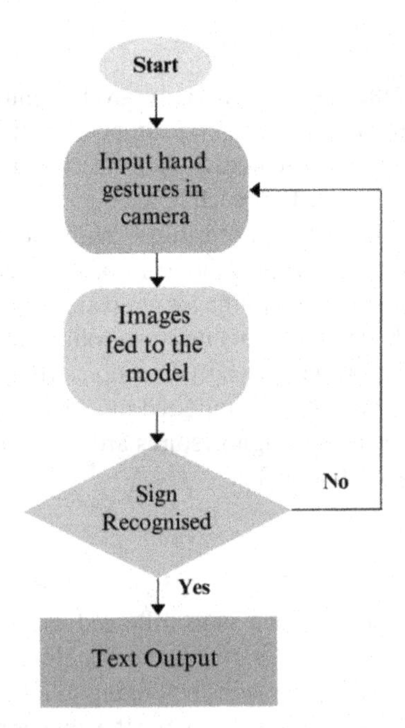

Fig. 2. Workflow of the proposed system

As seen in Fig. 2, initially the web camera will start and the users would project hand gestures in front of it. The image data will be fed into the trained model for recognition in real-time. The model would predict the sign and provide a text output if recognized.

4 Data and Implementation

This section talks about how the dataset collection is done and how the deep learning model is implemented. A self-created dataset has been used for training and testing, and a pre-trained MobileNet V1 model is used for increasing the accuracy. The model is implemented on Jetson Nano hardware KHD CPU which is powerful for embedded applications and AIoT.

4.1 Data Collection

The project involved two parts: data collection and testing for American Sign Language. Images were modified to a fixed size of 224x224 for consistency and organized into English alphabet folders within the 'images' directory. Data collection involved capturing images from various angles and shuffling for better training.

Then, some required libraries such as CVZone and MediaPipe were installed. Once the webcam was ready and the user could see their hands, the next step was to crop the images for which cv2 was imported. Using cap = cv2.VideoCapture(0), the webcam would start, with zero being the id number for the webcam. The images were shown using cv2.imshow. A Hand Detector was imported from cvzone.HandTrackingModule to get the skeleton of the hand as well as all the joints and the key points as this would help the classifier to better classify it rather than just having the shape of the hand. Next, the images were cropped by setting the desired dimensions of height and width. Also, to ensure all the images must be of the same size, an image with a white background was created and renamed as 'imgWhite'. This will also help in giving clarity to the hand key points which are being used for a particular sign gesture.

The last part was to save these images to the folders so that one by one data could be collected and used for training. The images were saved by clicking the key 'S' in the defined location. So, for example, for collecting a dataset for 'A', the location had to be changed to where the 'A' folder was, and so on. A counter was maintained to keep track of how many images were being saved. That is how the dataset for English alphabet from A to Y (excluding J) was collected.

4.2 Implementation

This paper describes an SLR system that uses Jetson Nano as its core processor and an HD Webcam for capturing hand gestures. The system relies on the OpenCV [21] and MediaPipe [22] libraries to process and analyze the hand gestures captured by the camera. The hand gesture recognition process begins by detecting the hand using the CVZone library's Hand Detector module, which maps 21 key points of the hand and captures the image of the sign with the marked points. The captured sign is then recognized in real-time using a trained deep-learning model. The size of the dataset captured is small. So, the model uses the pre-trained model MobileNet [23] to increase the accuracy. The MobileNet architecture as shown in Table 3 is a lightweight deep neural network as it significantly reduces the number of parameters. Thus, it utilizes less computational power. MobileNet uses special convolutional layers known as Depthwise Separable Convolution which performs two operations – Depthwise and Pointwise Convolution.

In a Depthwise convolution, each filter is applied independently to each channel of the input tensor, producing a set of channel-wise feature maps. This process does not mix information across channels, which allows the network to learn more fine-grained representations of each channel independently. The next step is feeding the resulting feature maps to a pointwise convolution. It mixes information across all channels by applying a single 1x1 filter to each pixel of the input tensor. This operation produces a set of output feature maps that represent a linear combination of the input channels at each pixel location. The resulting output tensor has the same spatial dimensions as the input but with a potentially different number of output channels.

Table 3. Model Architecture [23]

Layer Type	Stride	Shape of the filter	Size of input image
Standard convolution layer	2	3 × 3 × 3 × 32	224 × 224 × 3
Depthwise separable conv. Layer	1	3 × 3 × 32	112 × 112 × 32
Fixed Pre-trained MobileNet layers [24 fixed layers of Depthwise Separable and Standard Convolutions]			
Global Average Pooling	1	Pool 7 × 7	7 × 7 × 1024
Dense FC layer	1	1024 × 1024	1 × 1 × 1024
Dense FC layer	1	1024 × 512	1 × 1 × 1024
Dense FC layer	1	512 × 24	1 × 1 × 512
SoftMax activation function	1	Classifier	1 × 1 × 24

Table 3 represents the architecture of the model used in this system, summarizing the new layers added to the fixed pre-trained layers of the MobileNet model. The input shape of the image for the model is (224,224,3). The RGB picture is 224x224 pixels large and there is a matrix of pixel values for "red," "green," and "blue" since the RBG is three-channeled. All 26 layers of the MobileNet are added to the base model without any changes except the last one. The final layer of the architecture is replaced by a Global Average Pooling layer [25], which performs a down sampling of the image as it reduces the spatial resolution by computing height and width averages while maintaining a 2-D representation of the image. Finally, the model ends with three Dense ReLU fully connected layers and a SoftMax classifier. It uses an RMSprop optimizer and categorical cross-entropy loss function. The training metric is the accuracy of the model. The libraries used in model training are Keras and TensorFlow. The entire implementation of this American Sign Language Recognition (ASLR) system is performed on Jetson Nano – data collection, training, and real-time testing.

5 Results

The model has been trained for a batch size of 16 using 2400 images belonging to 24 classes. It achieved a training accuracy of 99.74% and a validation accuracy of 94.38%. The loss was calculated by the Categorical Cross Entropy Loss Function, and came out to be 01.18%. The results are promising considering the size of the dataset and time limitations. The accuracy was improved by data pre-processing and changing the number of epochs and batch size to find the perfect parameters.

The model accuracy and loss plotted in Fig. 3 and Fig. 4 respectively, are achieved by training the model for 3 epochs and a batch size of 16. Figure 5 represents the confusion matrix created by predicting new test images from the trained model. Figures 6, 7, 8 and 9 show the real-time detection of the ASL sign gestures of the letters- 'A', 'B', 'C', and 'X' respectively.

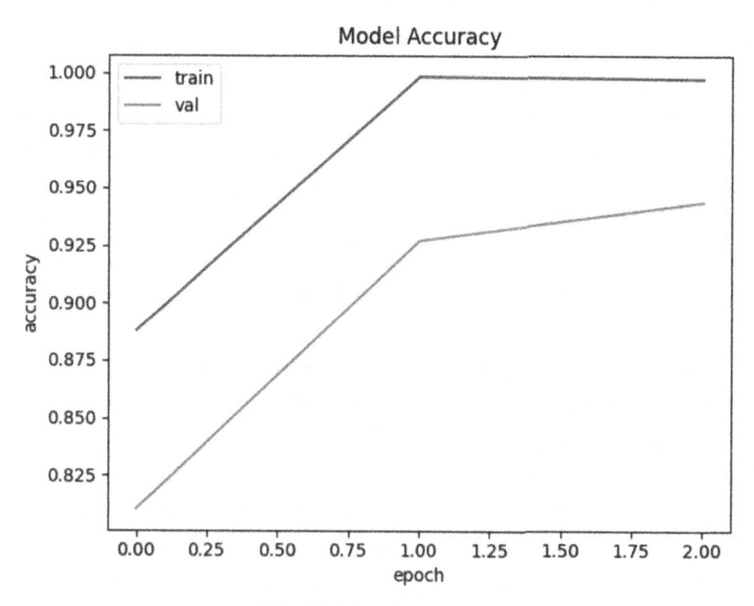

Fig. 3. Model Accuracy

Table 4 shows a comparative analysis of previous versus proposed research work done on Sign Language to Text Conversion using various Machine learning models.

The proposed work uses a transfer learning technique by utilizing a pre-trained MobileNet model as a starting point and fine-tuning it on a self-captured dataset for sign language recognition. The 27 convolutional layers used provide the depth required for better feature extraction and for achieving high accuracy, which is deeper than commonly used counterparts like VGG16 or Inception. Deeper models are generally able to capture more complex and abstract features in the data, which can improve their ability to discriminate between different sign language gestures.

Most of the previous research work includes a dataset that is self-captured and it lacks racial diversity which in turn affects the accuracy of the model's prediction. The dataset collection in the proposed work uses a hand tracking module which enables the model to learn the key points of the hand and accurately interpret the signer's hand gestures and convey their meaning in real-time.

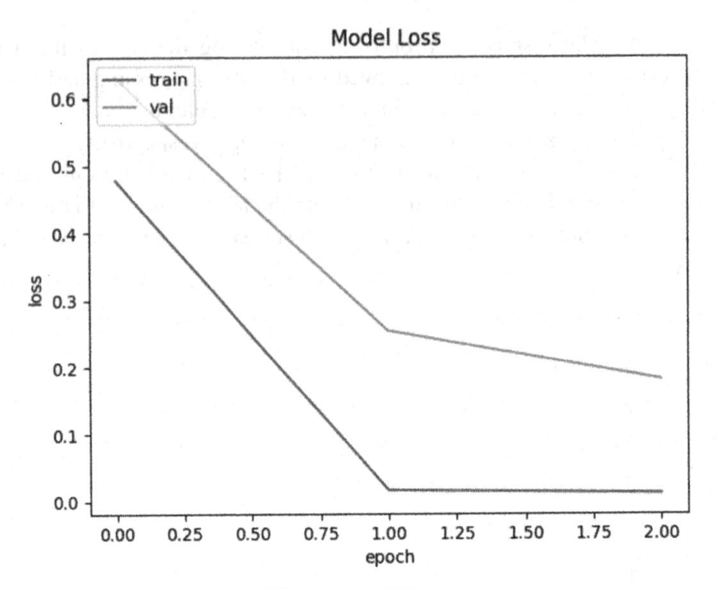

Fig. 4. Model Loss

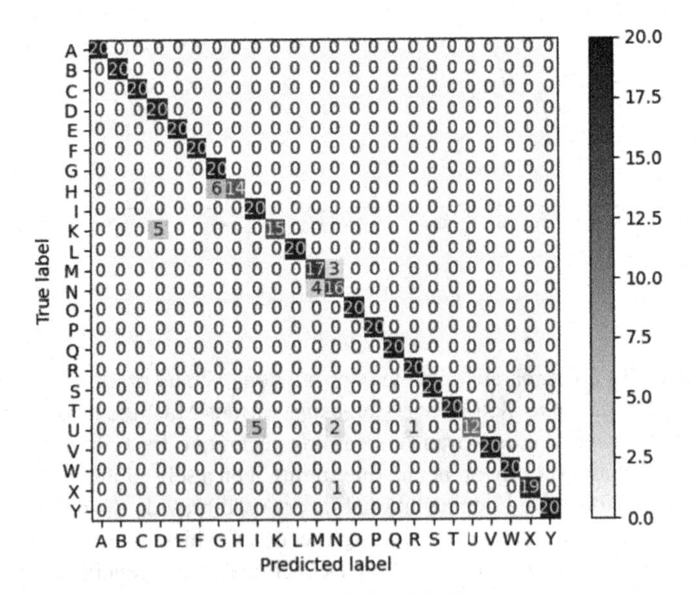

Fig. 5. Confusion Matrix

The proposed system works on Jetson Nano which is a portable embedded system making it ideal for sign language recognition applications. Jetson Nano acting as a core processor for sign language recognition provides real-time processing capabilities.

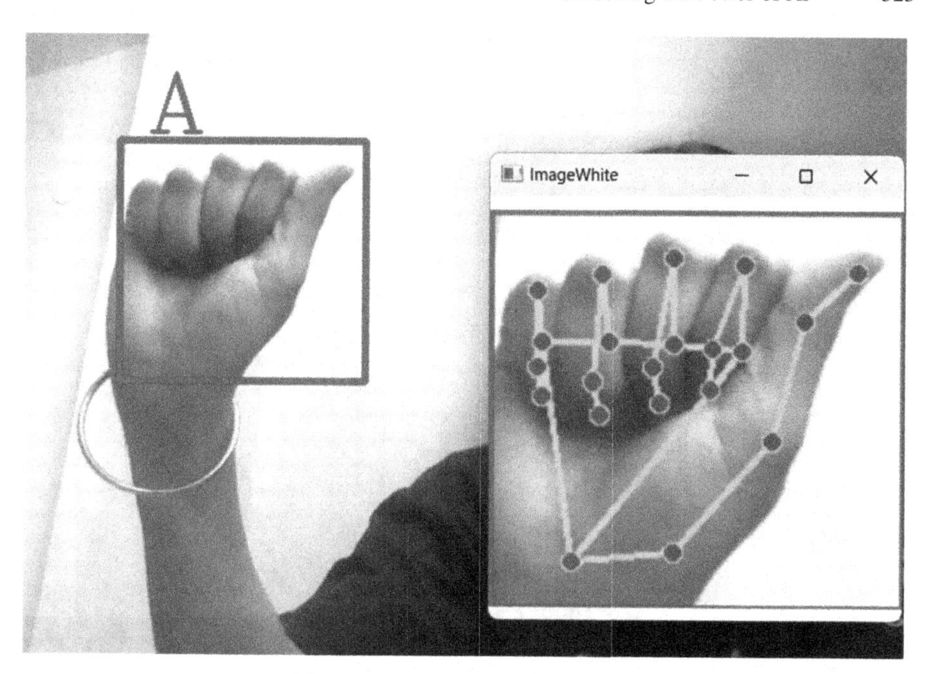

Fig. 6. Recognition of Alphabet 'A'

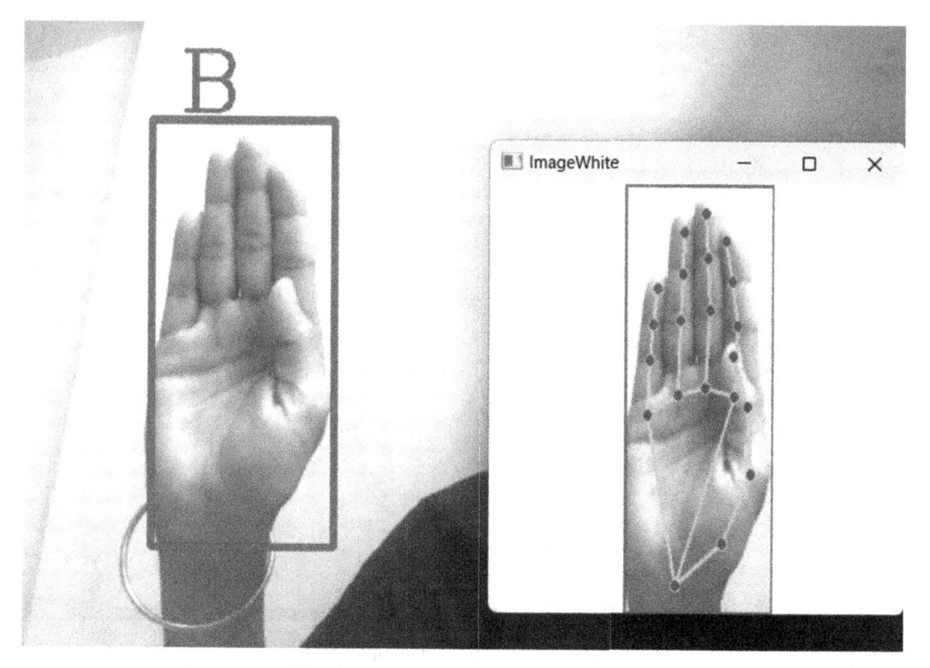

Fig. 7. Recognition of Alphabet 'B'

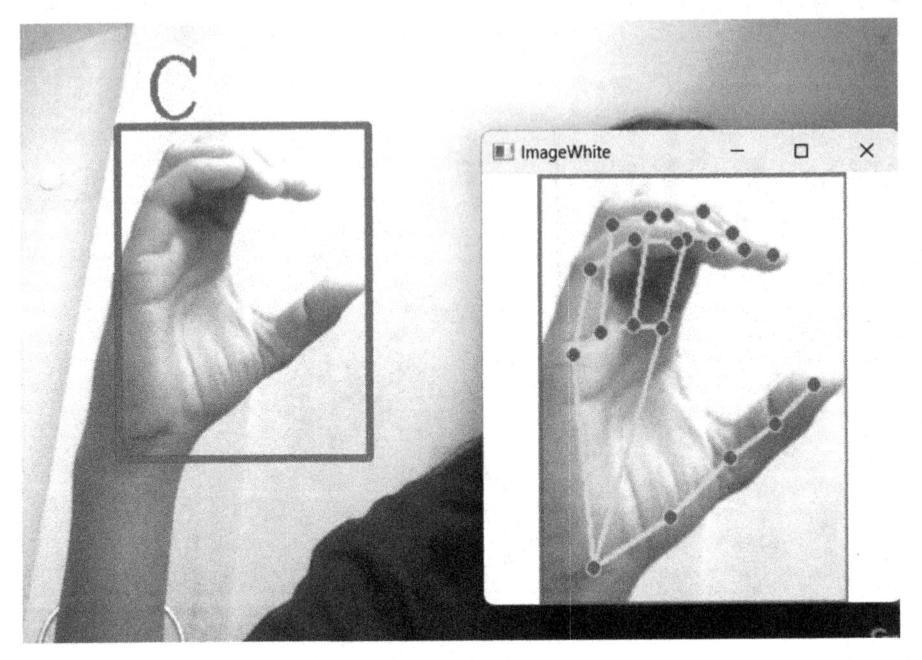

Fig. 8. Recognition of Alphabet 'C'

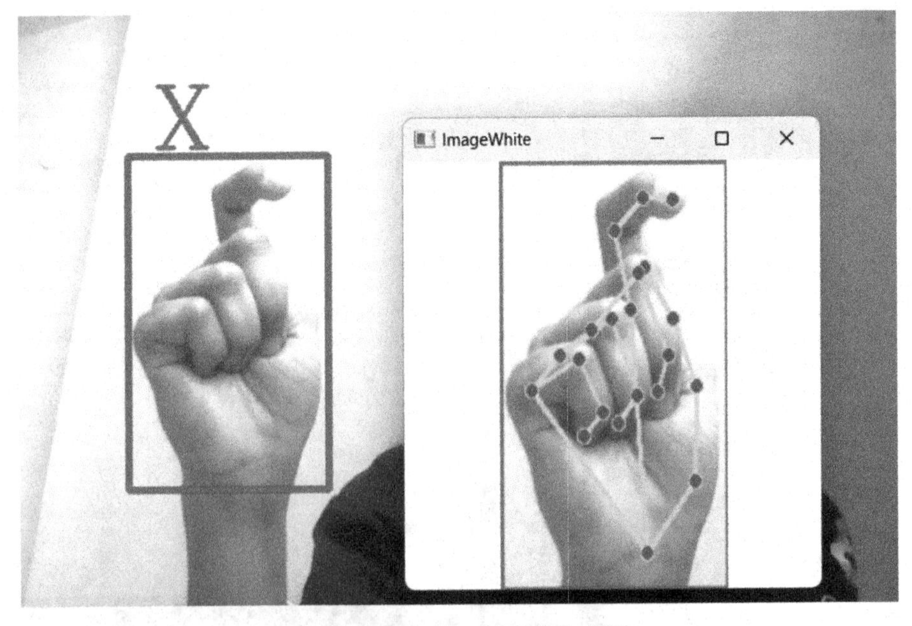

Fig. 9. Recognition of Alphabet 'X'

Table 4. Previous vs proposed research on Sign Language to Text Conversion in Real-Time using Deep Learning and Jetson Nano

Ref	Sign language	Algorithm	Pre-trained models	Dataset	Sample size	Accuracy
[12]	ASL	CNN	GoogLeNet	ASL Finger spelling Dataset from University of Surrey CVSSP	65,000	> 70%
[13]	ASL	CNN	VGG Net	Self-Captured	43,120	98.84%
[14]	Indian Sign Language	CNN, SVM	-	Self-Captured	36,000	99.64%
[15]	American, Indian, Italian, and Turkey Sign Language	SVM,KNN, Random forest,kNN, Decision Tree, Naïve Bayes,MLP	-	Multiple datasets including existing and self-captured	156000, 4972, 12856, and 4124	99.29%
[16]	Bengali Sign Language	CNN	-	Self-captured	1000	98.2%
[17]	ASL	ANN, K-convex hull	-	Self-captured	1850	94.32%
[18]	Turkish Sign Language	LR, KNN, RF, ANN	-	Self-captured	87000	98.97%
[19]	ASL	ANN	-	Self-captured	520	96.15%
Proposed	ASL	CNN	MobileNet	Self-captured	2400	94.38%

6 Conclusion

This paper proposes an approach to convert sign language to text using a deep learning model. The model pipeline is based on Convolutional Neural Network (CNN) architecture, which is designed to classify 24 alphabets (excluding the alphabet 'J' and 'Z'). The proposed approach has demonstrated an impressive training accuracy of 99.74% and validation accuracy of 94.38%, which suggests that it can be effectively used for real-world applications.

The deep learning model has been trained on a large dataset of sign language gestures, which has been carefully curated to include a wide range of variations and complexities. The proposed approach leverages the power of CNN to automatically extract relevant features from the input image and classify them into the appropriate category. This approach eliminates the need for human intervention in the feature extraction process, which is a time-consuming and error-prone task.

While the proposed approach has shown promising results, the research is still ongoing, and the focus is on expanding the sign database. The aim is to incorporate more variations and complexities in the dataset to improve the accuracy of the model further. The proposed approach has the potential to improve the communication and accessibility of hearing-impaired individuals and make a positive impact on their lives.

With Jetson Nano, models of varied complexity can also perform well. The proposed system works on Jetson Nano which is a portable embedded system making it easier for users to use it. The heterogeneous architecture of CPU-GPU helps to speed up complex machine-learning tasks. This system is integrated with a USB camera and in the future will be integrated with speakers to make real-time Sign language translation possible.

For the dataset collection, the hand tracking module of the CVZone package along with the OpenCV library ensures that cropped images having high-fidelity 3D hand key points are saved into separate classes. While reviewing previous research work performed in the field of sign language translation, various gaps were observed such as a drop in accuracy due to lighting variations in self-generated data and a lack of racial diversity in the dataset. Since the image dataset uses hand detection and finger tracking, the research gaps have been removed.

Worldwide, there are about 70 million deaf people who use sign language as their first language, and it is also the first language of many individuals with speech impairments. The proposed system can help in bridging a huge gap between the signers and non-signers.

Overall, the proposed Sign Language Recognition system has the potential to enhance the communication and accessibility of hearing-impaired individuals. The use of a lightweight deep learning model and advanced convolutional layers reduces the computational requirements of the system, making it more efficient and cost-effective. Further research could focus on expanding the dataset to improve the accuracy of the model and enable recognition of a broader range of sign language gestures.

References

1. Isaković, L., Kovačević, T., Srzić, M.: Sign languages: then and now. Зборник радова Филозофског факултета у Приштини. **50**, 293–314 (2020). https://doi.org/10.5937/zrffp50-28925
2. Shovkoplias, G., et al: Support for communication with deaf and dumb patients via few-shot machine learning. In: Proceedings 14th International Conference on ICT, Society and Human Beings (ICT 2021), the 18th International Conference Web Based Communities and Social Media (WBC 2021). (2021). doi: https://doi.org/10.33965/eh2021_202106c029
3. Rabiner, L., Juang, B.: An introduction to hidden markov models. IEEE ASSP Mag. **3**, 4–16 (1986). https://doi.org/10.1109/massp.1986.1165342
4. Zhu, H.-M., Pun, C.-M., Lin, C.: Robust hand gesture recognition using machine learning with positive and negative samples. Journal of Computers. 8, (2013). doi: https://doi.org/10.4304/jcp.8.7.1831-1835

5. Dominio, F., Donadeo, M., Marin, G., Zanuttigh, P., Cortelazzo, G.M.: Hand gesture recognition with depth data. Proceedings of the 4th ACM/IEEE international workshop on Analysis and retrieval of tracked events and motion in imagery stream. (2013). doi: https://doi.org/10.1145/2510650.2510651
6. Yang, M.-H., Ahuja, N.: Recognizing hand gesture using motion trajectories. Proceedings.In: 1999 IEEE Computer Society Conference on Computer Vision and Pattern Recognition (Cat. No PR00149). (1999). doi: https://doi.org/10.1109/cvpr.1999.786979
7. Mantowsky, S., Heuer, F., Bukhari, S.S., Keckeisen, M., Schneider, G.: ProAI: An efficient embedded AI hardware for automotive applications – a benchmark study.In: 2021 IEEE/CVF International Conference on Computer Vision Workshops (ICCVW). (2021). doi: https://doi.org/10.1109/iccvw54120.2021.00113
8. Feldbacher, C.: 6 Best microcontrollers: Hardware for embedded computing, https://blog.felgo.com/6-hardware-for-embedded-development#arduino
9. Suzen, A.A., Duman, B., Sen, B.: Benchmark analysis of jetson TX2, jetson nano and raspberry pi using deep-CNN. 2020 International Congress on Human-Computer Interaction, Optimization and Robotic Applications (HORA). (2020). doi: https://doi.org/10.1109/hora49412.2020.9152915
10. Verdin imx8M Mini - Toradex, https://docs.toradex.cn/108681-verdin-imx8m-mini-datasheet-v1.1
11. Bhatia, R.: Why convolutional neural networks are the go-to models in deep learning, https://analyticsindiamag.com/why-convolutional-neural-networks-are-the-go-to-models-in-deep-learning/
12. Garcia, B., Viesca, S.A.: Real-time american sign language recognition with convolutional neural networks. Convolutional Neural Networks for Visual Recognition. (2016)
13. Amer Kadhim, R., Khamees, M.: A real-time American sign language recognition system using convolutional neural network for real datasets. TEM Journal. 937–943 (2020). doi: https://doi.org/10.18421/tem93-14
14. Katoch, S., Singh, V., Tiwary, U.S.: Indian sign language recognition system using surf with SVM and CNN. Array. **14**, 100141 (2022). https://doi.org/10.1016/j.array.2022.100141
15. Haldera, A., Tayade, A.: Real-Time vernacular sign language recognition using mediapipe and machine learning. Int. J. Res. Publ. Rev. **2**, 9–17 (2021)
16. Hoque, O.B., Jubair, M.I., Islam, M.S., Akash, A.-F., Paulson, A.S.: Real time bangladeshi sign language detection using faster R-CNN. In: 2018 International Conference on Innovation in Engineering and Technology (ICIET). (2018). doi: https://doi.org/10.1109/CIET.2018.8660780
17. Islam, M.M., Siddiqua, S., Afnan, J.: Real time hand gesture recognition using different algorithms based on American Sign Language.In: 2017 IEEE International Conference on Imaging, Vision & Pattern Recognition (icIVPR). (2017). doi: https://doi.org/10.1109/ICIVPR.2017.7890854
18. Karaci, A., Akyol, K., Turut, M.U.: Real-time turkish sign language recognition using cascade voting approach with handcrafted features. Appl. Comput. Syst. **26**, 12–21 (2021). https://doi.org/10.2478/acss-2021-0002
19. Naglot, D., Kulkarni, M.: Real time sign language recognition using the leap motion controller.In: 2016 International Conference on Inventive Computation Technologies (ICICT). (2016). doi: https://doi.org/10.1109/INVENTIVE.2016.7830097
20. Creating a hand tracking module using python, OpenCV, and MediaPipe, https://www.section.io/engineering-education/creating-a-hand-tracking-module/
21. Bradski, G.: The OpenCV Library. Dr. Dobb's Journal of Software Tools. (2000)
22. Lugaresi, C., et al: MediaPipe: A framework for building perception pipelines. doi: https://doi.org/10.48550/arXiv.1906.08172

23. Howard , A.G., et al: MobileNets: efficient convolutional neural networks for mobile vision applications. arXiv preprint arXiv:1704.04861. (2017). doi: https://doi.org/10.48550/arXiv.1704.04861
24. Chng, Z.M.: Using Depthwise separable convolutions in TensorFlow, https://machinelearningmastery.com/using-depthwise-separable-convolutions-in-tensorflow
25. -D global average pooling layer, https://www.mathworks.com/help/deeplearning/ref/nnet.cnn.layer.globalaveragepooling2dlayer.html

AI-Assisted Geopolymer Concrete Mix Design: A Step Towards Sustainable Construction

Md Zia ul Haq[1(✉)], Hemant Sood[1], and Rajesh Kumar[2]

[1] Department of Civil Engineering NITTTR, Sector 26, Chandigarh 160019, India
ziazealous@gmail.com

[2] Department of Civil Engineering CCET, Sector 26, Chandigarh 160019, India

Abstract. The increasing interest in the utilization of Geopolymer Concrete as one of the sustainable alternatives to traditional Portland cement has been driven by its potential to mitigate the carbon footprint and its ability to be synthesized from industrial waste materials. However, the intricacies associated with the design mix process of Geopolymer has been identified as a limiting factor. Thus, in this research, an Artificial Intelligence (AI)-assisted approach for mix design was devised, with the aim of improving the efficiency and precision of the mix design process. The AI model was calibrated utilizing a dataset of experimental Geopolymer Concrete mixes, and subsequently employed to forecast the compressive strength and setting time of newly proposed mix designs. The outcome of the study revealed that the AI-assisted approach significantly enhanced the efficiency of the mix design process and also resulted in more precise predictions of the mechanical properties of the Geopolymer Concrete. Consequently, this study serves as a demonstration of the capability of AI to support sustainable construction by streamlining the process of Geopolymer Concrete mix design.

Keywords: Geopolymer Mix Design · Machine Learning Algorithms · AI · Neural Networks · Sustainability

1 Introduction

When it comes to AI-assisted geopolymer mix design, the possibilities are endless. One could imagine using a powerful neural network, trained on a vast dataset of experimental mixes, to predict the optimal combination of ingredients with precision and speed [1]. A decision tree model, on the other hand, could be employed to uncover the hidden relationships and factors that influence the final outcome, guiding the mix design process with clarity and insight. And let's not forget about the elegant simplicity of a support vector machine, which elegantly separates the data into distinct regions, revealing the most important characteristics of the mix design [2–4]. Regardless of the model chosen, what's important is that it should be able to capture the complexities of the mix design process and provide accurate predictions. The ultimate goal is to streamline the process and make it more efficient, while also improving the quality and sustainability of the final product [5–7]. With the right AI model in place, the future of geopolymer mix design is bright, and the path to sustainable construction that much clearer.

R. K. Challa et al. (Eds.): ICAIoT 2023, CCIS 1930, pp. 331–341, 2024.
https://doi.org/10.1007/978-3-031-48781-1_26

The potential of AI in geopolymer mix design is truly exciting. Imagine a system that can learn from past mix designs and adjust the proportions of ingredients in real-time to optimize the properties of the final product. This type of adaptive learning could revolutionize the way we approach mix design, making it faster, more accurate and more sustainable. Another exciting possibility is the use of AI for real-time monitoring of the concrete during the curing process [8–10]. By analyzing data from sensors embedded in the concrete, an AI model could predict the final properties of the concrete and make adjustments to the mix design as necessary. This would allow for more precise control over the final product and could lead to even more sustainable construction. In addition to these practical applications, AI-assisted geopolymer mix design could also open up new possibilities for research. With the ability to quickly and accurately generate thousands of mix designs, researchers would be able to explore a much wider range of possibilities and discover new, previously unknown properties of geopolymer concrete [11]. The integration of AI in the design of geopolymer mixtures holds significant promise for enhancing the efficiency, precision, and sustainability of the process. It is a thrilling time for the industry as we continually uncover the full capabilities of this advanced technology.

1.1 Geopolymer Mix Design

The design of geopolymer mixtures is a complicated procedure that encompasses the utilization of various elements, each possessing distinct characteristics and functions. The fundamental components of a geopolymer mix include:

Alkali Activator. The alkali activator plays a vital role in initiating the chemical reaction between the aluminosilicate materials and the alkali solution, thereby leading to the creation of the geopolymer matrix. Sodium hydroxide and potassium hydroxide are among the most widely used alkali activators.

Aluminosilicate Source Material. This is the primary component of the geopolymer matrix and can be derived from various sources such as fly ash, slag, or natural minerals like kaolinite, metakaolin, or rice husk ash.

Aggregate. This is the inert filler material that is used in the mix and can be made of natural or artificial materials like gravel, crushed stone, or recycled glass.

Water: Water is essential for activating the alkali activator and for providing the appropriate consistency for the mix.

Admixtures. By using admixtures such air-entraining agents, superplasticizers, and set retarders, the geopolymer concrete mixture's workability, compressive strength, and setting time may be improved. By using admixtures such air-entraining agents, superplasticizers, and set retarders, the geopolymer concrete mixture's workability, compressive strength, and setting time may be improved.

Reinforcement. The addition of reinforcement, in the form of steel bars or fibers, to the geopolymer concrete mix can improve its overall strength and durability.

The composition and the type of materials used for the components may vary based on the project requirements and the desired properties of the final geopolymer concrete product. The appropriate combination of components enables the customization of a geopolymer mix to fit the specific needs of any construction project, while promoting sustainability and minimizing its environmental impact.

1.2 Complexity and Importance of Mix Design in Geopolymer Concrete

Due to a number of issues, including the reliance on a chemical interaction between an alkali activator and aluminosilicate source materials, the design of geopolymer concrete mixes is complicated. In contrast to conventional Portland cement concrete, geopolymer concrete's source ingredients can come from a variety of places, such as fly ash, slag, or natural minerals. The intricacy of the geopolymer mix design is a result of this [12–14]. Numerous variables, such as the nature and ratios of the aluminosilicate source materials, the concentration of the alkali activator, and the curing conditions, affect this chemical reaction known as geopolymerization [15]. A thorough understanding of the chemical and physical characteristics of the various components and how they interact is required for the mix design of geopolymer concrete. This intricacy emphasizes the significance of in-depth subject expertise for the efficient design of geopolymer concrete mixes.

The mechanical qualities and the setting time must be balanced while designing the mix for geopolymer concrete. A few of the variables that might affect the mechanical characteristics of geopolymer are the water-to-binder solids ratio, the kind and size of aggregate, and the presence of reinforcement. These factors are taken into consideration in an efficient mix design to create concrete that has the best mechanical characteristics and setting time. On the other hand, the setting time can be influenced by the type and absorption of the alkali activator, the curing conditions, and the use of set retarders [16, 17]. Additionally, it's crucial to think about how the components will affect the environment while developing a mix design for geopolymer concrete. Since the geopolymerization process uses little energy, it is crucial to make sure that all of the mix design's components come from sustainable sources. Thus, a thorough understanding of the chemical and physical characteristics of its constituent parts and how they interact is required for mix design in geopolymer concrete. To create high-quality, long-lasting, and ecologically responsible geopolymer concrete, it is essential to strike a balance between mechanical qualities, setting time, and sustainability. The detailed chemical reaction of geopolymer is represented in Fig. 1 as shown below.

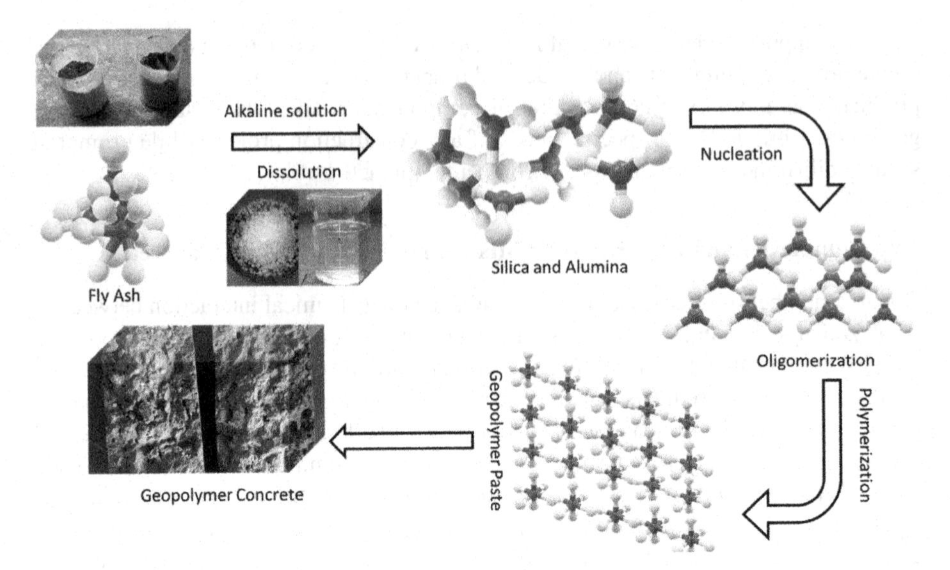

Fig. 1. Different Variables used in Geopolymer Concrete [18].

2 Goals and Parameters of the Study

In order to increase the accuracy and efficiency of the geopolymer concrete mix design process, this research aims to incorporate artificial intelligence into it. The goal of the work is to use a collection of experimental geopolymer concrete mix data to train a machine learning model, and then to use the model to effectively train data for compressive strength and setting time of new mix designs. The process will be streamlined, less time and money will be required, and the final product's quality and sustainability will be improved with the addition of AI to the mix creation process.

The Scope of the Research Includes the Following. Collection and preprocessing of a dataset of experimental geopolymer concrete mixes. Development and training of an AI model for mix design. Validation and testing of the AI-assisted approach using new mix designs. Comparison of the results with traditional mix design methods. Analysis of the implications of the results for the efficiency, accuracy, and sustainability of the mix design process.

The research is focused on the mix design process and the use of AI to assist in this process, rather than the production or testing of the final geopolymer concrete. The research also aims to show the AI-assisted approach's potential in the geopolymer concrete mix design, not to provide an exhaustive solution for all mix design scenarios.

2.1 Overview of Previous Research on Geopolymer Concrete Mix Design

Previous studies on geopolymer concrete mix design have delved into various aspects of the process, exploring everything from the perfect recipe for strength and setting time to the impact of different variables on the final product. Researchers have employed a plethora of methods, from experimental design and statistical analysis to artificial neural networks, all in the pursuit of the ultimate mix. Others have focused on uncovering the secrets of the different components, delving into the effect of source materials, alkali activators, water-to-solid ratios, and even curing conditions for the mechanical properties and setting time of the concrete. Innovative minds have also sought to push the boundaries of mix design, experimenting with a wide range of materials from industrial waste material which includes fly ash and slag to natural minerals like kaolinite and metakaolin, in the search for new and exciting mix designs.

Previous research has provided a solid foundation in understanding the intricacies of geopolymer mix design, but there is always room for improvement. The application of AI-assisted approach in the mix design process is expected to bring a new level of efficiency and accuracy, and also make the process more sustainable. It's an exciting time as we continue to uncover new possibilities in the realm of geopolymer mix design. However, there are still some gaps in the current knowledge that can be identified.

Efficiency of the Mix Design Process. Mix design in geopolymer concrete is a complex and time-consuming process. There is a need for more efficient methods that can reduce the time and effort required for mix design while maintaining the accuracy of the results.

Adaptability to Different Project Requirements. The mix design process should be able to adapt to the specific requirements of different projects and the desired properties of the final product. This requires a flexible approach that can easily be adjusted to different conditions.

Integration of Sustainability. Past studies have primarily concentrated on optimizing the mix proportions and evaluating the effect of various factors on the final product's properties. However, further research that assesses the sustainability and environmental impact of the mix components is necessary.

Real-Time Monitoring. There is a need for research that looks into the real-time monitoring of the concrete during the curing process and how it can be used to make adjustments to the mix design as necessary.

Combination of Different Approaches. Most of the previous research has used a single approach such as experimental design, statistical analysis, or artificial neural networks, there is a need for more research that combines different approaches and methods to improve the efficiency and accuracy of the mix design process.

The lack of research regarding the environmental impact of mix design materials in geopolymer concrete underscores the importance of further exploring the application of AI to enhance the efficiency, precision, and environmental sustainability of the mix design process.

3 Data Collection and Preparation for AI-Assisted Geopolymer Concrete Mix Design

A combination of simulated data and experimental data was used to create the dataset for the AI-assisted geopolymer concrete mix design research. The dataset consists of a total of 400 instances, each containing information about the mix design, such as the source material, alkali activator, water-to-solid ratio, aggregate, admixture, and curing condition, as well as the response parameters, compressive strength and setting time. A series of laboratory studies on geopolymer concrete mixtures made with various source materials, alkali activators, and mix proportions were used to gather the experimental results. The properties of the final products were measured, such as compressive strength and setting time. This experimental data was used as a basis to choose the range of the dataset. Simulation software was used to generate additional data, by simulating experiments using different variables and parameters, such as source materials, alkali activators, and mix proportions. The simulated data was validated against the experimental data to ensure its validity and results are shown in Fig. 2.

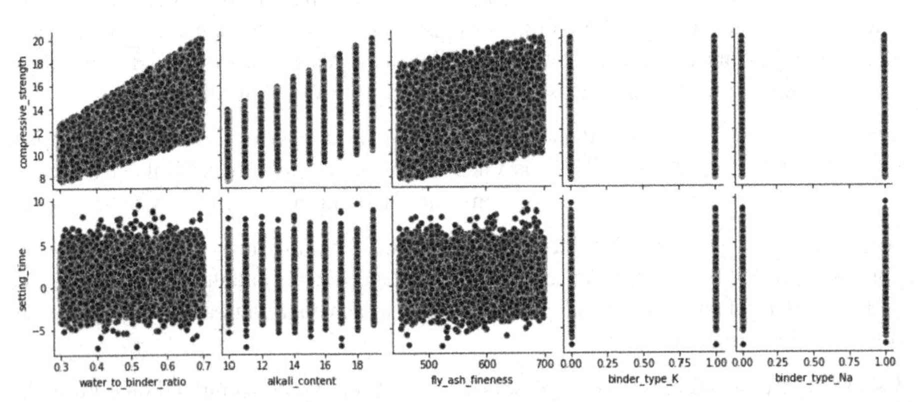

Fig. 2. Comparison of Different Dependent and Independent Variables used in Geopolymer Mix Design.

A dataset was created for the study of geopolymer materials in this research. The collection contains data regarding the characteristics of geopolymer materials, including the water-to-binder ratio, alkali concentration, binder type, fly ash fineness, compressive strength, and setting time. To order understand the relationship among independent variables and dependent variables in the geopolymer concrete dataset, a pair plot was generated using the seaborn library in Python. This visualization tool is beneficial in revealing correlations between multiple variables in a dataset. In this research, the pairplot was utilized to identify any strong associations between the independent variables, such as the water-to-binder ratio and alkali content, and the dependent variables, such as compressive strength and setting time. The diagonal plots in the pairplot represented in Fig. 2 that displayed the distribution of each variable. The off-diagonal plots, on the other hand, depicted scatterplots that showed the relationship between each pair of variables. The pair plot's warm color scheme makes it simpler to spot trends and patterns in the data. A strong association between several independent factors and dependent variables was found during pairplot analysis. Particularly, it was discovered that the compressive strength and setting time were closely connected to the water-to-binder ratio and alkali concentration. These findings suggest that these independent variables may have a significant impact on the properties of geopolymer materials. In this work, the connection between the independent and dependent variables in the dataset of geopolymer concrete was shown in Fig. 2 using a pairplot. The analysis's findings show a considerable connection between a few independent factors and dependent variables, pointing to a substantial influence on the characteristics of geopolymer materials. Future investigations into geopolymer materials will benefit greatly from these discoveries.

3.1 Machine Learning Algorithms Used for the Study

Linear Regression. A statistical method called linear regression examines the correlation between a dependent variable and one or more predictor variables. Finding the regression line or line with the greatest fit to this data is the goal in order to understand the relationship between these variables. The linear regression approach may be used to detect significant correlations between the variables and create predictions [19]. By reducing the difference between the anticipated and actual values, linear regression seeks to identify the link between a dependent variable and one or more independent variables. This is accomplished by fitting a straight line through the data points using the equation $Y = aX + b$. In this equation, the independent variable is X, the dependent variable is Y, and the coefficients are a and b [20].

K-Nearest Neighbors. A popular machine learning method for classification and regression problems is K-Nearest Neighbors (KNN). The approach locates a data point's K nearest neighbours, and then classifies or predicts the value of that data point based on the majority class or average value of these K neighbours. By using the KNN approach, one may calculate the majority class or average value of the K data points that are closest to a new sample [21].

Decision Tree. The Decision Tree algorithm models complex decision-making processes through the use of recursive splits in the dataset based on the most impactful feature. The resulting tree structure visually represents a series of if-else statements, with each node symbolizing a feature and each branch representing a decision [22, 23].

Random Forest. Random Forest is a sophisticated algorithm that leverages the collective strength of multiple decision trees to produce predictions. The methodology involves the creation of multiple decision trees and then averaging the predictions of each individual tree to form a final prediction. This technique enhances the prediction accuracy and reduces the risk of overfitting. The equation for Random Forest is a function of the average of all the decision trees predictions [24, 25] (Fig. 3).

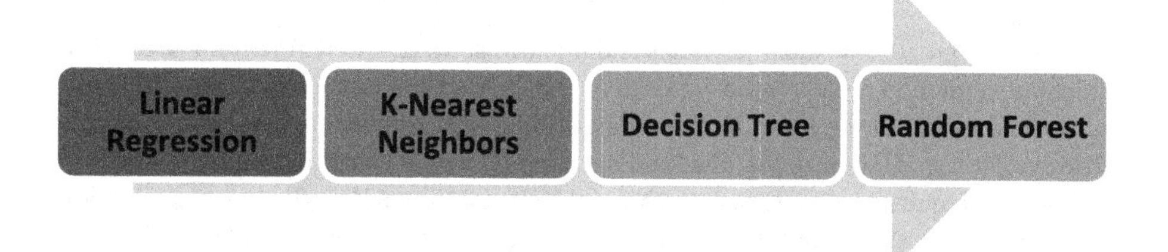

Fig. 3. Machine Learning Algorithm Used in Study.

Table 1. Performance of Different Machine Algorithm.

	R-squared	MAE	RMSE
Linear Regression	0.9791	0.26	0.1188
K-Nearest Neighbors	0.6295	1.1441	2.1083
Decision Tree	0.9964	0.1089	0.0207
Random Forest	0.9992	0.0497	0.0045

From Table 1, it can be seen that the Random Forest model has got the best performance with R-squared of 0.9992, MAE of 0.0497, and RMSE of 0.0045, which means that it has the highest accuracy in predicting the output. The Decision Tree model also has a good performance with R-squared of 0.9964, MAE of 0.1089, and RMSE of 0.0207. On the other hand, Linear Regression model has R-squared of 0.9791, MAE of 0.26, and RMSE of 0.1188, and K-Nearest Neighbors model has R-squared of 0.6295, MAE of 1.1441, and RMSE of 2.1083, showing lower accuracy in prediction.

The proper machine learning algorithm must be used in order to create the ideal mix for geopolymer concrete. After all, the precise proportions of the elements employed will determine the final product's strength and setting time. Choosing the right machine learning method is essential to achieving precise predictions for the compressive strength and setting time of geopolymer concrete mixtures. Four algorithms—Linear Regression, K-Nearest Neighbors, Decision Tree, and Random Forest—were examined in our study to see how well they performed. The algorithms' performance is assessed using performance measures including R-squared, Mean Absolute Error (MAE), and Root Mean Squared Error using a sample dataset of 300 observations (RMSE).

The amount of variation in the response variable that can be explained by the predictor variables is measured using the R-squared metric. The better the algorithm matches the data, the greater the R-squared score. The algorithms Random Forest and Decision Tree, which had remarkable R-squared values of 0.9992 for Random Forest and 0.9964 for Decision Tree, were determined to have the highest values. This indicates that a sizable amount of the variance in the response variable might be explained by these techniques. The average discrepancy between the response variable's expected and actual values is measured by the MAE and RMSE values. The better the algorithm matches the data, the lower the MAE and RMSE values. In this work the algorithm Random Forest has the lowest MAE value of 0.0497 and the lowest RMSE value of 0.0045. This means that the Random Forest algorithm makes the least amount of error in predicting the response variable.

With the greatest R-squared values and the lowest MAE and RMSE values, the Random Forest method is clearly the victor in our analysis. The Decision Tree algorithm also performed well, but the Random Forest algorithm proved to be slightly more accurate. However, it's worth noting that the final decision on which algorithm to use depends on the specific requirements of the project and the availability of data. But overall, Random Forest is the most suitable algorithm for predicting compressive strength and setting time in geopolymer concrete mix design (Fig. 4).

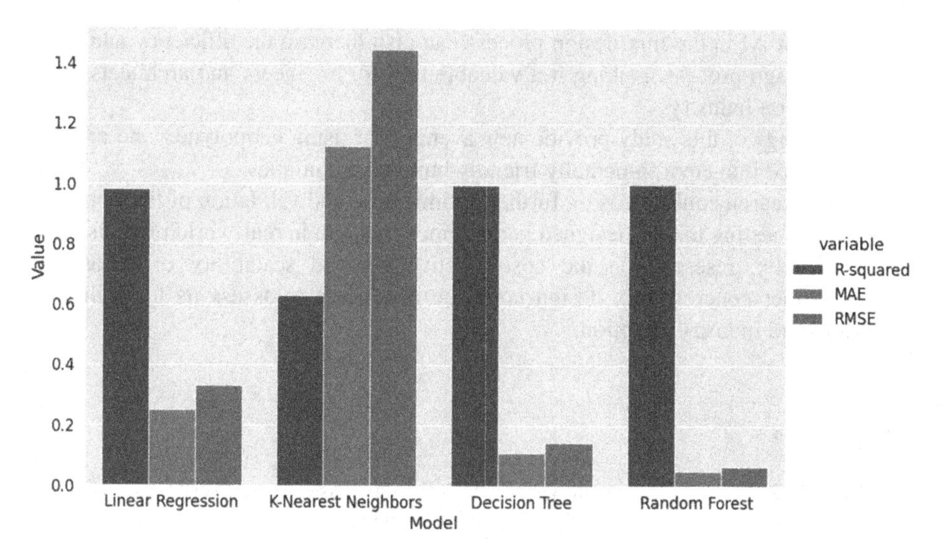

Fig. 4. Performance of Different Machine Algorithm.

4 Conclusion

The present work shows the potential of using artificial intelligence to the development of environmentally friendly geopolymer concrete. The R-squared, Mean Absolute Error (MAE), and Root Mean Squared Error (RMSE) values of the results show that the Decision Tree and Random Forest algorithms perform the best overall. This shows that AI-based methods might improve the efficient use of resources, reduce waste, and improve the durability and hardness of geopolymer concrete.

- The study used AI techniques to explore the potential of improving the sustainability of construction through geopolymer concrete mix design.
- Geopolymer concrete is a sustainable alternative to traditional concrete, as it requires less energy to produce and has a lower carbon footprint.
- In the study, four distinct AI algorithms were utilized: Linear Regression, K-Nearest Neighbors, Decision Tree, and Random Forest.
- The results showed that the decision tree and random forest models had the best performance in terms of R-squared, MAE, and RMSE values, suggesting a high potential for use in real-world applications.
- Linear regression had an R-squared value of 0.9791, a MAE of 0.26, and a RMSE of 0.1188.K-nearest neighbors had an R-squared value of 0.6295, a MAE of 1.1441, and a RMSE of 2.1083.Decision tree had an R-squared value of 0.9964, a MAE of 0.1089, and a RMSE of 0.0207.Random forest had an R-squared value of 0.9992, a MAE of 0.0497, and a RMSE of 0.0045.The decision tree and random forest models showed the best performance in terms of R-squared, MAE and RMSE values.
- The study highlights the potential of AI-assisted geopolymer concrete mix design to optimize the use of raw materials, reduce waste, and improve the strength and durability of concrete.

- The use of AI in the mix design process can also increase the efficiency and speed of the design process, making it a valuable tool for engineers and architects in the construction industry.
- The findings of this study provide new avenues for using geopolymer and artificial intelligence into environmentally friendly building techniques.
- Future research could focus on further optimization and validation of these models, as well as testing the AI-designed geopolymer concrete in real-world projects.
- Additionally, research on the cost-effectiveness and scalability of AI-assisted geopolymer concrete mix design could be conducted to assess its feasibility for widespread industry adoption.

References

1. Pazouki, G.: Fly ash-based geopolymer concrete's compressive strength estimation by applying artificial intelligence methods. Measurement **203** (2022). https://doi.org/10.1016/j.measurement.2022.111916
2. Raza, M.H., Zhong, R.Y.: A sustainable roadmap for additive manufacturing using geopolymers in construction industry. Resour. Conserv. Recycl. **186**, 106592 (2022). https://doi.org/10.1016/j.resconrec.2022.106592
3. Kanagaraj, B., Anand, N., Praveen, B., Kandasami, S., Lubloy, E., Naser, M.Z.: Physical characteristics and mechanical properties of a sustainable lightweight geopolymer based self-compacting concrete with expanded clay aggregates. Dev. Built Environ. **13** (2023). https://doi.org/10.1016/j.dibe.2022.100115
4. Huo, W., Zhu, Z., Sun, H., Ma, B., Yang, L.: Development of machine learning models for the prediction of the compressive strength of calcium-based geopolymers. J. Clean. Prod. **380**, 135159 (2022). https://doi.org/10.1016/j.jclepro.2022.135159
5. Turkey, F.A., Beddu, S.B., Ahmed, A.N., Al-Hubboubi, S.K.: Effect of high temperatures on the properties of lightweight geopolymer concrete based fly ash and glass powder mixtures, Case Stud. Constr. Mater. **17** (2022). https://doi.org/10.1016/j.cscm.2022.e01489
6. Shilar, F.A., Ganachari, S.V., Patil, V.B., Khan, T.M.Y., Dawood Abdul Khadar, S.: Molarity activity effect on mechanical and microstructure properties of geopolymer concrete: a review. Case Stud. Constr. Mater. **16** (2022). https://doi.org/10.1016/j.cscm.2022.e01014
7. Chan, C.L., Zhang, M.: Behaviour of strain hardening geopolymer composites at elevated temperatures, Cem. Concr. Compos. **132** (2022). https://doi.org/10.1016/j.cemconcomp.2022.104634
8. Zhu, Y., et al.: A novel aminated lignin/geopolymer supported with Fe nanoparticles for removing Cr(VI) and naphthalene: Intermediates promoting the reduction of Cr(VI). Sci. Total Environ. **866** (2023). https://doi.org/10.1016/j.scitotenv.2022.161379
9. Karuppaiyan, J., Mullaimalar, A., Jeyalakshmi, R.: Adsorption of dyestuff by nano copper oxide coated alkali metakaoline geopolymer in monolith and powder forms: kinetics, isotherms and microstructural analysis. Environ. Res. **218** (2023). https://doi.org/10.1016/j.envres.2022.115002
10. Raza, M.H., Zhong, R.Y., Khan, M.: Recent advances and productivity analysis of 3D printed geopolymers. Addit. Manuf. **52** (2022). https://doi.org/10.1016/j.addma.2022.102685
11. Liang, K., Wang, X.Q., Chow, C.L., Lau, D.: A review of geopolymer and its adsorption capacity with molecular insights: a promising adsorbent of heavy metal ions. J. Environ. Manage. **322** (2022). https://doi.org/10.1016/j.jenvman.2022.116066

12. Cai, J., Li, X., Tan, J., Vandevyvere, B.: Fly ash-based geopolymer with self-heating capacity for accelerated curing. J. Clean. Prod. **261** (2020). https://doi.org/10.1016/j.jclepro.2020.121119
13. Xie, T., Visintin, P., Zhao, X., Gravina, R.: Mix design and mechanical properties of geopolymer and alkali activated concrete: review of the state-of-the-art and the development of a new unified approach. Constr. Build. Mater. **256** (2020). https://doi.org/10.1016/j.conbuildmat.2020.119380
14. Mahmoodi, O., Siad, H., Lachemi, M., Dadsetan, S., Sahmaran, M.: Development of normal and very high strength geopolymer binders based on concrete waste at ambient environment. J. Clean. Prod. **279** (2021). https://doi.org/10.1016/j.jclepro.2020.123436
15. Zheng, Y., Rao, F., Yang, L., Zhong, S.: Comparison of ternary and dual combined waste-derived alkali activators on the durability of volcanic ash-based geopolymers. Cem. Concr. Compos. **136** (2023). https://doi.org/10.1016/j.cemconcomp.2022.104886
16. Shen, J., Li, Y., Lin, H., Feng, S., Ci, J.: Prediction of compressive strength of alkali-activated construction demolition waste geopolymers using ensemble machine learning. Constr. Build. Mater. 360 (2022). https://doi.org/10.1016/j.conbuildmat.2022.129600
17. Li, X., Bai, C., Qiao, Y., Wang, X., Yang, K.: Preparation, properties and applications of fly ash-based porous geopolymers: a review. J. Clean. Prod. 359 (2022). https://doi.org/10.1016/j.jclepro.2022.132043
18. Haq, M.Z. ul, Sood, H., Kumar, R.: Effect of using plastic waste on mechanical properties of fly ash based geopolymer concrete. Mater. Today Proc. **69**(Part 2), 147–152 (2022). https://doi.org/10.1016/j.matpr.2022.08.233
19. Ma, Q., Shi, Y., Xu, Z., Huang, K.: Research on a multivariate non-linear regression model of dynamic mechanical properties for the alkali-activated slag mortar with rubber tire crumb, Case Stud. Constr. Mater. **17** (2022). https://doi.org/10.1016/j.cscm.2022.e01371
20. Ali, A.A., Al-Attar, T.S., Abbas, W.A.: A statistical models to predict strength development of eight molarity geopolymer concrete. Case Stud. Constr. Mater. **17** (2022). https://doi.org/10.1016/j.cscm.2022.e01304
21. Tanyildizi, H.: Predicting the geopolymerization process of fly ash-based geopolymer using deep long short-term memory and machine learning. Cem. Concr. Compos. **123** (2021). https://doi.org/10.1016/j.cemconcomp.2021.104177
22. Ahmad, A., Ahmad, W., Aslam, F., Joyklad, P.: Compressive strength prediction of fly ash-based geopolymer concrete via advanced machine learning techniques. Case Stud. Constr. Mater. **16** (2022). https://doi.org/10.1016/j.cscm.2021.e00840
23. Hamie, H., Hoayek, A., El-Ghoul, B., Khalifeh, M.: Application of non-parametric statistical methods to predict pumpability of geopolymers for well cementing, J. Pet. Sci. Eng. **212** (2022). https://doi.org/10.1016/j.petrol.2022.110333
24. Kuang, F., Long, Z., Kuang, D., Guo, R.: Application of back propagation neural network to the modeling of slump and compressive strength of composite geopolymers. Comput. Mater. Sci. **206** (2022). https://doi.org/10.1016/j.commatsci.2022.111241
25. Li, Y., et al.: The data-driven research on bond strength between fly ash-based geopolymer concrete and reinforcing bars. Constr. Build. Mater. **357** (2022). https://doi.org/10.1016/j.conbuildmat.2022.129384

Deep Learning Mechanism for Region Based Urban Traffic Flow Forecasting

Nishu Bansal(✉) and Rasmeet Singh Bali

Department of Computer Science and Engineering, Chandigarh University, Chandigarh, India
nishubl@gmail.com

Abstract. Intelligent Vehicular Ad Hoc Networks which integrates deep learning techniques with modern vehicular communication networks can play a major role in prediction of vehicular traffic as well as efficient dissemination of critical information between vehicular nodes. An accurate traffic prediction mechanism is a key requirement for numerous applications of Intelligent Transportation System such as traffic management, accident prevention, route guidance and public safety. In this paper, a deep learning approach is proposed which is based on Convolutional Neural Network (CNN) combined with Temporal Convolutional Network (TCN), to predict the traffic patterns of vehicles. External factors like weather, weekend and holidays are considered along with internal factors such as location and time for analyzing their effect on vehicular traffic. Integration of CNN and TCN, captures spatio-temporal features, which are then merged with external factors to obtain a more accurate predicted traffic information. This predicted value is further disseminated within the vehicular network. Dataset of Indian cities is taken and converted to matrices of time vs space. Experimental results illustrate that our model outperforms other state-of-the-art techniques in regard to efficiency and accuracy.

Keywords: Vehicular flow forecasting · Intelligent Transportation Systems · Convolutional Neural Networks · Temporal Convolutional Network

1 Introduction

The Intelligent Transportation System (ITS) has evolved due to increase in demand of traffic management and road safety. The ITS offers different services assisted by other systems for sensing and gathering the vehicular information [1]. The sensing and gathering of different types of traffic information, is achieved by using devices such as loop detectors, video cameras, Radio-frequency identification (RFID) scanners, GPS etc. After processing, this information is communicated between various vehicles on roads to help them in traffic management. The network connectivity between vehicles is required for these applications to transfer the information at receiver end which helps in achieving the desired performance in terms of reducing traffic congestion and increasing convenience of road users. Some of the key issues in traffic management such as accidents and traffic jams can also be controlled in this way.

R. K. Challa et al. (Eds.): ICAIoT 2023, CCIS 1930, pp. 342–356, 2024.
https://doi.org/10.1007/978-3-031-48781-1_27

The above objectives are achieved by using vehicular ad hoc networks (VANETs). Due to their greater mobility than other nodes, vehicular nodes differ in their information dissemination [2, 3]. Vehicle-to-Infrastructure (V2I), Vehicle-to-Vehicle (V2V) and Vehicle-to-Pedestrian (V2P) type of communications are used for information exchange between the vehicles. Roadside units and On-Board Units are the means by which vehicles connect with one another. As a result, the development of VANETs as a component of ITS has promised to effectively facilitate a wide range of vehicular communication applications [4], such as the dynamic route planning [5], sharing of content, dissemination of safety messages, entertainment and gaming [6, 7].

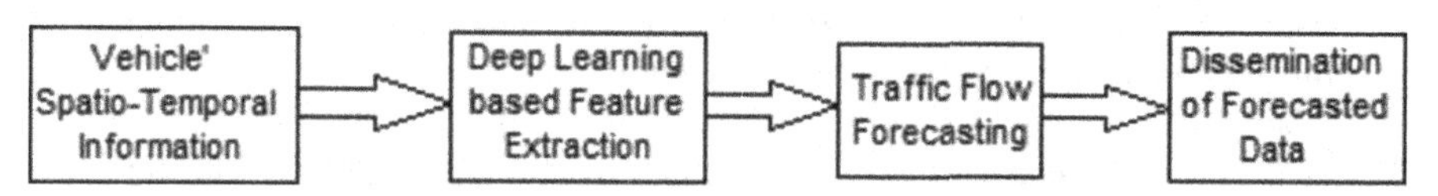

Fig. 1. Intelligent Vehicular Network Components

Due to communication, hardware and scaling constraints of VANET, Intelligent Vehicular Networks (IVN) are devised which integrates deep learning technique and modern vehicular communication networks. Although machine learning has shown great potential in predicting traffic patterns and helping to optimize traffic flow, but deep learning is particularly well-suited to solving problems involving large datasets. Compared to traditional machine learning algorithms, deep learning models can automatically extract relevant features from the input data. This makes them more effective at dealing with complex and high-dimensional data, such as traffic patterns. In the context of traffic prediction, deep learning models can be trained on large volumes of historical traffic data, allowing them to learn complex patterns and relationships between various factors, such as location, time of day, weather, road conditions, and traffic flow. These models enable more accurate vehicular analysis at different conditions and times.

Various components possessed by IVN are shown in Fig. 1. The spatial and temporal data of vehicles is passed to the model as input. Further feature extraction from this data takes place using different techniques of deep learning. Deep learning model helps to get the forecasting vehicular information for a given time span and at given location segment. The predicted data is forwarded to autonomous vehicles using vehicular communication networks [8]. This data is further used by vehicular network for decision-making and optimized planning of routes.

Vehicular prediction is mainly categorized into two main categories i.e. long term and short term forecasting. Short-term forecasting focuses on forecasting the traffic flow for small time spans while long-term forecasting focuses on predicting the vehicular flow for larger time spans. Vehicular flow prediction is helpful in attaining better performance for managing vehicular traffic. It also helps in decreasing congestion, pollution, accidents etc. Different prediction models have been described in literature. A traffic prediction model using fast forest quantile regression approach is described by Zahid et al. [9]. This approach makes use of part of single road data and is unable to apply the same for larger networks. A Conv-LSTM model [10] is proposed by Liu et al. for taking out spatial and temporal features from the traffic flow data combined with periodic features.

He et al. [11] described a vehicular flow prediction approach based on spatio-Temporal convolutional neural network which is based on encoder-decoder design. This approach extracts the spatial and temporal features from past data and predicts the traffic flow for longer span of time in advance.

The prediction models can be categorized into models based on statistics and neural network. Many statistical models like k-nearest neighbors [12], Auto-regressive integrated moving average [13], Support vector machines [14] etc. are being used for traffic flow forecasting. In models based on neural networks, Artificial neural network [15], Deep belief networks [16], De-noising stacked auto-encoders [17], Recurrent neural network [18], Deep learning models [19, 20] and Stack auto-encoder [21] are used for traffic prediction. But they may face scalability problem particularly in situations where real-time predictions are required for a large number of data.

In the proposed model, combined deep learning techniques are used to make IVN. This model first selects a set of intersection points in a region. The forecasting is performed on the selected points and is expanded for the complete city. For each point, the model extracts the spatial and temporal features of vehicle using Convolutional Neural Network (CNN) and Temporal Convolutional Network (TCN). Based on this fetched data, forecasting of traffic for different times can be computed. Using IVN, the forecasted data is transferred to other vehicular nodes thereby assisting them in deciding route information based on the existing traffic conditions and route scheduling. The sequence of paper has the hybrid model for vehicular forecasting at different span of time in Sect. 2. Section 3 explains the analysis of the simulation environment and the results. Finally, the last Sect. 4 concludes the work performed.

2 Proposed Methodology

The proposed model is used to forecast the traffic flow in a city. First the set of intersection points are selected among all the points available in the road topology of that city. For the selected points (locations), the traffic forecasting has been performed at a given time as per the model structure described in Fig. 2. The historical data is considered as a key parameter for providing traffic information at selected locations and at a particular time.

2.1 Intersection Point Selection

Let (x, y) indicates all the intersection nodes in the considered region. Among G nodes, $P(x, y)$ intersection points needs to be selected which will further undergo traffic forecasting. Selection of Pi points among G nodes is based on vehicular density at different points. For all (x, y), take 3 days historical data for a given time t. Three days include previous day, two days ago and one week ago (same day of the last week). Historical data of $(m-1)^{\text{th}}$, $(m-2)^{\text{th}}$ and $(m-7)^{\text{th}}$ day is considered to forecast the traffic on m^{th} day. For (x,y) points in a region as:

$$G(x, y) = \left[G_1(x, y)\ G_2(x, y) \cdots G_j(x, y)\right] \quad (1)$$

At G_j and at time t, traffic density d_j based on historical data is computed as the.

mean of previous days' data.

$$d_j = d(m-1) + d(m-2) + d(m-7) \tag{2}$$

Take H number of points for highest mean, M number of points for middle value and L number of points for lowest mean. While selecting,

$$H, M, L = P/3 \tag{3}$$

where

$$P = \begin{cases} G/10, & \text{if } G \geq 30 \\ G/5, & \text{if } 15 \leq G < 30 \\ G/2, & \text{if } 6 \leq G < 14 \\ 0, & \text{otherwise} \end{cases} \tag{4}$$

The total points selected For (x, y), are shown as:

$$P(x, y) = \left[P_1(x, y)\ P_2(x, y) \cdots P_i(x, y)\right] \tag{5}$$

It indicates the set of locations where traffic forecasting needs to be performed which is further divided into Data Input, Data forecasting and Data Communication.

The input data after pre-processing at different time scales are passed to CNN in parallel where spatial features are learnt and passed to TCN for extraction of temporal features shown in Fig. 2. Different modules of CNN are used for generalization. Both spatial and temporal correlations are extracted in whole process. After making the forecasting, the Data is communicated. The predicted data is passed on to all other neighbor nodes.

2.2 Data Input

The road traffic always shows a recurring behavior for consecutive workdays and same day of weeks. The input for the proposed model has been taken by considering the periodicity of data for the proposed model i.e. Three previous days' data have been taken as input for forecasting model. The first data item consists of traffic information for the previous day, the second data set is taken from the existing traffic two days before and the third data set includes traffic information of same day in the previous week. The model needs to make the prediction of the traffic patterns for a particular day for k consecutive segment locations at a particular time period.

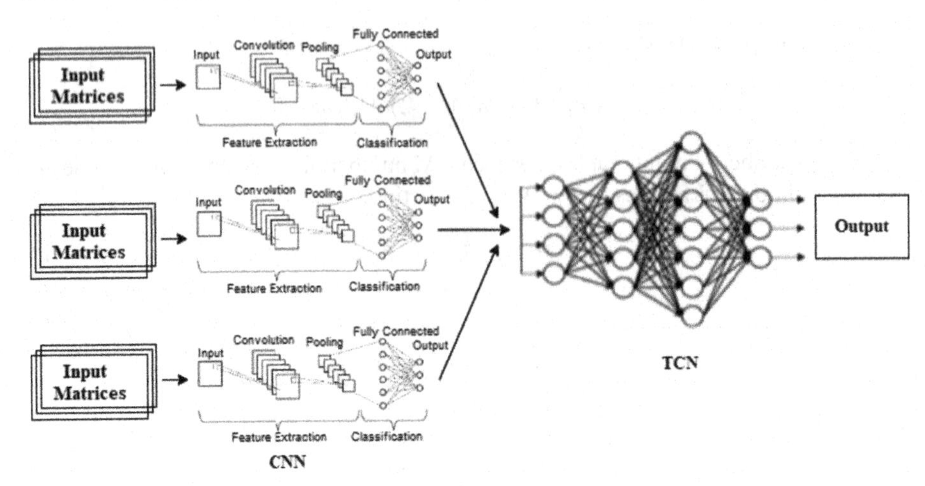

Fig. 2. Deep Learning Based Spatio-Temporal Traffic Forecasting

Historical data for k roads segments (s_1, s_2,..., s_k) at time (t_1, t_2,..., tn) is used in this model to make the forecast at time t where n is the total number of intervals taken in one day. The time vs road-segment matrix of type 2-dimensional is created as follows:

$$M = \begin{bmatrix} E_{s_1t_1} & E_{s_1t_2} & \cdots & E_{s_1t_{n-1}} & E_{s_1t_n} \\ E_{s_2t_1} & E_{s_2t_2} & \cdots & E_{s_2t_{n-1}} & E_{s_2t_n} \\ \vdots & \vdots & \ddots & \vdots & \vdots \\ E_{s_{k-1}t_1} & E_{s_{k-1}t_2} & \cdots & E_{s_{k-1}t_{n-1}} & E_{s_{k-1}t_n} \\ E_{s_kt_1} & E_{s_kt_2} & \cdots & E_{s_kt_{n-1}} & E_{s_kt_n} \end{bmatrix}$$

In the matrix shown, each row signifies a particular road segment and each column signifies a particular time period. The elements of matrix show the total vehicles present at a particular road segment at defined time.

Corresponding to the above representation, three matrices will be required as input depending on the day for which we want to make the prediction. Let say, if we want to predict the traffic for m^{th} day, then similar 2-D matrices will be required for $(m-1)^{\text{th}}$, $(m-2)^{\text{th}}$ and $(m-7)^{\text{th}}$ day. Let's take these matrices as M_1, M_2 and M_3 where M_1 is matrix representation for $(m-1)^{\text{th}}$ day, M_2 is matrix representation for $(m-2)^{\text{th}}$ day and M_3 is matrix representation for $(m-7)^{\text{th}}$ day.

Matrix M_1, M_2 and M_3 are shown in Fig. 3. These matrices are taken for a fixed time scale. Similarly at three different time scales, say, Ts_1, Ts_2 and Ts_3, all three matrices for $(m-1)^{\text{th}}$, $(m-2)^{\text{th}}$ and $(m-7)^{\text{th}}$ day are extracted from the historical data on cloud. Three days' matrices are taken as the input for the model for three different scales of time.

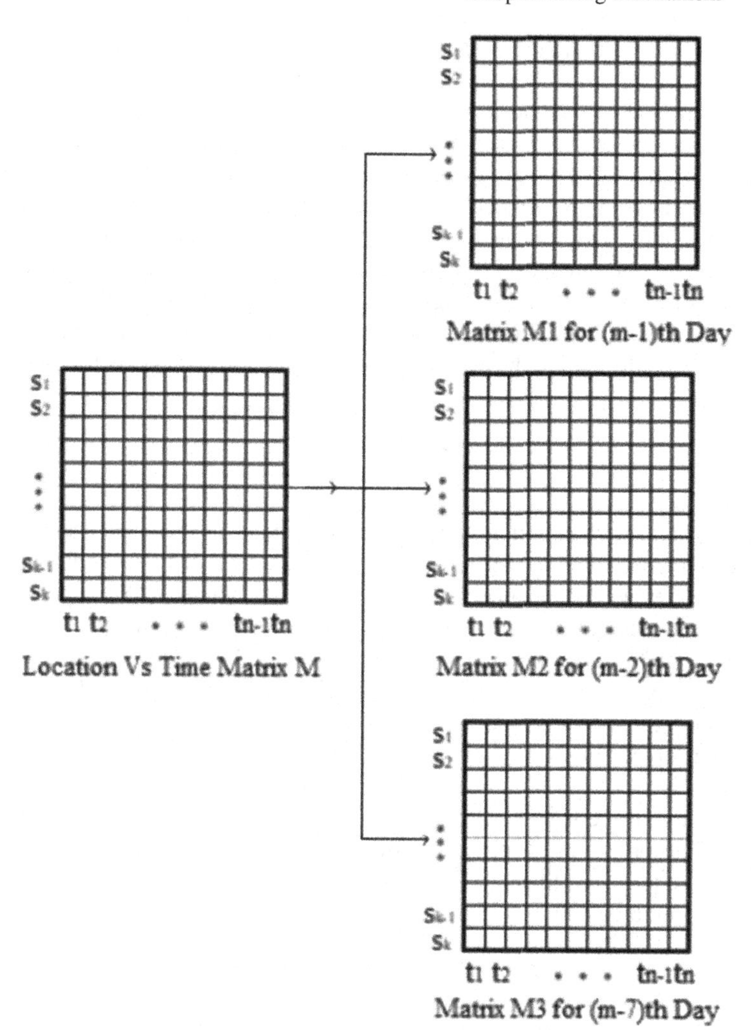

Fig. 3. Input Matrices for Convolutional Neural Network

2.3 Data Forecasting

After pre-processing, the input is converted to three different matrices taken for three different scales of time. The matrices are merged together for all three time scales Ts_1, Ts_2 and Ts_3 and final matrices are generated. The final generated matrices X_1, X_2 and X_3 at time scales Ts_1, Ts_2 and Ts_3 respectively are forwarded to *CNN* modules CNN_1, CNN_2 and CNN_3 in parallel as shown in Fig. 2. The spatial characteristics extraction is done at each *CNN* module and a tensor *TR* of that for time t is formulated as follows:

$$X_1(Ts_1) = M_1(Ts_1) + M_2(Ts_1) + M_3(Ts_1)$$
$$X_2(Ts_2) = M_1(Ts_2) + M_2(Ts_2) + M_3(Ts_2)$$
$$X_3(Ts_3) = M_1(Ts_3) + M_2(Ts_3) + M_3(Ts_3)$$

$$(TR)_t = X_1(Ts_1) + X_2(Ts_2) + X_3(Ts_3) \tag{6}$$

TCN is applied on the tensor for training and evaluation as a whole due to its easy applicability on time series based problems. Tensor *TR* formed in Eq. (6) includes the spatial characteristics which are converted to a long input sequence before sending to *TCN*. The filters on *TCN* are shared across layers with back propagation, so memory requirement is also less. Dialated convolutions are used in *TCN* where no pooling is required but field of perception is slowly increased using sequence of dialated convolutions. Output of each convolution gives rich information. Convolution operations of *TCN* help in extracting local as well as temporal information. Dialated convolutions are applied on long information dependent sequences which is defined as follows:

$$F(s) = \left(X * d_f\right)(s) = \sum_{i=0}^{m-1} f(i) X_{s-d_i} \tag{7}$$

where F is dialated convolution operation on element s of the sequence *TR*, d is the dialation factor, m is the filter size of f:{0,1,---,m-1}- > *TR* and s-d.i is direction of the past. For each input element, an output element will be obtained as $Y = \{y_1, y_2, ---, y_n\}$. The spatial correlations are extracted in *CNN* modules and the temporal correlations are extracted in *TCN*. So, the output obtained from *TCN* contains the combination of spatial and temporal features which are helpful for prediction. *CNN* and *TCN* modules are applied in sequence as written in algorithm 1 given in next section.

The final output of *TCN* is changed to a two-dimensional matrix and forwarded to next two layers of convolutional neural network. A generalization will be formed after extracting all the high level features. Procedures to reshape and convolution are also applied according to the algorithm shown in Sect. 3. The input undergoes n two-dimensional sequence of filters $\{F_1, F_2, ---, F_n\}$ via convolution operation (*) and feature maps are obtained. In general

$$P = f(Y * \{F_1, F_2, ---, F_n\}) \tag{8}$$

where Y is the input matrix and $\{F_1, F_2, ---, F_n\}$ are filter matrices respectively. After application of the activation function f, feature map P is obtained. Rectified linear unit (ReLU) used to serve the purpose of activation function which is defined as $f(x) = \max(0, x)$.

To avoid any loss of feature information, the input-output are made of the same size in a convolution operation. After application of the convolution operations, the high-level featured data in matrix form will be obtained. These are further flattened and merged with external factors of vector form and passed into a dense layer yields a linear vector for sections with k segments as shown below:

$$V = \left[V^1; V^2; \ldots; V^j; \ldots V^k\right] \tag{9}$$

The forecasting done for total number of vehicles at time t, is estimated as V. This model is proposed with an objective to minimize the mean squared error (MSE) between predicted value and ground truth. Optimization of the result is done using stochastic gradient decent method and the function for this can be written as:

$$\frac{1}{N}\sum_{n=1}^{N}(V' - V)^2 \tag{10}$$

where N signifies number of samples, V signifies the predicted value and V' signifies the test data for traffic i.e. ground truth.

2.4 Data Communication

Vehicular communication is very important aspects in traffic management. Its utilization may range from entertainment for passengers, to safety support, to local news delivery and advertisement, to Parking Space availability. All these applications use message dissemination as the key parameter. Combination of Road Side Unit (RSU) and the vehicles, considered as data carriers, creates a reliable data dissemination that is effective in reaching all nodes.

Autonomous vehicles are provided with information about the expected traffic at a specific time and location through vehicular network. The decision to take next steps to ease traffic congestion and eliminate gridlock on the roadways is being made in presence of this information. The vehicles will learn about congested roads and can reroute to avoid inconvenience. It offers the ability to adapt to different disruptions.

This way IVN helps in computation of the data using deep learning techniques and transferring the same to autonomous vehicles using vehicular network. Real-time traffic data will help in an estimated traffic scenario to properly plan the route, aid in traffic control, and promote public safety.

3 Experiments and Result Analysis

In this experiment using the proposed model, the data about total vehicles at a specific point and at a specific time on the roadways in Chandigarh, an Indian union territory, is collected using automatic traffic counter and classifier [22]. The description of dataset is given in Table 1.

For each day, the information gathered by loop detectors is transformed into spatio-temporal matrices. The data needs to be pre-processed before being made available for every hour. Data for a specific segment at a one-hour time interval is saved after the matrices are added column-by-column. Traffic density on a road section taken every one hour from 07:00 am–16:00 pm on May 8, 2019 is shown in Fig. 4. It is one day previous of the day of prediction. Figure 5 shows the density of traffic for same road segment for same time span for 07:00 am–16:00 pm on May 7, 2019 which shows the flow of traffic two days before the day of prediction. Figure 6 represents the flow of traffic of same road section for same time period on May 2, 2019. A total of 84,675 samples were produced after data pre-processing and were utilized to train the model. For the aim of the experiment, 20 road segments have been taken. 100 iterations of batch size 64 are utilized to train the model.

Table 1. Data Set Description

Dataset	**Chandigarh**
Time Span	1/3/2018–1/9/2019
Time Interval	10 min
Total road segments	20
Number of Samples	8,32,180
External Factors	**Code Values**
Weekend Holiday	0,1 0,1
Weather Wind power	0–9 0–3
Temperature	[−24, 34]

Three matrices are used as input, each with three different time scales, in accordance with the periodicity of the traffic data. Three matrices for $(m-1)^{\text{th}}$, $(m-2)^{\text{th}}$ and $(m-7)^{\text{th}}$ days at a time scale of one hour are combined to form the first input matrix. Similar to the first, the second input matrix is formed for time scale of 30 min and third input matrix is made taking time scale of 2 h. With a time-scale of 30 min, there are 48 time intervals in a day and 12 for every 2 h.

For primarily extracting the spatial characteristics, input is added to a CNN layer. For each CNN, a 3×3 kernel is employed, and a 32-level filter depth is used. After concatenating each CNN's output, a tensor is created. The sequence is created from the tensor and provided to the TCN module. For fully temporal feature extraction, 64 units of TCN are used, producing 64 sequences. To high level spatial properties ex-traction and generalization, this output is sent to CNN. It is once more reshaped into an 8×8 matrix. The first CNN produces a result output of 6×6 with deepness of 8 filters, and the second CNN produces a result output of 4×4 with deepness of 10 filters. The output of CNN is turned to a sequence of 160 by passing it via a flattening layer. The dense layer of a normal linear neurons, is used to obtain the vector form of result output, which is then combined with the vector of external factors to produce the final result output. All of the layers together contain 143,476 neurons. Figure 7 displays a layer-wise parameter representation. With a loss rate of 0.01, momentum of 0.9, Stochastic gradient decent optimizer is being used. Mean Absolute Error is the loss function used to enhance model learning.

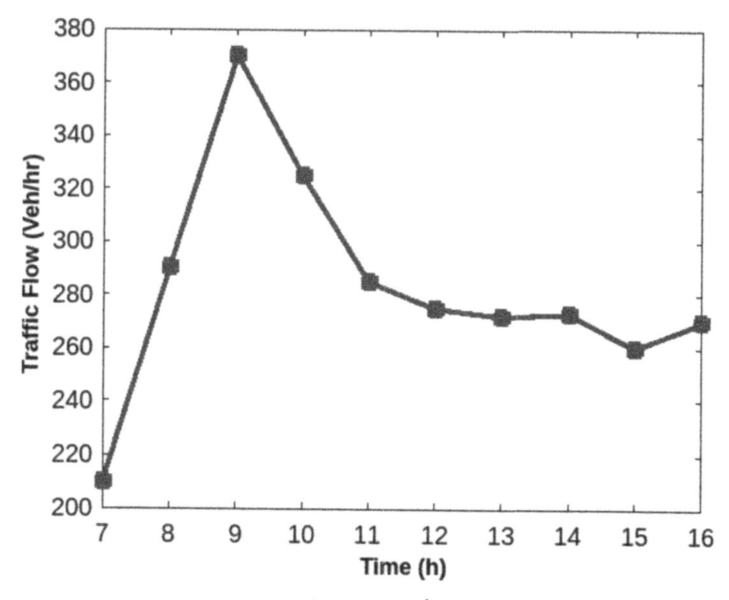

Fig. 4. Time vs vehicles on (m-1)th day i.e. May 8, 2019

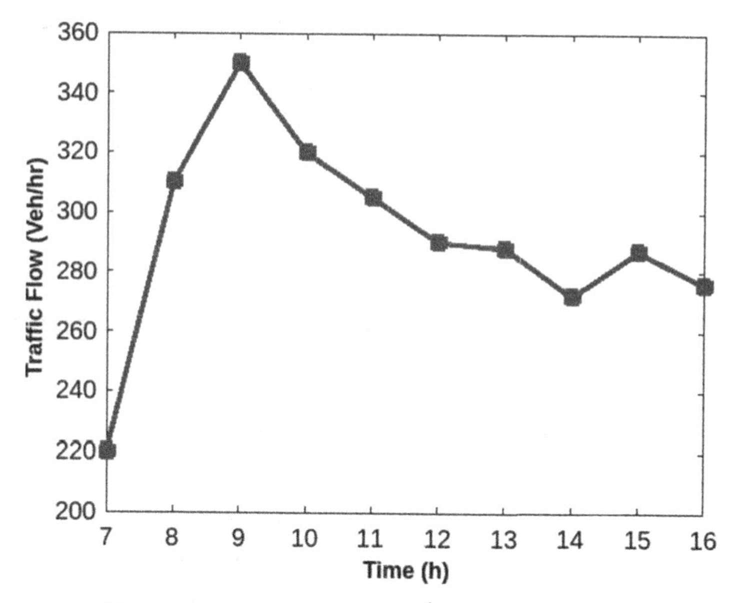

Fig. 5. Time vs vehicles on (m-2)th day i.e. May 7, 2019

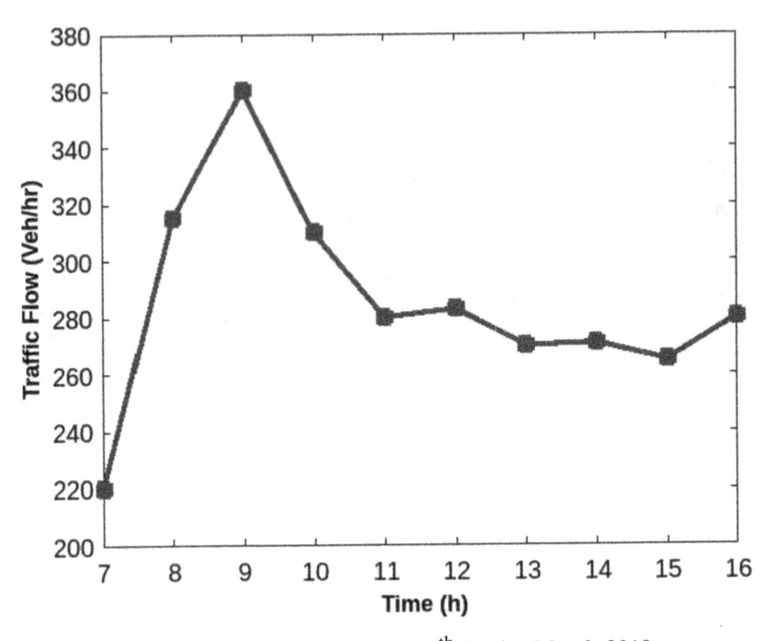

Fig. 6. Time vs vehicles on (m-7)th day i.e. May 2, 2019.

Layer (type)	Output Shape	Param #	Connected to
input_1 (InputLayer)	[(None, 32, 10, 1)]	0	
input_2 (InputLayer)	[(None, 56, 10, 1)]	0	
input_3 (InputLayer)	[(None, 20, 10, 1)]	0	
conv2d (Conv2D)	(None, 30, 8, 32)	320	input_1 [0][0]
conv2d_1 (Conv2D)	(None, 54, 8, 32)	320	input_2 [0][0]
conv2d_2 (Conv2D)	(None, 18, 8, 32)	320	input_3 [0][0]
concatenate (Concatenate)	(None, 102, 8, 32)	0	conv2d[0][0] conv2d_1[0][0] conv2d_2[0][0]
reshape (Reshape)	(None, 102, 256)	0	concatenate[0][0]
tcn (TCN)	(None, 64)	140096	reshape[0][0]
reshape_1 (Reshape)	(None, 8, 8, 1)	0	tcn[0][0]
conv2d_3 (Conv2D)	(None, 6, 6, 8)	80	reshape_1[0][0]
conv2d_4 (Conv2D)	(None, 4, 4, 10)	730	conv2d_3 [0][0]
flatten (flatten)	(None, 160)	0	conv2d_4 [0][0]
dense (Dense)	(None, 10)	1610	flatten[0][0]

Fig. 7. Model Demonstration

Algorithm 1 Vehicular Flow Forecasting

Inputs:

1. Two-dimensional matrix M_1 for (d-1)th day with location vs time t,
2. Two-dimensional matrix M_2 for (d-2)th day with location vs time t,
3. Two-dimensional matrix M_3 for (d-7)th day with location vs time t,
4. $M = 0$

Outputs: Predicted traffic Information at time t.

1: Begin

2: $X1{=}M1{+}M2{+}M3$ at time scale $T1$

3: $X2{=}M1{+}M2{+}M3$ at time scale $T2$

4: $X3{=}M1{+}M2{+}M3$ at time scale $T3$

5: for each matrix Xi do

6: Find the convolution matrix using

7: $Xi{=}conv2(Xi)$;

8: end for

9: for each matrix Mi do

10: $X = X + Xi$

11: end for //$tensor\ formation\ with\ concatenation$

12: $M = reshape(M)$; //$matrix\ reshaped\ into\ sequence$

13: $M = tcn(M)$;

14: $M = reshape(M)$; //$sequence\ reshaped\ into\ matix$

15: $M = conv(M)$; // $applied\ convolution\ for\ spatial\ feature\ extraction$

16: $M = flatten(M)$; //$merge\ the\ feature\ vectors\ with\ external\ factors\ vector$

17: $M = dense(M)$; // $final\ result\ as\ linear\ vector$

18: End

MSE for different models has been compared by Zang et al. [23] as given in Table 2. The methods are compared for Neihuan Elevated Highway. Out of them, the MSTFLN model yields the best results. Using our data set and the MSTFLN model, we obtained an MSE value of 189. Yet, the MSE achieved is 170 when the same data set is applied to our suggested model. Figure 8 displays a result comparison of both models with actual data.

Table 2. Result Comparison of Various Models

Model	MSE
MSTFLN	98.659
MSTFLN-FC	117.954
MSCNN-SC	104.904
MSCNN-FC	136.356
FCL-Net	140.877
DMVST-Net	147.552
ST-ResNet	129.954
CNN	142.145

While we obtained a lower MSE value for our model, it is clear that our methodology beats both conventional approaches and a number of cutting-edge models, including MSTFLN-FC, MSTFLN, ST-ResNet and FCL-Net.

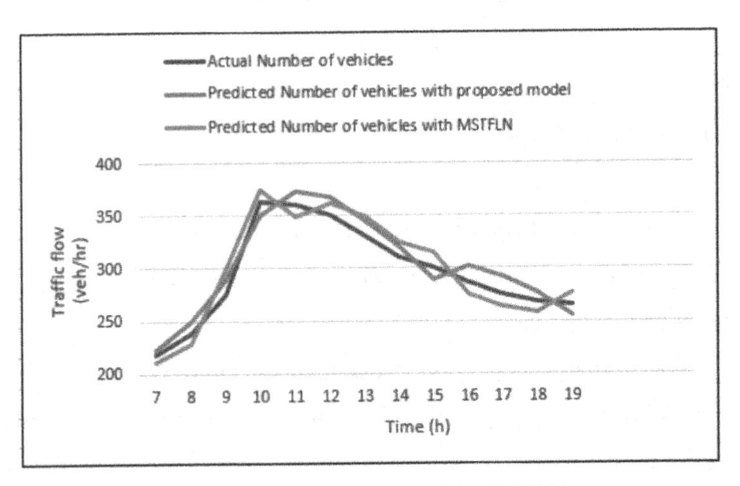

Fig. 8. Result comparison with MSTFLN

4 Conclusion

Traffic forecasting is a challenging task to perform because of several external factors (such as weekends, weather, holidays, etc.) as well as the dynamic nature of traffic. An efficient algorithm using TCN and CNN has been proposed in this paper for traffic prediction. The CNN and TCN are applied with regard to traffic in the proposed scheme using historical data. The vehicular network receives the predicted traffic information for dynamic route planning and decision making. The performance metric of the suggested model is assessed and determined to be satisfactory as compared to various other models.

References

1. Qureshi, K.N.: A survey on intelligent transportation systems. Middle-East J. Sci. Res. J. Sci. Res. **15**(5), 629–642 (2013)
2. Bali, R.S., Kumar, N.: An intelligent clustering algorithm for VANETs. In: 2014 international conference on connected vehicles and expo, ICCVE, pp. 974–979. IEEE (2014)
3. McDonald, A.B.: A mobility-based framework for adaptive clustering in wireless ad hoc networks. IEEE J. Sel. Areas Commun.Commun. **17**(8), 1466–1487 (1999)
4. Bali, R.S.: Learning automata-assisted predictive clustering approach for vehicular cyber-physical system. Comput. Electr. Eng.. Electr. Eng. **52**, 82–97 (2016)
5. Nzouonta, J.: VANET routing on city roads using real-time vehicular traffic information. IEEE Trans. Veh. Technol.Veh. Technol. **58**(7), 3609–3626 (2009)
6. Chen, R.: Broadcasting safety information in vehicular networks: issues and approaches. IEEE Network **24**(1), 20–25 (2010)
7. Chen, W.: Ad hoc peer-to-peer network architecture for vehicle safety communications. IEEE Commun. Mag.Commun. Mag. **43**(4), 100–107 (2005)
8. Garg, D., Kaur, A.: Truclu: trust based clustering mechanism in software defined vehicular networks. In: 2021 IEEE Global Communications Conference, GLOBECOM, pp. 1–6. IEEE, (2021)
9. Zahid, M.: Freeway short-term travel speed prediction based on data collection time-horizons: A fast forest quantile regression approach. Sustainability **12**(2), 646 (2020)
10. Liu, Y., Zheng, H.: Short-term traffic flow prediction with Conv-LSTM. In: 2017 9th International Conference on Wireless Communications and Signal Processing, WCSP, pp. 1–6. IEEE (2017)
11. He, Z., Chow, C.Y.: STCNN: a spatio-temporal convolutional neural network for long-term traffic prediction. In: 2019 20th IEEE International Conference on Mobile Data Management, MDM, pp. 226–233. IEEE (2019)
12. Xia, D.: A distributed spatial–temporal weighted model on MapReduce for short-term traffic flow forecasting. Neurocomputing **179**, 246–263 (2016)
13. Tran, Q.T.: A multiplicative seasonal ARIMA/GARCH model in EVN traffic prediction. Int. J. Commun. Netw. Syst. Sci.Commun. Netw. Syst. Sci. **8**(4), 43 (2015)
14. Wu, C.H.: Travel-time prediction with support vector regression. IEEE Trans. Intell. Transp. Syst.Intell. Transp. Syst. **5**(4), 276–281 (2004)
15. Zheng, W.: Short-term freeway traffic flow prediction: Bayesian combined neural network approach. J. Transp. Eng. **132**(2), 114–121 (2006)
16. Huang, W.: Deep architecture for traffic flow prediction: deep belief networks with multitask learning. IEEE Trans. Intell. Transp. Syst.Intell. Transp. Syst. **15**(5), 2191–2201 (2014)
17. Duan, Y., Lv, Y.: A deep learning based approach for traffic data imputation. In: 17th International IEEE conference on intelligent transportation systems, ITSC, pp. 912–917. IEEE, (2014)
18. Ma, X.: Large-scale transportation network congestion evolution prediction using deep learning theory. PLoS ONE **10**(3), e0119044 (2015)
19. Krizhevsky, A.: Imagenet classification with deep convolutional neural networks. Commun. ACM. ACM **60**(6), 84–90 (2017)
20. Bansal, N., Bali, R.S.: HTFM: hybrid traffic-flow forecasting model for intelligent vehicular ad hoc networks. In: ICC 2021-IEEE International Conference on Communications, pp. 1–6. IEEE, (2021)
21. Lv, Y.: Traffic flow prediction with big data: A deep learning approach. IEEE Trans. Intell. Transp. Syst.Intell. Transp. Syst. **16**(2), 865–873 (2014)

22. Sandhu, H.A.S., Bansal, A.: Study of traffic characteristics of major roads of Chandigarh using GIS–a case study
23. Zang, D.: Long-term traffic speed prediction based on multiscale spatio-temporal feature learning network. IEEE Trans. Intell. Transp. Syst.Intell. Transp. Syst. **20**(10), 3700–3709 (2018)

IoT Enabled Framework for Smart Home Automation Using Artificial Intelligence and Blockchain Technology

Abdul Mazid[1](✉), Sheeraz Kirmani[2], and Manaullah[1]

[1] Department of Electrical Engineering, Jamia Millia Islamia, New Delhi, India
abdulmazid1312@gmail.com

[2] Department of Electrical Engineering, Aligarh Muslim University, Aligarh, India

Abstract. Smart home automation is an emerging market trend that makes life simpler and easier to control. The term "home automation system" refers to the centralized management of lighting and electrical appliances, as well as the locking of doors and the activation of alarm systems, among other things, to improve the home's energy efficiency, safety, and comfort. To meet the smart home's aims of monitoring, protecting, and controlling, Internet of Things (IoT) technologies play a pivotal role by incorporating sensors and actuators that support various network operations and system automation. Consequently, people now have an assistant who manages their home and needs based on commands they provide. The limitations of existing home automation systems include a lack of IoT technology, Artificial Intelligence (AI) techniques, user-unfriendly interfaces, unauthorized access, short wireless connectivity ranges, cybersecurity issues, and excessive costs. As a result, this research proposes an affordable home automation system that integrates IoT, AI, and blockchain technology. In this regard, a framework is created to enable home monitoring and automated control of appliances via the Internet at any time and from any location. Our proposed work controls household appliances in an easy and effective manner over the internet, as well as identifies home intruders and provides support for autonomous home safety operations.

Keywords: Internet of Things · Smart Home · Blockchain Technology · Machine Learning · Artificial Intelligence · Automation

1 Introduction

The advancement of technology over time is making human life more and more dependent on it. There is an ever-increasing need for a luxurious lifestyle as technology progresses. The usage of smart technology in our homes makes it feasible for computers and people to interact with each other to regulate household appliances more efficiently. Due to the development of communication technology, home automation systems have garnered substantial interest [1]. A smart home, also known as an IoT application, enables users to remotely monitor and control their household appliances in real-time using the internet [2]. The goal in the creation of home automation systems was to reduce the

R. K. Challa et al. (Eds.): ICAIoT 2023, CCIS 1930, pp. 357–367, 2024.
https://doi.org/10.1007/978-3-031-48781-1_28

amount of effort expended by humans in the management of their household activities, such as operating washing machines, water heaters, air conditioners, and other similar devices [3]. The introduction of pervasive computing techniques into home automation allows for interaction with both the home's electrical appliances and the surroundings. The wireless network protocols, i.e., Z-Wave and ZigBee, are now the most advanced and widely used technologies for home automation. This technology is commonly known as domotics [4]. Devices like the Amazon Echo and the Nest Learning Thermostat are examples of the next generation of home automation technology since they use AI algorithms to control and report their functions. Robot buddies, also known as Rovio or Roomba, are considered part of the third generation of technology [5].

Recently, IoT technology has gained significant prominence. By interconnecting billions of IoT devices, the world is entering an IoT era. The primary objective of IoT is "to connect the disconnected." Over the course of a decade, it has drastically transformed human lifestyles and behaviors [6]. An IoT system consists of a collection of hardware devices, such as microprocessors, sensors, etc., responsible for transmitting data to and from the server and microcontroller [7]. Cloud computing is an additional component that will contribute to the success of IoT's performance. IoT has demonstrated its superiority because of cloud computing. Additionally, IoT enables users to complete typical computing operations at a low cost and with less memory. Also, user can manage data remotely. A notable example would be Google Photos, which allows users to transfer their photos from the internet back to their devices.

The power of these ideas has been magnified by advancements in AI techniques like Machine Learning (ML) and Deep Learning (DL). The goal of implementing these approaches in smart homes is to make them more efficient and responsive to the needs and routines of the people who live there. Instead of relying on explicit commands or predetermined scenarios, the homes will try to predict what the people will want or need [8, 9].

There are currently numerous home automation systems available, which employ various technologies such as Bluetooth, GSM, and Zigbee. However, when AI, Blockchain technology, and IoT technologies are utilized in the design and construction of a secure and smart home automation system, it functions very effectively. This work proposes a smart and secure home automation system that can control household appliances while also capturing images of potential intruders. In this model, all smart devices and users are authenticated by the blockchain network, making them secure by nature. To detect intruders, the MTCNN algorithm is employed, which achieves 92.6% accuracy. The proposed solution is both secure and cost-effective.

This work includes:

- Developed an AI-enabled remote-control application for electrical appliances.
- The proposed approach is a cloud-based system, that controls household appliances over the Web.
- Using blockchain technology, the proposed system guaranties the secure identification and verification of users and devices.
- Deployed a sensor-based approach for motion and intrusion detection.
- Implemented an MTCNN algorithm for identifying images within the house to prevent the security system from issuing false alarms.

2 Literature Review

Several studies have been conducted using various methods to achieve home automation. In paper [10], a smart home system is developed using Amazon Echo, cloud, and voice services. The author used Raspberry Pi3 to provide smart features to non-smart homes. Gladence et al. [11] implemented a client-server-based technique for smart home automation to enhance interaction between humans and machines, utilizing machine learning algorithms and natural language processing (NLP) techniques. Users can issue various commands to achieve tasks such as managing household devices, doors, and monitoring voice bed movement. The author also developed an NLP and AI module to help disabled people. Gambi proposed a home automation system architecture that utilizes LoRa and Message Queuing Telemetry Transport (MQTT) for effective communication [12]. The authors employed a low-cost, low-power, long-range communication device to investigate the development of IoT-based solutions utilizing such devices. Manu et al. introduced an LSTM deep learning method for conducting specified tasks based on the recognition of human behavior [13]. IoT technology and the LSTM algorithm are combined to forecast human activity toward user comfort, safety, and control over their homes. In paper [14], Reyas et al. developed an automatic smart home control platform that offers an overall customized automated control system and detects the pattern of resident behavior via ML and IoT, thereby enhancing comfort solutions for domestic systems. In [15], both machine learning and the HEMS-IOT system were used to help people take care of their homes' energy use in a smart way. Rule-ML and the Apache driver were used to improve the energy efficiency and safety of smart homes. To ensure the smart home's dependability and security, purpose experimentation was carried out as part of the validation process. An automation control system for SH that uses Bluetooth and a GSM module was proposed by Anandhavalli et al. [16]. This research is being conducted to assist disabled and elderly individuals in controlling household appliances from a remote location. People could operate their homes with wireless technologies such as Bluetooth and GSM. Indoor appliances were also controlled by Bluetooth, while GSM was used to control appliances outdoors. However, this method has limitations in both scenarios. Bluetooth suffers from significant limitations in both range and data transfer speed, while GSM is prohibitively expensive due to the cost of sending and receiving text messages. Bluetooth, GSM, Zigbee, and other emerging technologies are key components of the smart home automation system.

3 Proposed Methodology

In this work, a new intelligent-based secured home automation system is proposed that uses Artificial Intelligence and Blockchain Technology. The proposed methods enable users to verify and update the status of various connected smart devices in a smart home environment. To interpret natural language, we use natural language processing. To secure IoT device authentication and identification, blockchain technology is employed. Blockchain gives the IoT devices employed in our proposed home automation system a safe, reliable, and decentralized way to communicate with each other. The house has improved with increased security and occupant safety. A PIR motion sensor was utilized to monitor movement within the home and keep track of the graphs. Whenever any motion is detected, the user is notified with an alert through the Android mobile application. Before triggering the alarm, the MTCNN performs a background check to ensure the validity of the alert. The system employs intelligent decision-making capabilities for home surveillance and security based on the output of the MTCNN.

3.1 Application Interface

In this proposed model, Google Assistant and Dialogflow are employed as the AI platforms. Google Assistant enables voice commands for dialing, internet browsing, alarm setting, and opening certain mobile applications. In the Google Console action, a Google Assistant app is created and connected to a Dialogflow agent. The Dialogflow console imports all necessary data in the form of intent and agent files for agent training and working. Users can express their objectives through an intent, and the system will determine which actions correspond to those intentions. The intent file contains information about context, training phases, actions, and responses, while the agent file contains language and machine learning classification thresholds. Due to machine learning, the system can train itself to accurately respond to user input. To train the Dialogflow agent, we feed lots of user input into the agent file. The agent is trained using a set of user inputs. In our automation system, the user inputs are the actions that are performed by the Internet of Things devices. If we say 'turn on the television,' the system understands that the television must be turned on. If we tell it to 'turn off the bulb,' it knows that the bulb needs to be turned off. The Dialogflow agent has completed its training and is now prepared to receive user requests and give responses.

To integrate Google Assistant with mobile applications, an application programming interface (API) is developed. Incorporating the API into an Android application creates a more user-friendly environment and better understanding of the user's natural language input. After the device connects to the API, it is connected to the Google Cloud server. Once connected, Raspberry Pi is used to connect the Arduino to the server. The Google Cloud server allows access to the home environment from anywhere outside the house by receiving a request from an Android device and activating a pin on a Raspberry Pi based on that request.

3.2 Blockchain for Security

The primary goal of implementing blockchain technology in smart homes is to offer a safe and reliable platform for IoT device identification and authentication. Due to its advanced security and decentralization features, blockchain represents a major shift in the IT industry, particularly for IoT devices. The use of blockchain technology in home automation can reduce reliance on cloud servers. In the proposed method, blockchain is implemented in Python by defining the block's content. The hash block is computed using each block, and then SHA-256 is calculated using the hash block. A new block is formed after a connectivity request is granted, and whenever a new block is generated, we store it. To ensure that the generated hash of a block matches the hash calculated for that block, the blockchain validation process is carried out after the block is stored. If the validation is successful, blockchain allows access to the new block.

Through parent blocks, blocks on the blockchain are connected to one another. Each block header contains metadata, which includes version number, Merkle tree root hash, parent block hash, nBits, and a random number. The algorithm for user and device authentication is as follows:

```
Algorithm RegAuth:
Function Reg(device_id)
            Hashed_id = hash(device_id)
            stoneMapping(user_address, hash_id)
function Auth(device_id)
            hashed_id = hash(device_id)
            if (check mapping(user_id, hashed_id) = true)
                        dev_id = request ChianLink(device)
                        hashed_dev = hash(dev_id)
                        if (hashed_dev == hashed_id)
                                    authenticate()
                        else
                                    wrong_device id
            else
                        user not authorized
```

3.3 Home Environment

The primary elements of a smart environment include sensors, smart cameras, Wi-Fi modules, smart-gadgets (bulb, television, fans, AC), and smart smartphones. A level of smartness has been integrated into the home appliances, allowing them to respond intelligently to commands. In the home, sensors are employed to collect data, monitor, and detect. A smart camera can be used to detect intruders at the home; if an intruder is discovered, the camera will take a picture and communicate it to the user. The microcontroller is used to establish Wi-Fi connections between the various smart home devices and sensors. The user can remotely operate household appliances both inside and outside the house using a mobile application connected to a cloud server.

3.4 Implementation

The integration of IoT, AI, and blockchain technology results in a more user-friendly system. Figure 1 illustrates the integration of several IoT devices into the proposed home automation system. The design of the system encompasses the user, appliances, sensors, and a cloud platform. The GPIO pin of the Raspberry Pi is connected to the sensors using jumper wires, as shown in the configuration. Additionally, the Relay board module is linked to the Raspberry Pi, while household IoT devices are connected to the relay board module. The Android application functions as the client, and the Google Cloud IoT platform serves as the server.

In this proposed framework, the IoT devices in the home are accessed by giving commands to the Google Assistant AI. Once the device is connected to the cloud server,

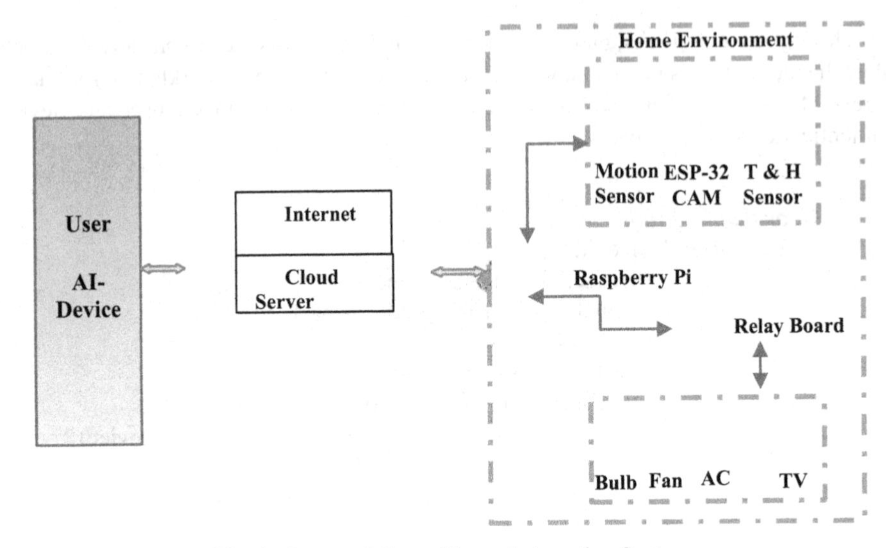

Fig. 1. Proposed Smart Home Automation System

the API sends the action to the server. The request is then received by the Raspberry Pi from the cloud server. Using the Google cloud server, we can connect to our home environment from anywhere in the world. It receives API requests and activates the Raspberry Pi pin in response to those requests. Sharing of data between an Android app and a cloud server is accomplished using JSON. Multiple hashing methods are used to protect APIs, and information about each request is stored in a database on the cloud server. Additionally, the user can review all their previous actions.

3.5 Hardware and Software Components

The proposed home automation system is designed to provide an appropriate level of intelligent home control, monitoring, and security by utilizing IoT hardware and software. The prototype of the smart home automation system was developed using the following hardware components: Raspberry Pi, Arduino, a Wi-Fi board, a 5-V 4 channel relay module, and ESP-32 CAM. A detailed description of each component is provided below.

Raspberry Pi. Raspberry Pi is an open-source hardware platform that integrates an IDE (Integrated Development Environment) and a programming language. It is an interesting innovation that is substantially more affordable than desktop, mobile or any smart device. The vast majority of Raspberry Pi's software and projects are available for free, and online user communities are enthusiastic about new software applications to ensure that they are maintained up to date. Python is a popular language for programming the Raspberry Pi as it is relatively easy to learn and requires less code than other languages, making it less complex. Raspberry Pi is not only low-cost, but it also has a low impact on the environment and does not need any kind of cooling system. Due to the various improvements introduced with respect to older versions, Raspberry Pi 4 was used for this work.

5 V Relay Module. The contacts of a switch are opened or closed by a relay, an electromechanical device. It is a type of automated switch that is often used to handle a large voltage with a low-voltage signal in automatic control circuits. 0 to 5 V is the range of the relay signal's input voltage.

ESP32-CAM. Built on the ESP32 architecture, the ESP32-CAM is a camera module that operates on low power and is small in size. It features an integrated OV2640 camera and a memory card reader, making it suitable for advanced IoT applications such as wireless video surveillance, Wi-Fi image uploading, and QR code identification. The ESP32-CAM is equipped with Wi-Fi and Bluetooth and is compatible with TF and microSD cards. The module can be programmed using Thonny IDE. In addition, the GPIO pins on the ESP32-CAM module allow it to be connected to external hardware. Since the ESP32-CAM does not have a USB port, a programmer such as FTDI is necessary to upload code onto the microcontroller.

PIR Motion Sensor Detector (HC-SR501). The PIR Motion Sensor Detector with Module HC-SR501 is capable of detecting motion and is primarily used to detect human movements within its range. The terms "PIR," "Pyroelectric," "Passive Infrared," and "IR Motion" are frequently used to describe these sensors. PIR motion sensors are commonly used in home automation applications, allowing us to program household equipment to automatically react to the presence of a person. When the PIR motion sensor generates an interrupt, any appliances connected to an ESP32 board will automatically respond according to the provided instructions. In this project, the PIR motion sensor will be connected to an ESP32 board and a 5mm LED. The LED will be controlled using the ESP32's interrupts and timers. When motion is detected, an interrupt will be triggered, a timer will be activated, and the LED will turn on for a predetermined amount of time. This will occur every time motion is detected.

The remaining hardware components consist of jumper wires, LED, capacitor, resistor, push button, a step-down transformer, sensor (humidity and temperature), voltage regulator LD-1117V33, buzzer module (Piezo speaker), and NPN bipolar transistor.

Software Components. Thonny IDE is an open-source integrated development environment (IDE) for Python programming. It can be used to write and upload Python applications to various development boards such as Raspberry Pi Pico, ESP32, and ESP8266, among others. Dialogflow is used to create intents under the agent. The system responds to the user's requests based on the training phrases that have been programmed into it. The inputs from the users are properly mapped in order to provide a response to the users. Thonny IDE is a technology that is based on interactions such as natural language processing (NLP) and natural language understanding (NLU).

4 Experimental Results and Discussion

Figure 2 depicts the proposed system's workflow. The Google Cloud server receives each user request, and depending on the request, APIs are called.

The Thonny IDE was used to program the microcontroller board and mobile app to interact. With the help of the Wi-Fi module, a wireless connection was established. To ensure that the current flowed properly, the home appliances were interfaced with the relay module. The system sends a signal to the mobile app via the Wi-Fi module.

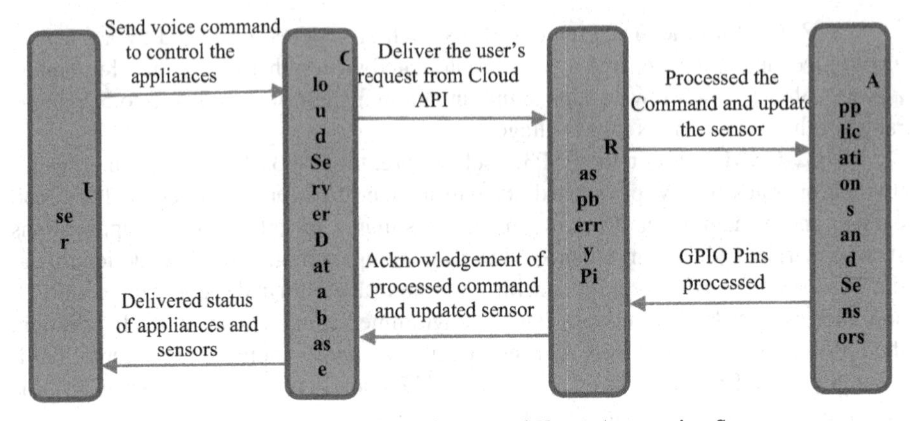

Fig. 2. Functional Behaviour of Proposed Home Automation System

When a command is issued via the mobile app, the relevant household appliance displays an ON/OFF indicator on the LEDs and relay board. The screenshots presented in Fig. 3(a), 3(b), and 3(c) demonstrate the expected results when the user interacts with the AI client device using voice commands. The device relays the action to the server application, which subsequently sends commands to the microcontroller for controlling the household appliances. The API incorporates machine learning technology to learn and improve its performance based on the training terms provided by the user earlier.

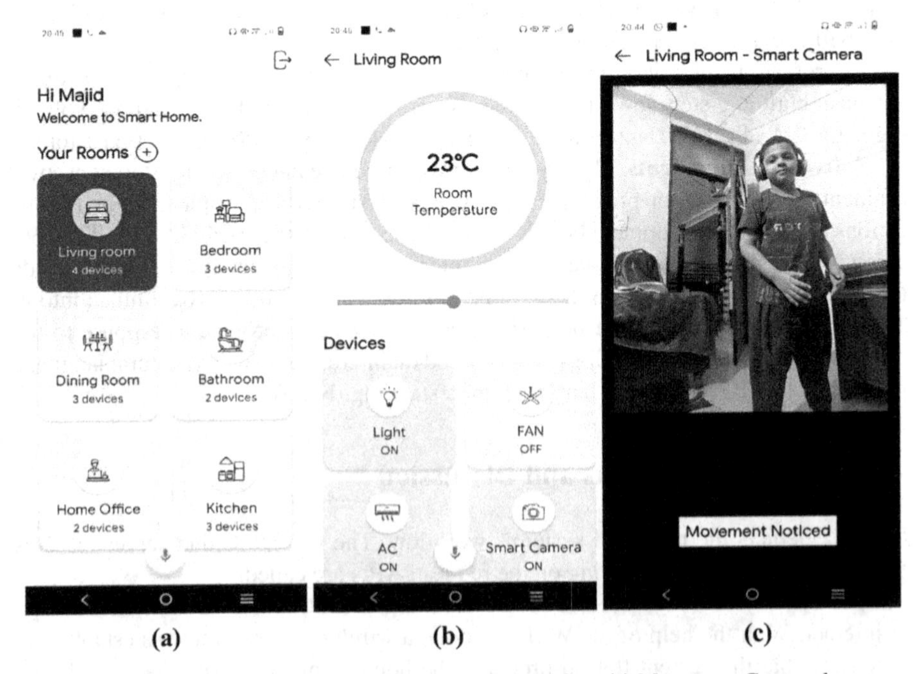

Fig. 3. (**a**) User Main Screen. (**b**) Living Room View. (**c**) Movement Captured

The mobile application's UI is presented in Figs. 3(a), 3(b), and 3(c). An example of the main menu display of the sitemap is shown in Fig. 3(a). Figure 3(b) shows the various devices deployed in the living room, and what happens when the user selects "Living Room"; the system retrieves and displays all the connected appliances in that room, along with their current states. Whenever the user makes a voice command through the smartphone app, the relevant home appliance displays an indicator on the LEDs and a relay board to indicate whether the device is on or off. When motion is detected by the sensor, the captured image of a person inside the house is displayed in Fig. 3(c).

In addition to controlling electrical home appliances such as light bulbs, fans, air conditioners, and televisions, the proposed system also monitors motion, takes pictures after detecting motion, and transmits those pictures to the user through a mobile app running on a smartphone at any time and from any location. The dataset used to train the deep learning algorithm for image classification is also presented.

A dataset titled "Real and Fake Face Detection" was downloaded from Kaggle [17] because the home environment that we are working on does not involve numerous images. This image dataset contains 1081 actual faces and 960 fake faces with varying degrees of difficulty in recognizing them. The photos are a composite of various faces, divided by the eyes, mouth, nose, or the entire face. The dataset was split into two different classes: intruders and home occupants. The images of real faces are stored in the occupants' class, while images of fake faces are kept in the intruders' class. To train and test the model, we split the dataset. The training set contains 1627 images, and the testing set contains 404 images.

In this dataset, we employed the MTCNN algorithm to evaluate its performance. MTCNN, also known as Multi-Task Cascaded Convolutional Neural Networks, is a type of neural network that can recognize faces and facial landmarks in images [18]. MTCNN is highly regarded as a popular and precise face detection method currently available. It possesses a high degree of accuracy and is extremely trustworthy. It can accurately identify faces despite variations in size, illumination, and dramatic rotations. Despite the fact that RGB images are used as the source of input, CNNs utilize color information [19]. The performance metrics for the proposed binary classification models include accuracy, recall, precision, and F1 score. The MTCNN model achieved an accuracy of 92.6%. We implemented the model using Python 3.9 programming language. The Keras API was running on top of the Tensorflow library on the backend. Figure 4 depicts the model's performance across multiple metrics.

In addition to the ease of use provided by smart home automation systems, another advantage of this technology is the ability to remotely control your home from anywhere. Therefore, the use of monitoring and control systems allows for remote supervision and regulation of the residence from any location. The surrounding environment of homes should also be monitored for safety. Existing systems that lack smart decision-making and analytical skills are no longer adequate. Moreover, the current technology does not provide secure communication among various IoT devices. The benefit of our proposed approach is that it enhances home security. To enable secure authentication and identification of IoT devices, we deployed blockchain technology. The MTCNN algorithm can eliminate false alerts and unwanted notifications. Although the system

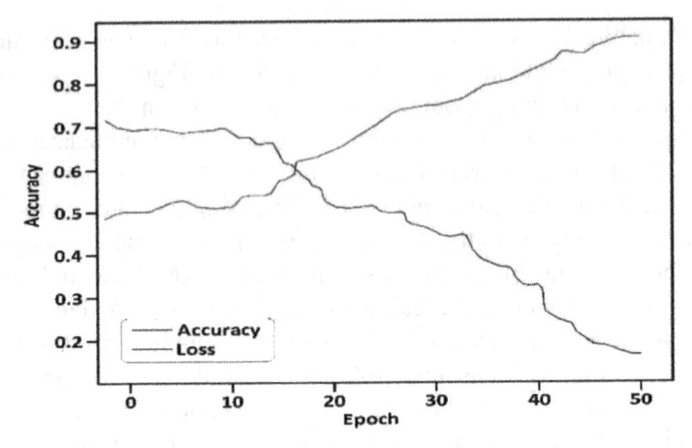

Fig. 4. Performance of MTCNN model

retains all images for future reference, the user is notified only when the system identifies an intruder in an image.

5 Conclusion

In conclusion, this study addresses pertinent areas of concern in the domains of enhanced living convenience, lifestyle, comfort, and home safety with the help of the introduction of a secure smart home automation system that harnesses the capabilities of emerging technologies such as the IoT, Blockchain, Cloud Server, and AI. The proposed system enables local and remote home control via an AI-integrated smartphone application, facilitating seamless communication between the user and the assistant, which comprehends user requests. In addition, the system integrates a motion sensor and an IoT camera to capture images when motion is detected. The use of the MTCNN machine learning concept enables the system to distinguish between images of typical home occupants and intruders, thus mitigating the risk of false alarms. To enhance security, our system adopts blockchain technology for user and device authentication and identification. Our proposed framework provides an affordable, low-cost, flexible, and dependable smart home automation system accessible to middle-class residents.

References

1. Daneshvar, M., Pesaran, M., Mohammadi, B.: Transactive energy in future smart homes. Energy Internet, 153–179 (2019)
2. Ning, H., Shi, F., Zhu, T., Li, Q., Chen, L.: A novel ontology consistent with acknowledged standards in smart homes. Comput. Netw. **148**, 101–107 (2019)
3. Batalla, J.M., Gonciarz, F.: Deployment of smart home management system at the edge: mechanisms and protocols. Neural Comput. Applic. **31**, 1301–1315 (2019)
4. Babun, L., Aksu, H., Ryan, L., Akkaya, K., Bentley, E.S., Uluagac, A.S.: Z-IoT: passive device-class fingerprinting of ZigBee and Z-wave IoT devices. In: ICC 2020 - 2020 IEEE International Conference on Communications (ICC), Dublin, Ireland, pp. 1–7 (2020)

5. Edu, J.S., Ferrer-Aran, X., Such, J., Suarez-Tangil, G.: SkillVet: automated traceability analysis of Amazon Alexa skills. IEEE Trans. Dependable Secure Comput. **20**(1), 161–175 (2023)
6. Mazon-Olivo, B., Pan, A.: Internet of things: state-of-the-art, computing paradigms and reference architectures. IEEE Lat. Am. Trans. **20**(1), 49–63 (2022)
7. Kirmani, S., Mazid, A., Khan, I.A., Abid, M.: A survey on IoT-enabled smart grids: technologies, architectures, applications, and challenges. Sustainability **15**(1), 717 (2023)
8. Sufyan, F., Banerjee, A.: Computation offloading for smart devices in fog-cloud queuing system. IETE J. Res., 1–13 (2021)
9. Kabir, S., Gope, P., Mohanty, S.P.: A security-enabled safety assurance framework for IoT-based smart homes. IEEE Trans. Ind. Appl. **59**(1), 6–14 (2023)
10. Achal, S.K., Nikhil, S.P., Atreyas., Srinivas, S., Kehav, V., Naveen, K.: Voice enabled home automation using Amazon Echo. Int. Res. J. Eng. Technol. **04**(08),682–684 (2017)
11. Gladence, M.V., Anu, M.V., Rathna, R., Brumancia, E.: Recommender system for home automation using IoT and artificial intelligence. J. Ambient Intell. Humanized Comput., 1–9 (2020)
12. Gambi, E.: A home automation architecture based on LoRa technology and message queue telemetry transfer protocol. Int. J. Distrib. Sens. Netw. **14**(10), 1550147718806837 (2017)
13. Manu, R.D., Kumar, S., Snehashish, S., Rekha, K.S.: Smart home automation using IoT and deep learning. Int. Res. J. Eng. Technol. **6**(4), 1–49 (2019)
14. Reyes-Campos, J., Alor-Hernández, G., Machorro-Cano, I., Olmedo-Aguirre, J.O., Sánchez-Cervantes, J.L., Rodríguez-Mazahua, L.: Discovery of resident behavior patterns using machine learning techniques and IoT paradigm. Mathematics **9**(3), 219 (2021)
15. Machorro-Cano, I., Alor-Hernández, G., Paredes-Valverde, M.A., Rodríguez-Mazahua, L., Sánchez-Cervantes, J.L., Olmedo-Aguirre, J.O.: HEMS-IoT: a big data and machine learning-based smart home system for energy saving. Energies **13**(5), 1097 (2020)
16. Anandhavalli, D., Mubina, N.S., Bharathi, P.: Smart home automation control using Bluetooth and GSM. Int. J. Informative Futuristic Res. **2**(8) (2015)
17. Real and Fake Face Image Detection dataset. https://www.kaggle.com/datasets/ciplab/real-and-fake-face-detection. Accessed Dec 2022
18. Ahmed, M., Masood, S., Ahmad, M., Abd El-Latif, A.A.: Intelligent driver drowsiness detection for traffic safety based on multi CNN deep model and facial subsampling. IEEE Trans. Intell. Transp. Syst. **23**(10), 19743–19752 (2022)
19. Caliwag, E.M.F., Caliwag, A., Baek, B.K., Jo, Y., Chung, H., Lim, W.: Distance estimation in thermal cameras using multi-task cascaded convolutional neural network. IEEE Sens. J. **21**(17), 18519–18525 (2021)

Big Data Techniques Utilization in Intelligent Transportation System Environment

Sumit Sharma[1(✉)], Gagangeet Singh Aujla[2], and Rasmeet Singh Bali[1]

[1] Chandigarh University, Mohali, Punjab, India
cu.sumitsharma@gmail.com
[2] Durham University, Durham, Great Britain

Abstract. Intelligent transportation systems (ITS) research is increasingly focusing on big data, as seen in numerous global projects. A lot of data will be generated by intelligent transportation systems. The creation and usage of intelligent transportation systems (ITS) will be significantly impacted by big data, which will lead to their safer use and increase of efficiency. The utilization of inferences drawn using analysis performed on big data in ITS is flourishing. Initial part of the paper includes discussion around parameters of Big Data in the context of ITS. Then a system model for Big Data analytics aimed at ITS along with different entities of different planes is discussed. The different phases of Big Data analytic process are discussed in detail along with existing work done in the domain. The final section of this paper discusses some unresolved issues with analysis of big data in ITS.

Keywords: Vehicles · Big Data · Traffic · Analytics · Machine Learning · Deep Learning · Intelligent transportation systems (ITS)

1 Introduction

Big Data has been an area of interest in all the domains around the world from last few years. It displays extensive and intricate data sets gathered from various sources. Big Data approaches are employed in many popular data processing techniques, like the field of exponential growth (Artificial Intelligence - Machine Learning – Deep Learning), and social networks [1]. Big Data analytics is widely used in many industries, with great success. For instance, some businesses use Big Data to more accurately understand consumer behavior in order to optimize product prices, enhance Boost operational effectiveness and cut labor costs [2]. In the social network space [2], the viewing trends of the users are being used to recommend the products and businesses which the user is interested in. The medical field as well also deploys its applications like doctors are able to assess the patient's condition in a detailed manner and able to provide better treatment plans and recommendations.

Grid operators in the field of smart grids can identify surge power load segments and by analyzing smart grid data, and they can even identify which lines are in a failed state. The electrical grid can be upgraded as well as renovated and maintained as a result of these data analysis findings [3].

R. K. Challa et al. (Eds.): ICAIoT 2023, CCIS 1930, pp. 368–383, 2024.
https://doi.org/10.1007/978-3-031-48781-1_29

ITS is being pondered and researched upon since the early 1970s. The transportation systems have huge scope of optimizations and thus is the future of research. The technologies have advanced in recent decades to include start of art sensing techniques, highly resilient, robust and ultra-low latency data transmission techniques, and multi aspect inference mechanisms. All these advancements are finding its way in the field of ITS with the aim of providing improved quality of service to passengers and drivers utilizing transportation networks. [4].

A number of sources, including sdCard, social platforms, RFID tags, GPS, sensors, and video cameras, can be used to collect data for ITS. Better service for ITS can be provided by accurate and efficient data analytics of data that appears to be disorganized [5]. The amount of data generated by ITS is increasing as it develops, going from trillion bytes to petabytes. Traditional data processing methods are not effective enough to handle the volume of data needed for data analytics. This is as a result of their failure to anticipate the exponential rise in data volume and complexity. ITS now has a new technical approach thanks to big data analytics. The following are some ways that big data analytics can help ITS.

i. Big Data analytics can handle the enormous amounts of varied and complex data produced by ITS. Data management, data analysis, and storage have all been solved by big data analytics. Massive amounts of data can be processed using big data platforms like Apache Hadoop and Spark, which are popular in both academia and business [6, 7].
ii. ITS operations can become more successful thanks to big data analytics. Many ITS subsystems must manage a lot of data in order to provide information or make decisions about how to manage traffic. The traffic management department is able to predict traffic flow in real time through quick data collection and analysis of current and historical massive traffic data. Transportation in general supports management in taking better informed decisions using Big Data analytics as they are able to understand how passengers use the transportation network, which can then be applied to better plan the provision of public transportation services. These applications also help end users by getting to their destination quickly and using the best route possible.
iii. The degree of ITS safety may be increased by big data analytics. The transportation happening 24 x 7 is a source of humungous data which can be obtained using state of art of sensor and detection technologies. Big Data analytics allows us to accurately forecast the likelihood of traffic accidents. When accidents happen or emergency rescue is needed, the Big Data analytics-based system's realtime reaction capabilities may considerably improve the emergency rescue capability. Big Data analytics may present fresh ways to spot issues with assets like

ageing ballast and pavement deterioration. It can assist in making maintenance decisions at the proper time and keep the infrastructure or vehicle from failing.

Despite the great potential of applications of Big Data analytics in ITS, there are still many pressing research problems and formidable obstacles that must be overcome. To the best of our knowledge, the work done till date in the field of structured analysis of utilization of big data analytics in the field of ITS still has certain elements to be touched upon which are explored in this paper. We first present a system model for Big Data Analytics which is aiming at ITS. Following which the acquisition of data from various sources in the ITS environment is discussed. It is discussed how to conduct big data analytics in ITS. We also provide an overview of ITS's platforms and data analytics techniques. Certain case studies are included to highlight the utilization of Big Data Analytics in the ITS domain.

The rest of the document is structured as follows. In Sect. 2, System model for the Big Data Analytics aimed at ITS is covered. The details about the analysis of Big Data are discussed in Sect. 3 where each subsection represent different phases of the ana lytics process. In Sect. 4, a few focus areas of research in the Big Data and ITS domain are mentioned. Finally, Sect. 5 serves as the paper's conclusion.

2 System Model for Big Data Analytics aimed at ITS

The data produced in an ITS environment has highly equivalent properties of that of Big Data. Intelligent transportation systems integrate cutting-edge technologies into the transportation systems, such as electronic sensor technologies, data transmission technologies, and intelligent control technologies. Better services for drivers and passengers in transportation systems are the goal of ITS. Advanced transportation management systems, advanced traveler information systems, advanced vehicle control systems, business vehicle management, advanced public transportation systems, and advanced urban transportation systems are the six fundamental parts of ITS, according to literature reviews through [4], most of the components of the above-mentioned fundamental parts are associated with the road transportation. Therefore, in this survey paper, we concentrate on ITS in-road transportation.

Intelligent transportation systems (ITS) collect data that is becoming more complex and has Big Data features. The three Vs—volume, variety, and velocity—have been proposed by major corporations like Gartner, IBM, and Microsoft, among others [8, 9].

Volume describes the amounts of data that are being produced by numerous sources and are still growing. The volume of data has increased tremendously in transportation due to the growth in traffic and detectors. Additionally, the upcoming vehicles and vehicle services install tracking responders for providing better services and thus more data is produced by people, things, and vehicles.

All types of data generated by plethora of detectors, user applications, CAN bus, infrastructure and sensors embedded in vehicles, and moreover the data generated from social media is an important focus area even social media are the main focus of variety.

Data on transportation now come in a much wider range. Modern vehicles, for instance, are able to report internal system telemetry and passenger and crew information in real time.

Due to better communications technologies, faster monitoring and processing, and increased processing power, the third v of data i.e., velocity in transportation has increased. For instance, smart card or tag-based ticketing and tolling transactions are being captured instantaneously with the help of technology are now immediately reported, whereas the manual processes being followed earlier would require the allotment of manpower to infer useful information from the manual registers/records.

The proposed system model for the Big Data Analytics comprises of three planes namely, physical plane, communication plane and application plane. The pictorial representation of the system model is presented in Fig. 1 and elaborated as follows:

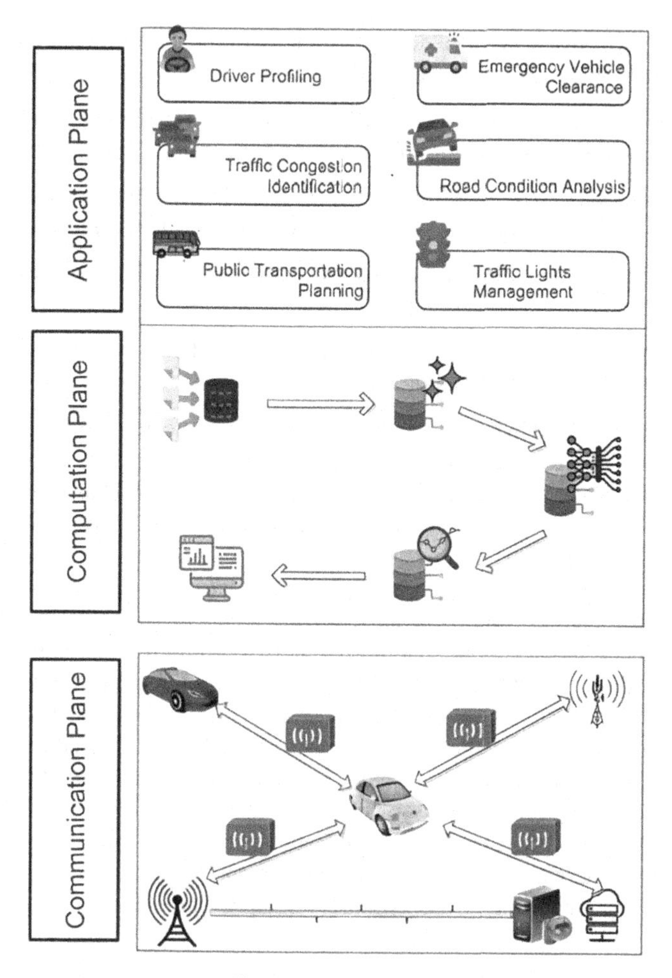

Fig. 1. System Model

2.1 Communication Plane

Various entities involved in the Intelligent Transportation System model in a smart city are listed below.

Vehicles. Vehicles are machines which are used by humans for transportation from one location to another location. Vehicles could be classified on multiple categories like public/private, n-wheeler, freight carriage, human carriage etc.

Drivers. The vehicles are required to be driven by humans which are referred to as Drivers. Drivers are to go through a testing process before being handed over by a driving license by the government.

Road Side Unit. The road side units are infrastructure entities which are being installed on the edge of road with the data collection, transmission capabilities and/or computa tional capabilities.

Edge Server. An edge server is a device installed on either road side unit or connected to road side unit at one end and connected to Cloud Server on another end.

Cloud Server. Cloud Server is a device which acts as centralized source of data storage, information processing and global inferences generation for a system.

Sensors. Sensors are small hardware which are capable of reading/sensing values of an attribute using electric, magnetic, pressure based and acoustic techniques.

Cellular Towers. Cellular towers are giant structures fitted with hardware for providing cellular services to a defined geographical area.

Satellites. Satellites are machines orbiting the earth and supporting large diaspora of services like GPS, Network, Satellite telephony etc.

The entities involved in the physical plane are heterogeneous in nature, some entities like Road Side Unit, Edge Server, Cloud Server, Cellular Towers have a fixed geographical positioning whereas some entities like vehicles behave mobile as well as stationary from time to time. The sensor entities are deployed on vehicles as well as some are deployed on the road infrastructure.

The proposed ecosystem uses a hybrid network for providing connectivity among the entities. The entities communicate with each other as per the applications using either Wireless or Wired communication channels. The entities are required to register themselves with the ecosystem before being able to use the Smart City services. The communication model being followed in the ecosystem is multi-level and bi-directional i.e., the information generated at the end devices(vehicles) is required to be processed at End Device/ Edge server and/or Cloud Server as well as information from both Edge Server and Cloud Server are also transmitted to end device from time to time. The transmission between the Edge Server and End Device; Cloud Server and End Device is conducted via the wireless channels as the end devices are mobile in nature whereas all the communication between the Edge Server and Cloud Server is conducted via the backbone network.

2.2 Computation Plane

The computation plane focuses on the various phases related to the data collection and preprocessing steps. The phases of the computation plane are explained below:

Data Collection. The data generation is rooted to various sources in the Intelligent Transportation Systems like Cabs, Inductive Loop Detector Sensors, Speed Cameras, Traffic Light Cameras, Road Side Units, CAN bus data, In-vehicle camera data, Smartphone sensors data, pressure sensitive steering wheels data etc. The type of data being generated from multiple type of hardware is heterogeneous in nature as well as the frequency of the data generation from multiple type of hardware are also varying in nature. The first step is to create a policy for collection of data from various devices.

Data Cleaning. The data collected in the Intelligent Transportation Systems data is sometimes sparsely generated or may have outliers due to hardware glitches and in some cases have duplicate values for elongated periods of time which needs to be removed to reduce the volume of data.

Machine Learning. The data after transformation is analyzed using different types of machine learning algorithms CNN, LSTM, Vision Transformers, RNN, Decision Trees etc. for pattern identification, Weightage identification with different objectives like Traffic Forecasting, Intelligent Traffic Control, Public Transportation review from Social Media data, Road Condition Analysis, Driver Behavior Profiling etc.

2.3 Application Plane

The Smart City ecosystem is capable for providing multiple Intelligent Transportation System Services like Public Transportation Planning, Traffic Lights Management, Road Condition Analysis, Emergency Vehicle Clearance, Driver Profiling and Traffic Congestion Identification.

Public Transportation Planning. The number of vehicles required for public transportation, departure time of vehicles, departure venues of vehicles, extra vehicles deployment during rush hours/festivals all these are part of public transportation planning which aims at creating an optimized plan for supporting the population needs.

Traffic Lights Management. The concept of traffic light management is very important as the waiting time at the traffic lights not only impacts the travel time of passengers but also impacts the environment.

Road Condition Analysis. The data from the smartphone sensors of the passengers/drivers of the vehicles acts as a crowdsourcing data for continuous monitoring of the conditions of the road. The data can provide information about potholes, unevenness index and impact of road patches.

Emergency Vehicle Clearance. The emergency vehicles like Ambulance, Fire Brigades, Police car are strictly bound by the timeline for reaching the destination from the source and any delay in timeline may result to be fatal. Thus, this application area requires route selection, rerouting, traffic lights clearance for ensuring the Emergency vehicles clearance.

Traffic Congestion Identification. The segments of the roads where traffic congestion occurs are an area of work for the urban town planning as well as the traffic police. The traffic congestion could be because of the infrastructure attributes at the geo-location or the unavailability of traffic management measures.

Driver Profiling. Another important application area is driver profiling which acts as a contributor to many of the other applications. The Driver profiling not only helps in providing corrective feedback to the driver himself but also provides preventive feedback to all the other entities getting affected by the driver.

Each of the applications are vast research sub-domains in the field of Intelligent Transportation System. One example application is taken ahead to showcase the various con texts which are required for holistic evaluation of a scenario in a research area. Each context further requires data to be stored from a plethora of sources and with varying frequencies of data record generation. Driver Profiling domain with the aim of providing attentiveness alerts to the fellow drivers in case an erratic driver is in their vicinity. The Driver profiling can be categorized on basis of various parameters like time span, profiling hardware, location of computation etc. as shown in Fig. 2.

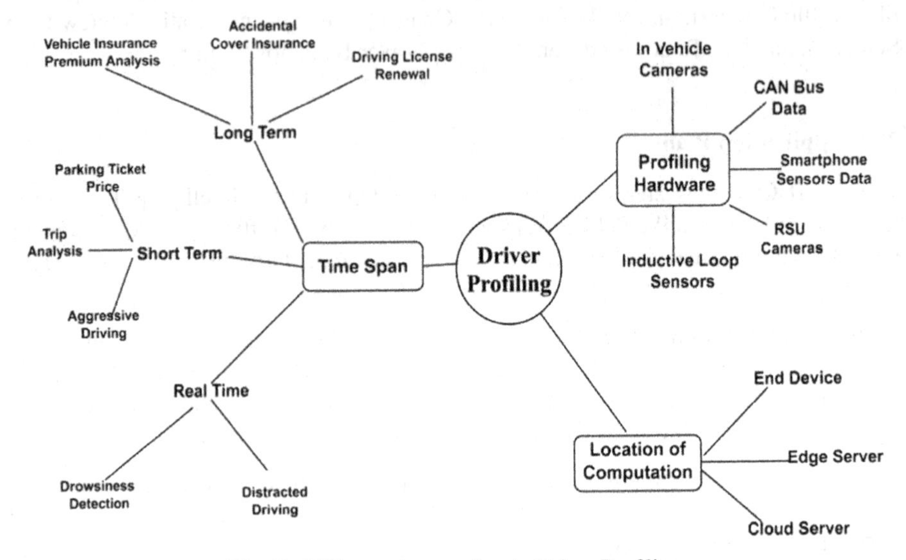

Fig. 2. Different Approaches to Driver Profiling

The driver profiling approaches requires to ensure that the latency of the emergency information passing is below the threshold, the frequency of data transmission from the MobileEntitiesVehicles (MEv) is optimized, the type of data transmission from the MEv is optimized and the computational tasks are distributed in a balanced way over end devices, edge server and cloud server.

The MEv have a large battery backup, thus there are no constraint off power consumption but the computational hardware capabilities are comparatively smaller than the Cloud Server. As the vehicles are manufactured to meet the needs of the masses, thus manufacturers cannot provide large computational capabilities due to substantial increase in the price of the vehicle with it. Therefore, the data being produced at the End Device cannot be processed at the End Device and need to be transmitted to Cloud Server for processing. The application of Driver Profiling requires the data transmission from the end device to Cloud Server in very low latency due to near real time processing requirements. Due to the fact that the cloud servers are located at geographically far off locations, this brings the need of an intermediate layer which can provide processing capabilities like Cloud Server but is located comparatively nearer to the end device. Edge Servers are the processing units which fits well into the requirement which can help in offloading of complex tasks from End Device to itself or serving cloud as well as can support in the analytical framework for data processing at the different computational devices.

Thus, it could be seen that the impact of data in the field of ITS is significant and there is a large scope of improvisation and utilization of state of art techniques to achieve maximum efficiency out of the existing transportation systems.

3 Analysis of Big Data in ITS environment

The computation plane of the proposed model is responsible for ingesting all the data from the communication plane and act as a source for the application plane. The data being generated from communication plane has all the attributes of Big Data and thus needs to pass through a cycle of activities as explained below.

3.1 Big Data Aggregation

The major components of ITS are systems focusing on traveler information, optimization of public transport services for maximum efficiency and utility, business vehicle/ platoons' system and advanced vehicle control system.[10] The above-mentioned components are supposed to perform analytics (real-time as well as long term) on the huge amount of data generated by various ITS environment devices. Therefore, there is a need of understanding the ways of collecting ITS generated big data to achieve above mentioned agenda. The sources of acquisition of data are on road sensors, on board sensors, smart cards, smartphone sensors, VANET and passive mediums.

In the context of advanced public transportation management system, there is a need to perform spatial-temporal analysis of the public transport usage i.e., buses, trams, trains, monorails etc. The data could either be collected via a traditional paper-based feedback form mechanism or with the usage of smart cards. The usage of smart cards has been prominent in many countries and various attempts has been made to draw inferences out of these data [11, 12]. Nishiuchi et al. showcased performance analysis of various data mining algorithms on the data collected via smart cards. Comparative analysis of the performance of rough set-based algorithm, C4.5, K-NN, Neural Network and Naive Bayes was performed by taking accuracy and time as parameters. The results showcased that proposed rough set-based algorithm was able to achieve nearly equal accuracy to C4.5 in only half of the time [13]. Rough set-based algorithm is hence found to perform better in ITS environment for finding travel pattern regularities.

Deployment of high-speed cameras, CCTV cameras, stop line cameras, on-board cam-eras in ITS environment is done to acquire the visual feed. The visual feed could be in-vehicle outside environment view, in-vehicle inside vehicle view, outside-vehicle vehicle view, outside-vehicle vehicle surroundings view. The video feeds could be used for vehicle identification, automatic challenging systems, automatic parking fare collection system etc. Other than these utilities, this data could also put to use in safety enhancement by evaluating the state of driver for unwanted symptoms like dizziness, yawning, pose of passengers etc. which will further lead to decreasing the number of on road fatalities. Another important research segment is the pedestrian intention prediction using the video feed data [14].

GPS sensor is one of the most commonly utilized sensor in research related to vehicle tracking and further calculating the traffic density as well as traffic intensity. GPS sensors are now a days embedded into almost all of the vehicles except bicycles, wheelchairs. The GPS data collected can be used for various purposes. Gong et al. proposed an algorithm for finding the travel mode of commuter with the help of GPS data. Different classes of travel mode used were bus, subway, rail, car and walk. The cumulative results of the proposed approach showcased an accuracy of 82.6% [15]. GPS data is not only limited to finding the travel mode of the user but can also be used to identify the trip segments as well trip purpose imputation. Accuracy of trip segment identification or tour identification directly impacts the performance of travel mode detection as well as trip purpose imputation. Shen et al. performed an analysis of various techniques used for trip purpose imputation and showcased their accuracy for undertaken instances [16]. VANET, which is an indispensable part of ITS environment generates voluminous real time data. The data produced also has lot of variety because of the fact that elements of the network are heterogenous in nature. Alongside the variety, due to the variability of entities' density the data produced is sometimes sparse in nature. The data generated from VANET is of Big Data nature. Analysis of aggregated VANET Big Data can also be performed for finding an efficient route between nodes [17]. The data collection from various heterogeneous devices leads to large volumes. The type of data format used for communication plays huge role in the size of packets transmitted and hence as well as leads to latency due to larger size. In [18] authors proposed a new data inter change format PON which reduced the parsing time for large data as well as for gener ating. The sources whose objective of data collection is different from what is being used for are called passive sources.

Other than above mentioned sources, there are many passive sources of data as well which could be utilized for ITS. The location information fetched from GPS devices could be incorrect in some instances due to issues like urban canyon. To support GPS based positioning service, passive mobile handset generated dataset can also be used. This data is produced because of the fact that whensoever a user communicates with the network, there is a need to know the location of user. User's location help network provider to select the channel for event handling. The Call Detail Record (CDR) contains information like positional coordinates of the user which is based on position of tower from where the call was channeled for the first time which could be converted to GPS coordinates. Passive data is not limited to network carriers' data only, but also to wearable devices data, social media posts. Users post the status of various events like heavy rainfall, empty roads, traffic jams, check-ins in real-time on social media. The different types of analytics applicable on social transportation can be categorized into statistical analysis, data mining and visual analytics. The data collected from such sources is highly unstructured and requires a lot of cleaning but has a lot of significant information which can act as a major secondary source of information. This has led to another field of study known as Cyber Physical Social Systems (CPSS) which includes study about human factors, driving skills, behavior analysis, frequency of visits, timing of visits and duration of visits [19].

3.2 Analytics Paradigms of Big Data in ITS

Out of various data analytics techniques available, machine learning is most prominently being used in big data ecosystems. The various sub categories based on the type of data available in which machine learning approaches could be divided are supervised, unsupervised and reinforcement learning. All these approaches have been applied in the field of ITS for various objectives. With the widespread adoption of AI, deep learning models are now also being used in ITS environment.

Supervised Learning. The machine learning approach where labeled data is available is called Supervised Learning. The dataset includes input variables as well as the output variables. The approach tries to infer the mapping function which generates the output from the given set of input. Henceforth, once a mapping function is inferred; the same could be applied to new set of inputs to forecast the output. The different types of supervised learning models include naive bayes, logistic regression, linear regression, linear discriminant analysis, Support Vector machines and Neural Networks. The representative sample of various work done along with their details is shown in Table 1.

Unsupervised Learning. It is the branch of machine learning which is used to draw inferences from unlabeled data. The most common application area of this branch is to perform cluster analysis. K-means is an unsupervised learning model which has been used by researchers to predict travel time and perform highway transportation planning. Meng et al. [20] discussed the usage of ACAK algorithm, which is an ensemble of Ant Colony Optimization and K-means algorithm for highway transportation planning. Applying ant colony optimization helps in eradicating the initial dependence of K-means algorithm with cluster centers and numbers. Similarly optimized K-means algorithm is being used in the research conducted by Nath et al. [21] for predicting travel time with the usage of historical traffic data. The research focused on predicting with more accuracy in uncertain conditions.

Reinforcement Learning. Reinforcement learning is different from supervised learning in the context that the latter works on given sample data which contains inputs as well as output whereas the latter works by interacting with the environment. The agent involved receives awards for the actions performed. The rewards could be either positive or negative. The agent aims at maximization of the expected cumulative reward. Reinforcement learning is highly applicable in the areas of control and optimization. Therefore, reinforcement learning has been applied for by Li et al. [28] for formulating the traffic signal timing plan. The approach used is capable of learning the optimal control plan and traffic system dynamics simultaneously by modeling control actions and state transitions with the help of Q-Learning model. The usage of deep reinforcement learning has a huge scope in the field of ITS.

Deep Learning. Deep learning is the class of machine learning which is based on ANN and where the number of layers is increased to progressively infer higher level information from raw data. Credit Assignment Path (CAP) depth is high in case of Deep Learning based systems. The CAP depth of Feedforward Neural Network is equal to number of hidden layers plus one whereas in case of Recurrent Neural Network this could be infinite. Deep learning algorithms could be applicable on both supervised learning tasks as well as unsupervised learning tasks. Hu et al. [29] used Deep Neural Networks for performing fault diagnosis of bogies of high-speed trains. The deep neural networks are capable of generating insights about faults in self-adaptive mode from the signal frequency spectrums. The results showcased that deep neural network's diagnostics approaches 100% accuracy whereas single hidden layer neural network and genetic algorithm based neural network approaches 81% and 91% respectively. Polson et al. [30] created a deep learning architecture for predicting short term traffic flow. The architecture created was compared with single hidden layer and VARM8L approach for the predicted results. It

Table 1. Supervised Learning in ITS

Authors	Learning Model Used	Application Area	Key Finding
Sun et al.[22]	LocalLinear Regression	Short Term Traffic Forecasting	Performance local linear is betterthan k-NN and kernel smoothing in traffic speed analysis
Shanet al. [23]	Multiple Linear Regression	Estimating missing data for traffic speed estimation	Spatial-temporal approach in multiple linear regression method help reduce the RMSE by 73.3%
Abellan et al. [24]	Decision Tree	Accident Severity Detection	Considering each input variable as separate decision tree helps to infer larger amount of knowledge about the accident
Vlahogianni et al. [25]	ANN	Short-term flow prediction	Traffic flow prediction ca be obtained to a satisfactory level by neural networks with parameters like step size, hidden layer units and momentum generated genetically
Lint et al. [26]	SSNN	Accurate Travel time prediction	Incorrect and missing data is an inherent property of the existing data collection systems, where SSNN helps to provide robust travel time prediction
Xiao et al. [27]	SVM	Traffic incident detection	Multiple kernels based SVM needs only one time training and also has robust performance for binary classification of incident free traffic pat tern and incident traffic pattern

was found that MSE of deep learning architecture was 4% and 14% more respectively. The real time data collected in Guangzhou city's selected road segment along with the historical data of the road segment is fed into the platform created for analytics of big data which acts as an input to the Deep Belief Network to build an optimized model for significantly improved congestion evacuation performance [31]. The analytics model's applicability in the real-life scenarios is dependent on yet another important factor of user authentication. The data collected for analytics acts an input to the analytics model and thus ensuring that the on-board sensor data is only accessed by authenticated user is very important. In [32], authors proposed and demonstrated the application of three factor user authentication scheme which enhanced the security along with multiple functionality attributes as well. Deep learning computation models find difficulty in effectively extracting features in multi-modal data. Moreover, it also leads to exponential increase in the learning parameters which requires huge number of computational resources as well as lead to increased processing time, thus making them unfit for deployment on edge devices. In [33], authors proposed and experimented a lightweight tensor based Deep learning computation model on multiple real datasets which was able to achieve constant network model performance when compressing the learning parameters along with reduction in the model's training time as well as computational complexity. The data analytics model can further also be improved by identifying the important traffic nodes in the complete network. This approach requires ranking of the available traffic nodes which could be then filtered using certain thresholding techniques. The authors in [34] highlighted that the commonly used metrics of degree and betweenness are prone to inducing bias in the system and the traditional approach of iterative node deletion is not feasible in large scale network. Authors proposed and demonstrated the effectiveness of their machine learning based clustering by utilizing the traffic flow of the nodes.

3.3 Prominent Tools for Big Data Analytics

As it is visible from the preceding sections that the data being generated in the CV environment is of Big Data in nature i.e., it has four V's Volume, Variety, Velocity and Veracity. To extract the fifth V i.e., value out of this data collected from the environment there is a need of software which will help to analytic operations. There are some open sources as well as big data analytics platforms available. Big data platforms utilize the amalgamation of parallel processing and distributed processing concept for enabling faster processing of data. The most common platform being used is Apache Hadoop. The vital components of this platform are [35] listed below.

Hadoop Common - This component comprises of all the utilities which are required for supporting the other Hadoop modules.
HDFS - It is a distributed file system that ensures high-throughput access to the data.
Hadoop YARN - This component is responsible for resource management of clusters and job scheduling.
Hadoop MapReduce - This component is dependent on YARN and is responsible for parallel processing of large data sets.
Hadoop Ozone - This component is responsible for storing the objects created in Hadoop.
Hadoop Submarine - It is an engine which enables machine learning for Hadoop.

Alongside the Apache Hadoop project, there are plenty of other projects which are applicable in the area of Big Data Analytics and are based on Apache Hadoop. Apache

Ambari is a web-based tool for monitoring and controlling the clusters of Hadoop. Ambari also provides features for viewing the cluster health in the form of heatmap as well as to analyze the performance of Pig, MapReduce and Hive applications. Apache Avro is a data serialization system which relies on pre-defined schemas. While data is read, the availability of schema helps in no per value overheads. Whensoever Avro data is exported, the schema is also saved alongside the data. This feature helps in easy access of the data by any other tool. Similarly, there are other projects like Cassandra, Spark, Hive, Mahout, Pig and Tez. Research carried in the area of Big Data analytics revolves around the usage of such tools but with different configurations which are tailored as per the application requirements. Main et al. [36] proposed an architecture for handling the big data generated by Greater Toronto Area. The platform proposed is an ensemble of different types of analytics engines and processing which ranges from lower computational analytics like text processing to complex computational analytics like video processing. The various engines involved include Storage engines, Workflow engines, Graph engine, MapReduce engine, Text engine, Mining engine. Gang Zeng [37] discussed the key technologies on the data analysis layer i.e., Traffic flow, average duration of distance covered for a road, vehicle trajectory, fake vehicles identification. The key of map function is a combination of bayonet Id and direction id. The value of map function is passing time. Similarly, the other applications utilize HBase, Cassandra, Apache Spark. The updated versions of Cross Industry Standard Process for Data Mining (CRISP-DM) have been utilized for data mining projects. CRISP-DM architecture has been taken as base for creating an ETL architecture of the CV environment for the dynamic toll charging for highway by Guerreiro et al. [38]. The performance of Apache Spark was compared with other traditional approaches and it was found that effectiveness of Spark Cluster is significantly higher.

4 Focus Areas of Future Research

After the detailed analysis of the aforementioned existing proposals with respect to current status of the ITS, the following focus areas of research have been identified.

Lightweight and scalable data interchange format. The nodes which are included in ITS environment contains nodes which may contain very less processing power, storage capacity and battery constrained. So, there is need of a data format for communication which is lightweight in nature and shall support scalability [17].

Algorithms and Protocols which are less sensitive to the number of nodes.
The algorithms and protocols used in ITS environment are capable to perform well for a certain range of devices. The performance of chosen paradigms fluctuates with the. spike in device count. Therefore, there is need of paradigms which may perform considerably well for all the scenarios.

Maintain reliability throughout the heterogeneous nature of the ITS network. The ITS environment is directly related to the real-life application of travel and commute practices. The availability of services provided for applications of ITS is highly required as they are related to the life of the users as well in many cases which is not achieved due to highly heterogeneous nature and different speeds of Vehicles [39].

Social connection analysis and Trust mechanism for real-time and long-term application scenarios. There is a need of standard framework for vehicle's driver social connection analysis for transportation services usage prediction. The decisions related to probability of vehicles entering the commute process is also dependent on certain social connections and their geospatial locations [40]. Devices which are communicating with each other in ad hoc network structure need to build a trust mechanism with the constraint of real-time inference need as well as long-term trust assurance to create the concepts of buddy evaluators [41].

5 Conclusion

More studies that seek to address the drawbacks of the current Big Data algorithms in order to explore new facets of the field are being conducted as a result of the continuous growth of ITS, its different elements, and the increase in volume, diversity, velocity, and truthfulness of ITS data. The majority of currently published review articles in the literature either do extensive mathematical modelling research or narrowly concentrate on a single use of Big Data methods in ITS. The most important ITS applications, in which big data methods play a key part, are reviewed in this paper, but it also sheds light on how these applications relate to the models that are used in them. As a result, one of the consequences of this study is its clear categorization of big data techniques and approaches in the context of ITS which could be put to use by scholars working in this area.

The prediction concept is the ITS use case that has been documented in the literature most frequently. Traffic flow prediction is the most popular study topic among the several aspects of prediction in the transportation system. Detection and recognition of driver behavior are two more popular ITS applications that draw academic interest.

References

1. Bello-Orgaz, G., Jung, J.J., Camacho, D.: Social big data: recent achievements and new challenges. Inf. Fusion. **28**, 45–59 (2016). https://doi.org/10.1016/j.inffus.2015.08.005
2. Chen, C.: Storey: business intelligence and analytics: from big data to big impact. MIS Q. **36**, 1165 (2012). https://doi.org/10.2307/41703503
3. Mayilvaganan, M., Sabitha, M.: A cloud-based architecture for Big-Data analytics in smart grid: a proposal. In: 2013 IEEE International Conference on Computational Intelligence and Computing Research. IEEE (2013)
4. Zhang, J., Wang, F.-Y., Wang, K., Lin, W.-H., Xu, X., Chen, C.: Data-driven intelligent transportation systems: a survey. IEEE Trans. Intell. Transp. Syst. **12**, 1624–1639 (2011). https://doi.org/10.1109/tits.2011.2158001
5. Shi, Q., Abdel-Aty, M.: Big Data applications in real-time traffic operation and safety monitoring and improvement on urban expressways. Transp. Res. Part C Emerg. Technol. **58**, 380–394 (2015). https://doi.org/10.1016/j.trc.2015.02.022
6. Lin, X., Wang, P., Wu, B.: Log analysis in cloud computing environment with Hadoop and Spark. In: 2013 5th IEEE International Conference on Broadband Network & Multimedia Technology, pp. 273–276. IEEE (2013)

7. Zaharia, M., et al.: Fast and interactive analytics over Hadoop data with spark. https://www.usenix.org/system/files/login/articles/zaharia.pdf?spm=5176.100239.blogcont37396.229.RGRYGj&file=zaharia.pdf. Accessed 18 Apr 2023
8. Zikopoulos, P., deRoos, D., Parasuraman, K., Deutsch, T.A., Giles, J., Corrigan, D.: Harness the Power of Big Data the IBM Big Data Platform. Osborne/McGraw-Hill, New York (2012)
9. Basche, L.: Says solving 'big data' challenge involves more than just managing volumes of data. Bus. Wire, San Francisco, CA, USA, Technical report (2011)
10. An, S.-H., Lee, B.-H., Shin, D.-R.: A survey of intelligent transportation systems. In: 2011 Third International Conference on Computational Intelligence, Communication Systems and Networks, pp. 332–337. IEEE (2011)
11. Ma, X., Wu, Y.-J., Wang, Y., Chen, F., Liu, J.: Mining smart card data for transit riders' travel patterns. Transp. Res. Part C Emerg. Technol. **36**, 1–12 (2013). https://doi.org/10.1016/j.trc.2013.07.010
12. Nishiuchi, H., King, J., Todoroki, T.: Spatial-temporal daily frequent trip pattern of public transport passengers using smart card data. Int. J. Intell. Transp. Syst. Res. **11**, 1–10 (2013). https://doi.org/10.1007/s13177-012-0051-7
13. Quintero Minguez, R., Parra Alonso, I., Fernandez-Llorca, D., Sotelo, M.A.: Pedestrian path, pose, and intention prediction through Gaussian process dynamical models and pedestrian activity recognition. IEEE Trans. Intell. Transp. Syst. **20**, 1803–1814 (2019). https://doi.org/10.1109/tits.2018.2836305
14. Gong, H., Chen, C., Bialostozky, E., Lawson, C.T.: A GPS/GIS method for travel mode detection in New York City. Comput. Environ. Urban Syst. **36**, 131–139 (2012). https://doi.org/10.1016/j.compenvurbsys.2011.05.003
15. Shen, L., Stopher, P.R.: Review of GPS travel survey and GPS data-processing methods. Transp. Rev. **34**, 316–334 (2014). https://doi.org/10.1080/01441647.2014.903530
16. Bedi, P., Jindal, V.: Use of big data technology in vehicular ad-hoc networks. In: 2014 International Conference on Advances in Computing, Communications and Informatics (ICACCI). IEEE (2014)
17. Nordahl, M., Magnusson, B.: A lightweight data interchange format for internet of things with applications in the PalCom middleware framework. J. Ambient. Intell. Humaniz. Comput. **7**, 523–532 (2016). https://doi.org/10.1007/s12652-016-0382-3
18. Chen, C., Ma, J., Susilo, Y., Liu, Y., Wang, M.: The promises of big data and small data for travel behavior (aka human mobility) analysis. Transp. Res. Part C Emerg. Technol. **68**, 285–299 (2016). https://doi.org/10.1016/j.trc.2016.04.005
19. Zheng, X., et al.: Big data for social transportation. IEEE Trans. Intell. Transp. Syst. **17**, 620–630 (2016). https://doi.org/10.1109/tits.2015.2480157
20. Meng, Y., Liu, X.: Application of K-means algorithm based on ant clustering algorithm in macroscopic planning of highway transportation hub. In: 2007 First IEEE International Symposium on Information Technologies and Applications in Education. IEEE (2007)
21. DebNath, R.P., Lee, H.-J., Chowdhury, N.K., Chang, J.-W.: Modified K-means clustering for travel time prediction based on historical traffic data. In: Setchi, R., Jordanov, I., Howlett, R.J., Jain, L.C. (eds.) Knowledge-Based and Intelligent Information and Engineering Systems. LNCS (LNAI), vol. 6276, pp. 511–521. Springer, Heidelberg (2010). https://doi.org/10.1007/978-3-642-15387-7_55
22. Sun, H., Liu, H.X., Xiao, H., He, R.R., Ran, B.: Use of local linear regression model for short-term traffic forecasting. Transp. Res. Rec. **1836**, 143–150 (2003). https://doi.org/10.3141/1836-18
23. Shan, Z., Zhao, D., Xia, Y.: Urban road traffic speed estimation for missing probe vehicle data based on multiple linear regression model. In: 16th International IEEE Conference on Intelligent Transportation Systems (ITSC 2013), pp. 118–123. IEEE (2013)

24. Abellán, J., López, G., de Oña, J.: Analysis of traffic accident severity using Decision Rules via Decision Trees. Expert Syst. Appl. **40**, 6047–6054 (2013). https://doi.org/10.1016/j.eswa.2013.05.027
25. Vlahogianni, E.I., Karlaftis, M.G., Golias, J.C.: Optimized and meta-optimized neural networks for short-term traffic flow prediction: a genetic approach. Transp. Res. Part C Emerg. Technol. **13**, 211–234 (2005). https://doi.org/10.1016/j.trc.2005.04.007
26. van Lint, J.W.C., Hoogendoorn, S.P., van Zuylen, H.J.: Accurate freeway travel time prediction with state-space neural networks under missing data. Transp. Res. Part C Emerg. Technol. **13**, 347–369 (2005). https://doi.org/10.1016/j.trc.2005.03.001
27. Xiao, J., Liu, Y.: Traffic incident detection using multiple-kernel support vector machine. Transp. Res. Rec. **2324**, 44–52 (2012). https://doi.org/10.3141/2324-06
28. Li, L., Lv, Y., Wang, F.-Y.: Traffic signal timing via deep reinforcement learning. IEEE/CAA J. Autom. Sin. **3**, 247–254 (2016). https://doi.org/10.1109/jas.2016.7508798
29. Hu, H., Tang, B., Gong, X., Wei, W., Wang, H.: Intelligent fault diagnosis of the high-speed train with big data based on deep neural networks. IEEE Trans. Industr. Inform. **13**, 2106–2116 (2017). https://doi.org/10.1109/tii.2017.2683528
30. Polson, N.G., Sokolov, V.O.: Deep learning for short-term traffic flow prediction. Transp. Res. Part C Emerg. Technol. **79**, 1–17 (2017). https://doi.org/10.1016/j.trc.2017.02.024
31. Liu, Y., Zhang, Q., Lv, Z.: Real-time intelligent automatic transportation safety based on big data management. IEEE Trans. Intell. Transp. Syst. **23**, 9702–9711 (2022). https://doi.org/10.1109/tits.2021.3106388
32. Srinivas, J., Das, A.K., Wazid, M., Vasilakos, A.V.: Designing secure user authentication protocol for big data collection in IoT-based intelligent transportation system. IEEE Internet Things J. **8**, 7727–7744 (2021). https://doi.org/10.1109/jiot.2020.3040938
33. Liu, D., Yang, L.T., Zhao, R., Wang, J., Xie, X.: Lightweight tensor deep computation model with its application in intelligent transportation systems. IEEE Trans. Intell. Transp. Syst. **23**, 2678–2687 (2022). https://doi.org/10.1109/tits.2022.3143861
34. Huang, X., Chen, J., Cai, M., Wang, W., Hu, X.: Traffic node importance evaluation based on clustering in represented transportation networks. IEEE Trans. Intell. Transp. Syst. **23**, 16622–16631 (2022). https://doi.org/10.1109/tits.2022.3163756
35. Apache Hadoop. https://hadoop.apache.org/. Accessed 18 Apr 2023
36. Mian, R., Ghanbari, H., Zareian, S., Shtern, M., Litoiu, M.: A data platform for the highway traffic data. In: 2014 IEEE 8th International Symposium on the Maintenance and Evolution of Service-Oriented and Cloud-Based Systems, pp. 47–52. IEEE (2014)
37. Zeng, G.: Application of big data in intelligent traffic system. IOSR J. Comput. Eng. **17**(1), 01–04 (2015)
38. Guerreiro, G., Figueiras, P., Silva, R., Costa, R., Jardim-Goncalves, R.: An architecture for big data processing on intelligent transportation systems. an application scenario on highway traffic flows. In: 2016 IEEE 8th International Conference on Intelligent Systems (IS), pp. 65–72. IEEE (2016)
39. Ang, L.-M., Seng, K.P., Ijemaru, G.K., Zungeru, A.M.: Deployment of IoV for smart cities: applications, architecture, and challenges. IEEE Access **7**, 6473–6492 (2019). https://doi.org/10.1109/access.2018.2887076
40. Butt, T.A., Iqbal, R., Shah, S.C., Umar, T.: Social internet of vehicles: architecture and enabling technologies. Comput. Electr. Eng. **69**, 68–84 (2018). https://doi.org/10.1016/j.compeleceng.2018.05.023
41. Darwish, T.S.J., Bakar, K.A.: Fog based intelligent transportation big data analytics in the internet of vehicles environment: motivations, architecture, challenges, and critical issues. IEEE Access **6**, 15679–15701 (2018). https://doi.org/10.1109/ACCESS.2018.2815989

Correction to: A Very Deep Adaptive Convolutional Neural Network (VDACNN) for Image Dehazing

Balla Pavan Kumar, Arvind Kumar, and Rajoo Pandey

Correction to:
Chapter 4 in: R. K. Challa et al. (Eds.):
***Artificial Intelligence of Things*, CCIS 1930,**
https://doi.org/10.1007/978-3-031-48781-1_4

In the originally published version of chapter 4, one of the references had been incorrect. This has been corrected.

The updated version of this chapter can be found at
https://doi.org/10.1007/978-3-031-48781-1_4

R. K. Challa et al. (Eds.): ICAIoT 2023, CCIS 1930, p. C1, 2024.
https://doi.org/10.1007/978-3-031-48781-1_30

Author Index

R. K. Challa et al. (Eds.): ICAIoT 2023, CCIS 1930, pp. 385–387, 2024.
https://doi.org/10.1007/978-3-031-48781-1

Printed in the United States
by Baker & Taylor Publisher Services